Advances in Fuzzy Decision Theory and Applications

Advances in Fuzzy Decision Theory and Applications

Editors

Jun Ye

Yanhui Guo

MDPI

Basel • Beijing • Wuhan • Barcelona • Belgrade • Novi Sad • Cluj • Manchester

Editors
Jun Ye
School of Civil and
Environmental Engineering
Ningbo University
Ningbo, China

Yanhui Guo
Department of
Computer Science
University of Illinois
Springfield
Springfield, IL, USA

Editorial Office
MDPI
St. Alban-Anlage 66
4052 Basel, Switzerland

This is a reprint of articles from the Special Issue published online in the open access journal *Mathematics* (ISSN 2227-7390) (available at: https://www.mdpi.com/journal/mathematics/special_issues/advances_fuzzy_decision_theory_applications).

For citation purposes, cite each article independently as indicated on the article page online and as indicated below:

Lastname, A.A.; Lastname, B.B. Article Title. *Journal Name* **Year**, *Volume Number*, Page Range.

ISBN 978-3-0365-9720-1 (Hbk)
ISBN 978-3-0365-9721-8 (PDF)
doi.org/10.3390/books978-3-0365-9721-8

Contents

About the Editors

Jun Ye

Jun Ye is a Professor in the School of Civil and Environmental Engineering and Deputy Director of the Institute of Rock Mechanics at Ningbo University. Since 2019, he has served as the Vice President of the China Branch of the Neutrosophic Science International Association (NSIA). He is currently Editor-in-Chief of the international journals *Current Computer Science and Decision Making and Analysis*. He is Academic Editor of the *Journal of Classification, Journal of Mathematics, Mathematical Problems in Engineering*, and *Neutrosophic Sets and Systems*. He is also an Editorial Board Member for other journals. He hosted or participated in three projects with the National Natural Science Foundation and two projects with the Zhejiang Natural Science Foundation. He has published more than 300 papers in domestic and foreign journals. From 2019 to 2022, he was selected as the "Elsevier China Highly Cited Scholar". He was selected as one of the World's Top Computer Scientists by Research.com in 2022 and 2023.

Yanhui Guo

Yanhui Guo received his Ph.D. in the Department of Computer Science, Utah State University, USA. He was a Research Fellow in the Department of Radiology at the University of Michigan and an Assistant Professor at St. Thomas University. Dr. Guo is currently an Associate Professor in the Department of Computer Science at the University of Illinois Springfield. Dr. Guo's research area includes computer vision, machine learning, data analytics, neutrosophic sets, computer-aided detection/diagnosis, and computer-assisted surgery. He has published three books, more than 110 journal papers, and 40 conference papers, completed more than 10 grant-funded research projects, has two patents, and has worked as an Associate Editor for different international journals and as a reviewer for top journals and conferences. Dr. Guo successfully applied neutrosophic sets to image processing in 2008 and has published much research work in this area. Dr. Guo was a Co-Founder and Chief Scientist of MedSights Tech Inc., a high technology company focusing on a computer-assisted surgery system. Dr. Guo was awarded a University Scholar in 2019, the university system's highest faculty honor, recognizing outstanding teaching and scholarship.

Preface

In the realm of complex decision making, characterized by inherent incompleteness and uncertainty, the foundational work of Lotfi A. Zadeh on fuzzy set theory has been instrumental. The efficacy of classical fuzzy sets in addressing vagueness has prompted an exploration of various extensions, each catering to the intricacies of real-world decision-making problems. This book delves into an array of advanced fuzzy theories, including type-2 fuzzy sets, hesitant fuzzy sets, multivalued fuzzy sets, cubic sets, intuitionistic fuzzy sets, Pythagorean fuzzy sets, spherical fuzzy sets, neutrosophic sets, and more. The richness of these extensions reflects the dynamism of fuzzy theories in diverse decision-making applications.

Comprising ten research papers, this book presents a synthesis of the latest progress and achievements in fuzzy decision theories. The contributions span theoretical developments and practical applications, demonstrating the versatility of advanced fuzzy theories across domains such as finance, healthcare, engineering, and beyond. The collective knowledge presented here serves as a testament to the growing significance of fuzzy decision theories in addressing the complexities of contemporary decision science.

In the paper entitled "A Hybrid MCDM Approach Based on Fuzzy MEREC-G and Fuzzy RATMI" by Anas A. Makki and Reda M. S. Abdulaal, the authors address the critical realm of multi-criteria decision making (MCDM), which plays a pivotal role in navigating complex problems where diverse alternatives must be evaluated against conflicting criteria. Traditional MCDM methods have been integral in this regard, yet the increasing prevalence of uncertain and ambiguous decision-maker inputs in real-world scenarios necessitates the application of fuzzy logic. The authors introduce a novel hybrid fuzzy MCDM approach that combines the strengths of two recent methodologies: fuzzy MEREC-G, designed to handle linguistic input terms from multiple decision makers and generate consistent fuzzy weights, and fuzzy RATMI, which ranks alternatives based on their fuzzy performance scores on each criterion. Notably, this paper presents the first fuzzy extension of both MEREC-G and RATMI methods. The study provides detailed algorithms for fuzzy MEREC-G and fuzzy RATMI and demonstrates their application in solving real-world problems. Through correlation and scenario analyses, the authors validate the new approach's accuracy, consistency, and sensitivity, highlighting its potential to deliver robust and reliable decision-making outcomes in the face of uncertain and dynamic decision contexts.

The paper titled "MemConFuzz: Memory Consumption Guided Fuzzing with Data Flow Analysis", by Chunlai Du, Zhijian Cui, Yanhui Guo, Guizhi Xu, and Zhongru Wang, tackles the critical issue of uncontrolled heap memory consumption—a software vulnerability exploited by attackers to consume significant amounts of heap memory, leading to system crashes. Existing efforts in vulnerability fuzzing of heap consumption, such as MemLock and PerfFuzz, often fall short in considering the impact of data flow. In response, the authors present MemConFuzz, a novel heap memory consumption-guided fuzzing model. MemConFuzz leverages static data flow analysis to extract the locations of heap operations and data-dependent functions. Notably, the paper introduces a seed selection algorithm based on data dependency, allocating more energy to samples with higher priority scores during the fuzzing process. Experimental results demonstrate that MemConFuzz outperforms existing approaches like AFL, MemLock, and PerfFuzz, showcasing superior efficiency in both quantity and time consumption when exploiting vulnerabilities related to heap memory consumption. This innovative contribution enhances our understanding and approach to addressing critical software vulnerabilities associated with uncontrolled heap memory consumption.

In the scholarly work titled "Some Operations and Properties of the Cubic Intuitionistic Set with Application in Multi-Criteria Decision-Making" authored by Shahzad Faizi, Heorhii Svitenko, Tabasam Rashid, Sohail Zafar, and Wojciech Sałabun, the authors present a comprehensive exploration of operations and properties associated with the cubic intuitionistic set, offering valuable insights with potential applications in multi-criteria decision making (MCDM). The introduced concepts include the internal cubic intuitionistic set (ICIS), external cubic intuitionistic set (ECIS), P-order, R-order (P-(R-) order), P-union, R-union (P-(R-) union), and P-intersection, R-intersection (P-(R-) intersection). The paper delves into the investigation of various properties related to the P-(R-) union and P-(R-) intersection of ICISs and ECISs, accompanied by illustrative examples to elucidate these theoretical constructs. Additionally, the authors put forth significant theorems pertaining to ICISs and ECISs, supported by rigorous proofs. The practical applicability of these operations is demonstrated through a real-world scenario, applying the proposed concepts to solve a multi-criteria decision-making problem. This work not only contributes to the theoretical foundation of cubic intuitionistic sets but also showcases their relevance in addressing complex decision-making challenges, thereby enriching the toolkit available for researchers and practitioners in the field.

In the paper titled "Study on Chaotic Multi-Attribute Group Decision Making Based on Weighted Neutrosophic Fuzzy Soft Rough Sets" authored by Fu Zhang and Weimin Ma, the authors delve into the realm of multi-attribute group decision making (MAGDM) with a distinctive dimension termed Chaotic MAGDM. This novel scenario incorporates considerations not only for the weights of decision makers (DMs) and decision attributes but also for the familiarity of DMs with these attributes. The authors leverage the weighted neutrosophic fuzzy soft rough set theory to address the complexities inherent in Chaotic MAGDM, presenting a new algorithm tailored for MAGDM applications. A notable contribution lies in the integration of familiarity into MAGDM within the framework of neutrosophic fuzzy soft rough sets. Furthermore, the paper introduces a novel MAGDM model grounded in neutrosophic fuzzy soft rough sets and develops a sorting/ranking algorithm based on the same set of theories. To illustrate the practical utility of the proposed algorithm, a case study is provided, showcasing the application of the devised model. This work not only advances the theoretical foundations of decision making under chaotic conditions but also provides a valuable methodology for handling real-world MAGDM challenges through the fusion of weighted neutrosophic fuzzy soft rough sets and chaos theory.

The paper titled "A Novel Driving-Strategy Generating Method of Collision Avoidance for Unmanned Ships Based on Extensive-Form Game Model with Fuzzy Credibility Numbers," authored by Haotian Cui, Fangwei Zhang, Mingjie Li, Yang Cui, and Rui Wang, addresses the crucial issue of intelligent collision avoidance for unmanned ships at sea. The study introduces an innovative approach by proposing a novel driving strategy generation method rooted in an extensive-form game model employing fuzzy credibility numbers. The key contribution lies in formulating an extensive-form game model that accounts for the two-sided clamping situation of unmanned ships, validated through a fuzzy credibility assessment. The research quantitatively divides the head-on situations of ships at sea, facilitating targeted collision avoidance decisions for unmanned ships. The utilization of an extensive-form game model, particularly in scenarios involving two-sided clamping, is a notable aspect of the study. The integration of fuzzy credibility degrees into the game model allows for the assessment of whether the collision avoidance decisions made by unmanned ships achieve optimal results. Through case analysis and simulation, the effectiveness of the introduced game model is confirmed, demonstrating its practical utility in real-time collision avoidance decision making for unmanned ships in scenarios involving two-sided clamping. The proposed mathematical model, as illustrated in an example, stands as a promising tool for enhancing the ability of unmanned ships to navigate safely and make informed decisions in complex maritime environments.

The paper titled "Medical Diagnosis and Pattern Recognition Based on Generalized Dice Similarity Measures for Managing Intuitionistic Hesitant Fuzzy Information" by Majed Albaity and Tahir Mahmood addresses the intersection of medical diagnosis, pattern recognition, and the representation of intuitionistic hesitant fuzzy (IHF) information. Pattern recognition, a fundamental aspect of computer science, has wide-ranging applications, including machine learning, information compression, signal processing, and bioinformatics. The authors introduce the theory of generalized dice similarity (GDS) measures to establish relationships between pieces of IHF information, making valuable contributions to real-life problem solving. The GDS measures offer versatility by allowing the derivation of various measures through parameter variations, known as DGS measures. The paper leverages this theory to extend the well-established dice similarity measures (DSMs) to the context of IHF sets, which encompass both membership and non-membership grades within the finite subset [0, 1]. Pioneering the theory of generalized DSMs (GDSMs) computed based on IHF sets, the authors introduce the IHF dice similarity measure, IHF weighted dice similarity measure, IHF GDS measure, and IHF weighted GDS measure. The application of these measures is demonstrated through medical diagnosis and pattern recognition problems, showcasing their proficiency and capability. The authors conduct a comparative analysis with existing measures to enhance the practical value of the proposed measures, thereby contributing to the advancement of computational methodologies in the management of IHF information for medical diagnosis and pattern recognition.

In the paper titled "A Ship Fire Escape Speed Correction Method Considering the Influence of Crowd Interaction" authored by Jingyuan Li, Weile Liu, Fangwei Zhang, Taiyang Li, and Rui Wang, the authors delve into the critical issue of passenger ship fire evacuation and investigate the impact of various personnel attributes and interactions on evacuation efficiency. The study introduces a novel speed correction method designed to account for human attributes and interactions among different populations during evacuation scenarios. Initially, hesitant fuzzy sets and hesitant fuzzy average operators are employed to quantify four distinct personnel attributes. Subsequently, the study extracts a formula for acceleration that considers the interactive influence of different groups of people. Leveraging the first-order linear relationship between velocity and acceleration, the authors propose an interactive velocity correction method for ship personnel evacuation. To validate the effectiveness of the method, the study employs personnel evacuation simulation software, Pathfinder, conducting experiments with both corrected and uncorrected speeds introduced into the evacuation simulation process. The results demonstrate that the simulation outcomes of the revised speed plan align more closely with real-world scenarios, emphasizing the practical significance of considering personnel attributes and interactions in refining ship fire escape speed strategies for enhanced evacuation efficiency and safety.

The paper titled "A Hybrid Intuitionistic Fuzzy Group Decision Framework and Its Application in Urban Rail Transit System Selection" by Bing Yan, Yuan Rong, Liying Yu, and Yuting Huang addresses the complex decision-making process of selecting an urban rail transit system, focusing on green and low-carbon perspectives to promote sustainable urban development. Acknowledging the uncertainty arising from conflicting criteria and the inherent fuzziness in decision-makers cognition, the authors present a hybrid intuitionistic fuzzy multi-criteria group decision making (MCGDM) framework. The proposed methodology addresses various aspects of the decision process. Firstly, the weights of experts are determined using an improved similarity method. Subsequently, the subjective and objective weights of criteria are calculated using DEMATEL and CRITIC methods, and a comprehensive weight is obtained through linear integration. Considering the experts' regret degree and risk preference, the COPRAS method based on regret theory is introduced to determine the prioritization of the urban rail transit system ranking. The practicality and effectiveness of the developed method are demonstrated through a case study of urban rail transit system selection for

City N. The results reveal that a metro system (P1) is the most suitable option for City N's urban rail transit system construction, followed by a municipal railway system (P7). A sensitivity analysis, a comparative analysis, and a thorough case study validate the robustness, stability, and practicality of the proposed decision-making framework, showcasing its efficacy in supporting informed decisions in the context of urban rail transit system selection.

The paper titled "Group Decision-Making Problems Based on Mixed Aggregation Operations of Interval-Valued Fuzzy and Entropy Elements in Single- and Interval-Valued Fuzzy Environments" by Weiming Li and Jun Ye addresses the intricate challenges of operational problems in group decision making (GDM) scenarios involving single- and interval-valued fuzzy multivalued hybrid information expressions. Fuzzy sets and interval-valued fuzzy sets serve as crucial tools for representing uncertain and vague information in real-world contexts. To tackle the complexity of these mixed multivalued information expressions and operational challenges, the study introduces the concept of single- and interval-valued fuzzy multivalued set/element (SIVFMS/SIVFME) with identical and/or different fuzzy values. The conversion of SIVFMS/SIVFME into the interval-valued fuzzy and entropy set/element (IVFES/IVFEE) is presented, relying on the mean and information entropy of SIVFME to address operational problems with varying lengths. The study defines the operational relationships of IVFEEs, introduces expected value functions and sorting rules, and proposes IVFEE-weighted averaging and geometric operators, along with their mixed-weighted-averaging operation. Leveraging these operations and functions, a GDM method is developed for multicriteria GDM problems within the SIVFMS environment. The proposed method is applied to a supplier selection problem in a supply chain as a practical example, demonstrating the rationality and efficiency of SIVFMSs. A comparative analysis with other decision-making methods highlights the superiority of the developed GDM method, providing a more reasonable and flexible approach that compensates for existing methodological deficiencies in the GDM process.

The paper titled "An Intelligent Expert Combination Weighting Scheme for Group Decision Making in Railway Reconstruction" by Lihua Zeng, Haiping Ren, Tonghua Yang, and Neal Xiong introduces an intelligent approach to expert combination weighting for group decision making in the context of railway reconstruction. The study addresses the limitations of existing intuitionistic fuzzy entropies by proposing an improved version based on the cotangent function. This enhanced entropy not only considers the deviation between membership and non-membership but also incorporates the hesitancy degree of decision makers, providing a more comprehensive measure of uncertainty for intuitionistic fuzzy sets. Furthermore, the paper introduces a novel intuitionistic fuzzy (IF) similarity measure, whose values are IF numbers. The improved entropy and similarity measures are then applied to the determination of expert weights in group decision making. The study presents an intelligent expert combination weighting scheme, leveraging the new intuitionistic fuzzy similarity to transform the decision matrix into a similarity matrix. Through the analysis of threshold change rates and the design of risk parameters, the scheme achieves reasonable expert clustering results. In this scheme, each category is weighted, and experts within each category are weighted using entropy weight theory. The total weight of experts is then determined by synthesizing these two weights. This comprehensive approach provides a new method for objectively and reasonably determining expert weights in group decision-making scenarios. The proposed scheme is applied to the evaluation of a railway reconstruction scheme, demonstrating its feasibility and showcasing its potential in real-world applications.

As the editors responsible for curating this Special Issue, we extend our gratitude to the authors whose dedicated research has enriched this collection. Their insights and expertise have contributed to the depth and breadth of our exploration of fuzzy decision theories. We also express appreciation to the reviewers whose meticulous assessments have ensured the scholarly rigor and quality of each contribution.

We would like to acknowledge the support and collaboration of the editorial team and the publisher in bringing this book to fruition. Their commitment to academic excellence has been integral to the success of this endeavor.

This book is not just a static representation of the current state of fuzzy decision theory; it is an invitation to researchers, practitioners, and students to engage with the evolving landscape of decision science. We hope that the insights shared within these pages inspire further inquiry, spark innovative ideas, and pave the way for continued advancements in the fascinating and ever-expanding field of fuzzy decision theory.

Jun Ye and Yanhui Guo
Editors

mathematics

Article

An Intelligent Expert Combination Weighting Scheme for Group Decision Making in Railway Reconstruction

Lihua Zeng [1], Haiping Ren [1,*], Tonghua Yang [2] and Neal Xiong [3]

1 School of Software, Jiangxi University of Science and Technology, Nanchang 330013, China; 9520080011@jxust.edu.cn
2 School of Vocational Education and Technology, Jiangxi Agricultural University, Nanchang 330045, China; yangth883@jxau.edu.cn
3 Department of Mathematics and Computer Science, Northeastern State University, Tahlequah, OK 74133, USA; neal.xiong@sulross.edu
* Correspondence: 9520060004@jxust.edu.cn; Tel.: +86-15979014223

Abstract: The intuitionistic fuzzy entropy has been widely used in measuring the uncertainty of intuitionistic fuzzy sets. In view of some counterintuitive phenomena of the existing intuitionistic fuzzy entropies, this article proposes an improved intuitionistic fuzzy entropy based on the cotangent function, which not only considers the deviation between membership and non-membership, but also expresses the hesitancy degree of decision makers. The analyses and comparison of the data show that the improved entropy is reasonable. Then, a new IF similarity measure whose value is an IF number is proposed. The intuitionistic fuzzy entropy and similarity measure are applied to the study of the expert weight in group decision making. Based on the research of the existing expert clustering and weighting methods, we summarize an intelligent expert combination weighting scheme. Through the new intuitionistic fuzzy similarity, the decision matrix is transformed into a similarity matrix, and through the analysis of threshold change rate and the design of risk parameters, reasonable expert clustering results are obtained. On this basis, each category is weighted; the experts in the category are weighted by entropy weight theory, and the total weight of experts is determined by synthesizing the two weights. This scheme provides a new method in determining the weight of experts objectively and reasonably. Finally, the method is applied to the evaluation of railway reconstruction scheme, and an example shows the feasibility of the method.

Keywords: intuitionistic fuzzy entropy; hesitant degree information; intuitionistic fuzzy group decision making; clustering; intuitionistic fuzzy similarity

Citation: Zeng, L.; Ren, H.; Yang, T.; Xiong, N. An Intelligent Expert Combination Weighting Scheme for Group Decision Making in Railway Reconstruction. *Mathematics* **2022**, *10*, 549. https://doi.org/10.3390/math10040549

Academic Editor: Vassilis C. Gerogiannis

Received: 10 December 2021
Accepted: 8 February 2022
Published: 10 February 2022

1. Introduction

With the characteristics of high speed, large volume, low energy consumption, little pollution, safety and reliability, railway transportation has become the main transportation mode in the modern transportation system in China (see Figures 1 and 2) [1–3] and plays an important role in the development of the national economy.

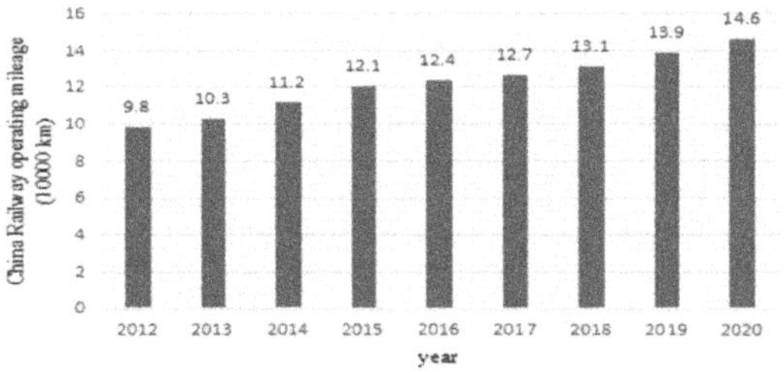

Figure 1. Business mileage of China's railways.

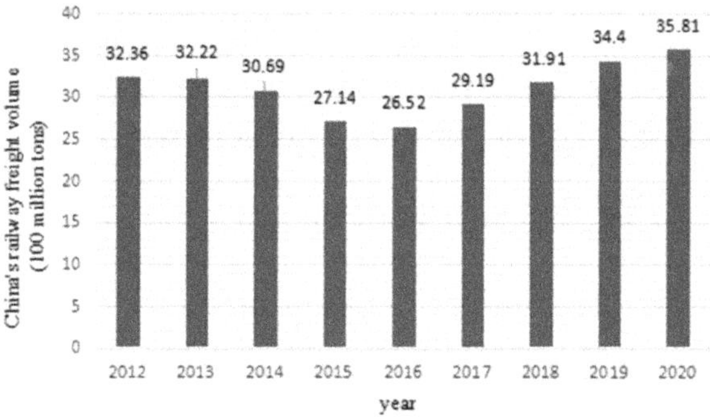

Figure 2. Total railway freight volume in China.

As an important national infrastructure and popular means of transportation, railway is the backbone of China's comprehensive transportation system. With the continuous acceleration of China's urbanization process and the urban expansion, railway construction has entered a period of rapid development, and the railway plays an increasingly important role in people's choice of travel mode (see Figure 3) [1,4].

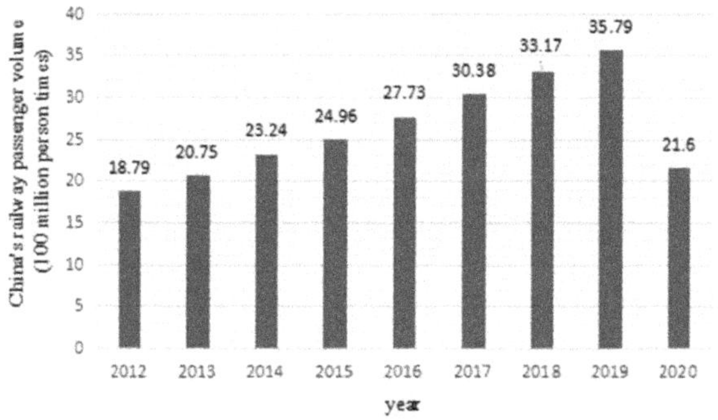

Figure 3. China railway passenger volume.

With regard to railway reconstruction, due to the huge investment and complex factors [5–7], it is necessary to compare and select various construction schemes in order to optimize the scheme with more reasonable technology and economy. Therefore, the use of scientific evaluation methods is very important. At present, the method of expert scoring and evaluation with the help of fuzzy theory has been more common, but the expert scoring is more or less subjective. This paper proposes an intelligent expert combination weighting method to optimize the scheme.

The rest of this paper is structured as follows. Section 2 introduces the related work of this study. Section 3 introduces the preparatory knowledge. Section 4 puts forward the weighted scheme of intelligent expert combination. Section 5 introduces the risk factors of the railway reconstruction project and uses the method proposed in the fourth section to optimize the railway reconstruction scheme. Finally, Section 6 summarizes the whole paper.

2. The Related Work

Fuzziness, as developed in [8], is a kind of uncertainty that often appears in human decision-making problems. Fuzzy set theory deals with uncertainties happening in daily life successfully. The membership degrees can be effectively decided by a fuzzy set. However, in real-life situations, the non-membership degrees should be considered in many cases as well. Thus, Atanassov [9] introduced the concept of an intuitionistic fuzzy (IF) set that considers both membership and non-membership degrees. IF set has been implemented in numerous areas due to its ability to handle uncertain information more effectively [10–24]. Tao et al. [10] provided an insight with an alternative queuing method and intuitionistic fuzzy set into dynamic group MCDM, which ranked the alternatives based on preference relation. Intuitionistic fuzzy sets based on the weighted average were adopted for aggregating individual suggestions of decision makers by Singh et al. [11]. Chaira [12] suggested a novel clustering approach for segmenting lesions/tumors in mammogram images using Atanassov's intuitionistic fuzzy set theory. Jiang et al. [13] studied a novel three-way group investment decision model under an intuitionistic fuzzy multi-attribute group decision-making environment. Wang et al. [14] put forward a novel three-way multi-attribute decision-making model in light of a probabilistic dominance relation with intuitionistic fuzzy sets. Wan and Dong [15] developed a new intuitionistic fuzzy best-worst method for multi-criteria decision making. Kumar et al. [16] formulated an intuitionistic fuzzy set theory-based, bias-corrected intuitionistic fuzzy c-means with spatial neighborhood information method for MRI image segmentation. In addition, intuitionistic fuzzy sets are extended to various forms and applied to practical problems. Senapati and Yager [17–19] proposed Fermatean fuzzy sets and introduced four new weighted aggregated operators, as well as defined basic operations over the Fermatean fuzzy sets. Ashraf et al. [20] introduced a new version of the picture fuzzy set, so-called spherical fuzzy sets (SFS), and discussed some operational rules. Khan et al. [21] introduced a method to solve decision-making problems using an adjustable weighted soft discernibility matrix in a generalized picture fuzzy soft set. Riaz and Hashmi [22] introduced the novel concept of the linear Diophantine fuzzy set (LDFS) with the addition of reference parameters.

Shannon used probability theory as a mathematical tool to measure information. He defined information as something that eliminates uncertainty, thus connecting information with uncertainty. Taking entropy as a measure of the uncertainty of information state, Shannon put forward the concept of information entropy. De Luca and Termini [25] studied the measurement of fuzziness of fuzzy sets, extended probabilistic information entropy to non-probabilistic information entropy and proposed axioms that fuzzy information entropy must satisfy. Szmidt and Kacprzyk [26] extended the axioms of De Luca and Termini and extended fuzzy information entropy to IF information entropy. Some scholars have conducted in-depth research in this aspect and constructed IF entropy formulae from different angles and applied it to the fields of multi-attribute decision making and pattern recognition [27–34]. Whether these entropy formulas can reasonably measure the uncertainty of IF sets is directly related to the rationality of their application. In this paper, some entropy formulas in existing literature are classified, and their advantages and disadvantages are analyzed with data. On this basis, a new IF entropy is constructed that not only considers the deviation between membership and non-membership but also includes the hesitancy in the entropy measure. The rationality of entropy is fully explained by data analysis and comparison.

In recent years, the decision-making problem with IF information has attracted many scholars' attention [35]. Due to the complexity and uncertainty of pragmatic problems, expert group decision-making method is commonly used in decision-making problems. Expert group decision making can fully gather the experience and knowledge of various experts, making the decision-making results more scientific and reasonable. However, in the actual evaluation, experts in group decision making are influenced by numerous factors, such as knowledge structure, understanding of scheme, interest correlation and so on. They often hold different views and attitudes. How to determine the weight of experts and effectively aggregate the decision-making information of experts with different preferences has become the focus of scholars [36–41].

In traditional group decision making, the expert weighting method usually uses the consistency ratio of the judgment matrix to construct the weight coefficient, which lacks the attention to the overall consistency of group decision-making objectives. In order to surmount the shortcomings of the traditional method, a cluster analysis method is often used to realize the expert weighting in group decision making. The basic principle of expert cluster analysis is to measure the similarity degree of expert evaluation opinions according to certain standards and cluster experts based on the similarity degree. He and Lei [36] extended fuzzy C-means clustering to IF C-means clustering and proposed a clustering algorithm based on IF sets. Zhang et al. [37] and He et al. [38] proposed the concept of IF similarity, whose value is an IF number; they also constructed the IF similarity matrix, the IF equivalent matrix and its λ- cut matrix and gave a clustering method based on the IF similarity matrix. Wang et al. [39] proposed a new method of an IF similar matrix, avoided the tedious process of calculating an IF equivalent matrix and used the membership degrees of elements in an IF similar matrix to cluster. Zhou et al. [40] conducted cluster analysis on experts according to the principle of entropy, used information similarity coefficients to measure the similarity degrees of expert opinions and then classified the experts.

The above clustering methods have the following problems when clustering IF information.

(1) In reference [36], the clustering results of IF sets are expressed in real numbers, which does not accord with the characteristics of IF sets.
(2) After obtaining the IF similarity matrix, the method proposed in reference [37,38] also needs to test whether it is an IF equivalent matrix. If not, it needs a lot of iterative operations until it becomes an IF equivalent matrix, which requires a large amount of calculation.
(3) Reference [39] reduced the amount of calculation, but after obtaining the IF similarity matrix, only membership degree is used for clustering, ignoring non-membership degree and hesitation degree, which will inevitably cause the loss of information.
(4) In literature [40], there is no analysis on the value of the clustering threshold. The value directly affects the clustering results, so the rationality of the value is particularly important.

Considering the above situation, this paper proposes a method of clustering and weighting experts based on IF entropy. According to the evaluation information of IF numbers given by experts, a new IF similarity measure is constructed, whose value is an IF number. Then the decision matrix is transformed into a similar matrix. By analyzing the change rate of the threshold and designing the risk parameters, the decision maker can choose the appropriate clustering threshold and risk parameters so as to obtain the reasonable expert clustering results, and based on this result, experts are weighted between categories. It can make more experts in a category, so that the weight of the category is greater, which reflects the important principle of the minority obeying the majority in group decision making. Using the new IF entropy proposed in this paper, the experts in the same category with clear logic and an accurate evaluation can get a larger weight. The total weight of experts is determined by synthesizing the weight between categories and within categories. Finally, the IF weighted aggregation operator is used to aggregate weighted experts and their IF information, and the alternatives are optimized and sorted.

3. Preliminaries

In the following part, we introduce some basic concepts, which will be used in the next sections.

Definition 1 ([9]). *Let X be a given universal set. An IF set is an object having the form $A = \{< x_i, \mu_A(x_i), \nu_A(x_i) > | x_i \in X\}$ where the function $\mu_A : X \to [0,1]$ defines the degree of membership, and $\nu_A : X \to [0,1]$ defines the degree of non-membership of the element $x_i \in X$, respectively, and for every $x_i \in X$, it holds that $0 \le \mu_A(x_i) + \nu_A(x_i) \le 1$. Furthermore, for any IF set A and $x_i \in X$, $\pi_A(x_i) = 1 - \mu_A(x_i) - \nu_A(x_i)$ is called the hesitancy degree of x_i. All IF sets on X are denoted as $IFSs(X)$.*

For simplicity, Xu and Chen [41] denoted $\alpha = (\mu_\alpha, \nu_\alpha)$ as an IF number (IFN), where μ_α and ν_α are the degree of membership and the degree of non-membership of the element $\alpha \in X$ to A, respectively.

The basic operational laws of IF set defined by Atanassov [9] are introduced as follows:

Definition 2 ([9]). *Let* $A = \{< x_i, \mu_A(x_i), \nu_A(x_i) > |x_i \in X\}$ *and* $B = \{ < x_i, \mu_B(x_i), \nu_B(x_i) > |x_i \in X\}$ *be two IF sets; then,*

(1) $A \subseteq B$ *if and only if* $\mu_A(x_i) \leq \mu_B(x_i)$ *and* $\nu_A(x_i) \geq \nu_B(x_i)$ *for all* $x_i \in X$;
(2) $A = B$ *if and only if* $A \subseteq B$ *and* $B \subseteq A$;
(3) *The complementary set of* A, *denoted by* A^C, *is* $A^C = \{< x_i, \nu_A(x_i), \ >|x_i \in X\}$
(4) $A^n = \{< x_i, [\mu_A(x_i)]^n, [\nu_A(x_i)]^n > |x_i \in X\}$;
(5) $A \prec B$ *called* A *less fuzzy than* B, *i.e., for* $\forall x_i \in X$

if $\mu_B(x_i) \leq \nu_B(x_i)$, *then* $\mu_A(x_i) \leq \mu_B(x_i), \nu_A(x_i) \geq \nu_B(x_i)$;
if $\mu_B(x_i) \geq \nu_B(x_i)$, *then* $\mu_A(x_i) \geq \mu_B(x_i), \nu_A(x_i) \leq \nu_B(x_i)$.

Definition 3 ([9]). *Let* $A = \{< x_i, \mu_A(x_i), \nu_A(x_i) > |x_i \in X\}$ *and* $B = \{ < x_i, \mu_B(x_i), \nu_B(x_i) > |x_i \in X\}$ *be two IF sets and* $\omega = (\omega_1, \omega_2, \cdots, \omega_n)^T$ *be the weight vector of the element* $x_i (i = 1, 2, \cdots, n)$, *where* $\omega_j \geq 0$ *and* $\sum_{j=1}^{n} \omega_j = 1$. *The weighted Hamming distance for* A *and* B *is defined as follows:*

$$d(A, B) = \frac{1}{2}\sum_{i=1}^{n} \omega_i(|\mu_A(x_i) - \mu_B(x_i)| + |\nu_A(x_i) - \nu_B(x_i)| + |\pi_A(x_i) - \pi_B(x_i)|).$$

Definition 4 ([26]). *A map* $E : IFSs(X) \to [0, 1]$ *is called the IF entropy if it satisfies the following properties:*

(1) $E(A) = 0$ *if and only if* A *is a crisp set;*
(2) $E(A) = 1$ *if and only if* $\mu_A(x_i) = \nu_A(x_i), \forall x_i \in X$;
(3) $E(A) = E(A^C)$;
(4) *If* $A \prec B$, *then* $E(A) \prec E(B)$.

Definition 5 ([37]). *Let* $z_{ij} (i = 1, 2, \cdots, m; j = 1, 2, \cdots, n)$ *be a collection of IFNs, and the matrix* $Z = (z_{ij})_{m \times n}$ *is called an IF matrix.*

Definition 6 ([37]). *Let* $\psi : IFSs(X) \times IFSs(X) \to IFNs$ *and* C_1, C_2, C_3 *be three IF sets.* $\psi(C_1, C_2)$ *is called an IF similarity measure of* C_1 *and* C_2 *if it satisfies the following properties:*

(1) $\psi(C_1, C_2)$ *is an IFN;*
(2) $\psi(C_1, C_2) =< 1, 0 >$ *if and only if* $C_1 = C_2$;
(3) $\psi(C_1, C_2) = \psi(C_2, C_1)$;
(4) *If* $C_1 \subseteq C_2 \subseteq C_3$, *then* $\psi(C_1, C_3) \subseteq \psi(C_1, C_2)$, *and* $\psi(C_1, C_3) \subseteq \psi(C_2, C_3)$.

Definition 7 ([42]). *The membership degree* $\mu_i(x_j)$ *is expressed as* μ_{ij}, *and the non-membership degree* $v_i(x_j)$ *is expressed as* v_{ij}. *If an IF matrix* $Z = (a_{ij})_{m \times n}$ *where* $a_{ij} =< \mu_{ij}, v_{ij} >$ *satisfies the following conditions:*

(1) *Reflexivity:* $a_{ii} =< 1, 0 >, i = 1, 2, \cdots, m$.
(2) *Symmetry:* $a_{ij} = a_{ji}, i = 1, 2, \cdots, m, j = 1, 2, \cdots, n$.

then Z *is called an IF similarity matrix.*

In order to compare the magnitudes of two IF sets, Xu and Yager [43] introduced the score and accuracy functions for IF sets and gave a simple comparison law as follows:

Definition 8 ([43]). *Let $A =< \mu, \nu >$ be an IFN; the score function $M(A)$ and accuracy function $\Delta(A)$ of A can be defined, respectively, as follows:*

$$\begin{cases} M(A) = \mu - \nu \\ \Delta(A) = \mu + \nu \end{cases} \quad (1)$$

Obviously, $M(A) \in [-1, 1], \Delta(A) \in [0, 1]$.

Based on the score and accuracy functions, a comparison law for IF set is introduced as below:

Let A_j and A_k be two IF sets, $M(A_j)$ and $M(A_k)$ be the scores of A_j and A_k, respectively, and $\Delta(A_j)$ and $\Delta(A_k)$ be the accuracy degrees of A_j and A_k, respectively; then,

(1) If $M(A_j) > M(A_k)$, then $A_j > A_k$.

(2) If $M(A_j) = M(A_k)$, then $\begin{cases} \Delta(A_j) = \Delta(A_k) \Rightarrow A_j = A_k \\ \Delta(A_j) < \Delta(A_k) \Rightarrow A_j < A_k \\ \Delta(A_j) > \Delta(A_k) \Rightarrow A_j > A_k \end{cases}$.

The weighted aggregation operator for an IF set developed by Xu and Yager [43] is presented as follows:

Definition 9 ([43]). *Let $A_j =< \mu_j, \nu_j > (j = 1, 2, \cdots, n)$ be a collection of IF sets, and $\omega = (\omega_1, \omega_2, \cdots, \omega_n)^T$ be the weight vector of $A_j (j = 1, 2, \cdots, n)$, where ω_j indicates the importance degree of A_j, satisfying $\omega_j \geq 0 (j = 1, 2, \cdots, n)$ and $\sum_{j=1}^{n} \omega_j = 1$, and let $f_\omega^A : F^n \to F$. If*

$$f_\omega^A(A_1, A_2, \cdots, A_n) = \sum_{j=1}^{n} \omega_j A_j =< 1 - \prod_{j=1}^{n} (1 - \mu_j)^{\omega_j}, \prod_{j=1}^{n} \nu_j^{\omega_j} > \quad (2)$$

then the function f_ω^A is called the IF weighted aggregation operator.

4. Our Proposed Intelligent Expert Combination Weighting Scheme

4.1. A New IF Entropy

The uncertainty of IF sets is embodied in fuzziness and intuitionism. Fuzziness is determined by the difference between membership and non-membership. Intuitionism is determined by its hesitation. Therefore, entropy is used as a tool to describe the uncertainty of IF sets; the difference between membership and non-membership and their hesitation should be considered at the same time. Only in this way can the degree of uncertainty be reflected more fully. Next, we will classify the existing entropy formulas according to whether they describe the fuzziness and intuitiveness of IF sets. In addition, the motivation behind the origination of fuzzy and non-standard fuzzy models is their intimacy with human thinking. Therefore, if an entropy measure does not meet some cognitive aspect, we call it a counterintuitive case.

In this section, suppose that $A = \{< x_i, \mu_A(x_i), \nu_A(x_i) > | x_i \in X, i = 1, 2, \cdots, n\}$ is an IF set.

(1) The entropy measure only describes the fuzziness of IF sets. For example, the IF entropy measure of Ye [27] is

$$E_Y(A) = \frac{1}{n} \sum_{i=1}^{n} [(\sqrt{2} \cos \frac{\mu_A(x_i) - \nu_A(x_i)}{4} \pi - 1) \times \frac{1}{\sqrt{2} - 1}]$$

The IF entropy measure of Zeng and Li [28] is

$$E_Z(A) = 1 - \frac{1}{n} \sum_{i=1}^{n} \left| \mu_A(x_i) - \nu_A(x_i) \right|.$$

The IF entropy measure of Zhang and Jiang [29] is

$$E_{ZJ}(A) = -\frac{1}{n}\sum_{i=1}^{n} \left[\frac{\mu_A(x_i)+1-\nu_A(x_i)}{2}\log_2\left(\frac{\mu_A(x_i)+1-\nu_A(x_i)}{2}\right) + \frac{\nu_A(x_i)+1-\mu_A(x_i)}{2}\log_2\left(\frac{\nu_A(x_i)+1-\mu_A(x_i)}{2}\right)\right].$$

The exponential IF entropy measure of Verma and Sharma [30] is

$$E_{VS}(A) = \frac{1}{n(\sqrt{e}-1)}\sum_{i=1}^{n}\left[\left(\frac{\mu_A(x_i)+1-\nu_A(x_i)}{2}e^{1-\frac{\mu_A(x_i)+1-\nu_A(x_i)}{2}} + \frac{\nu_A(x_i)+1-\mu_A(x_i)}{2}e^{1-\frac{\nu_A(x_i)+1-\mu_A(x_i)}{2}} - 1\right)\right].$$

Example 1. *Let* $A_1 = \{< x, 0.3, 0.4 > | x \in X\}$ *and* $A_2 = \{< x, 0.2, 0.3 > | x \in X\}$ *be two IF sets. Calculate the entropy of* A_1 *and* A_2 *with the entropy formulae* E_Y, E_Z, E_{ZJ} *and* E_{VS}.

According to the above formulae, the results are as follows:

$E_Y(A_1) = E_Y(A_2) = 0.9895$, $E_Z(A_1) = E_Z(A_2) = 0.9$,
$E_{ZJ}(A_1) = E_{ZJ}(A_2) = 0.9928$, $E_{VS}(A_1) = E_{VS}(A_2) = 0.9905$.

It can be seen that x belongs to IF sets A_1 *and* A_2*; the absolute value of deviation between membership and non-membership is equal; and the hesitation degree increases, so the uncertainty of* A_1 *is smaller than* A_2*. However, the entropy formulae* E_Y, E_Z, E_{ZJ} *and* E_{VS} *calculated the entropy of two IF sets as equal. In fact, for any IF sets* $\widetilde{A} = \{< x_i, \mu_{\widetilde{A}}(x_i), \nu_{\widetilde{A}}(x_i) > | x_i \in X\}$ *and* $\widetilde{B} = \{< x_i, \mu_{\widetilde{B}}(x_i), \nu_{\widetilde{B}}(x_i) > | x_i \in X\}$ *if* $\mu_{\widetilde{A}}(x_i) - \nu_{\widetilde{A}}(x_i)$ *for all* $x_i \in X$*, then any entropy formula E above is adopted, and all of them have* $E(\widetilde{A}) = E(\widetilde{B})$*. These are counterintuitive situations.*

(2) The entropy measure only describes the intuitionism of IF sets.

For example, we show the IF entropy measure of Burillo and Bustince [31]:

$$E_{B_1}(A) = \sum_{i=1}^{n}[1-(\mu_A(x_i)+\nu_A(x_i))] = \sum_{i=1}^{n}\pi_A(x_i)$$

$$E_{B_2}(A) = \sum_{i=1}^{n}[1-(\mu_A(x_i)+\nu_A(x_i))^{\lambda}], \lambda = 2, 3, \cdots, \infty;$$

$$E_{B_3}(A) = \sum_{i=1}^{n}[1-(\mu_A(x_i)+\nu_A(x_i))]e^{[1-(\mu_A(x_i)+\nu_A(x_i))]};$$

$$E_{B_4}(A) = \sum_{i=1}^{n}[1-(\mu_A(x_i)+\nu_A(x_i))]\sin\left(\frac{\pi}{2}(\mu_A(x_i)+\nu_A(x_i))\right);$$

Example 2. *Let* $A_3 = \{< x, 0.09, 0.41 > | x \in X\}$ *and* $A_4 = \{< x, 0.18, 0.32 > | x \in X\}$ *be two IF sets. Calculate the entropy of* A_3 *and* A_4 *with the entropy formula* E_{B_1}.

From Formula E_{B_1}*, we can get the following results:* $E_{B_1}(A_3) = E_{B_1}(A_4) = 0.5$*. For IF sets* A_3 *and* A_4*, the hesitancy degree of element x is equal, but the absolute value of the deviation between the membership degree and non-membership degree of* A_3 *is greater than that of* A_4*, so the uncertainty of* A_3 *is obviously smaller than that of* A_4*. However, the entropy formulae* E_{B_1}, E_{B_2}, E_{B_3} *and* E_{B_4} *calculated the entropy of two IF sets as equal, which is inconsistent with people's intuition. In fact, for any IF sets* $\widetilde{A} = \{< x_i, \mu_{\widetilde{A}}(x_i), \nu_{\widetilde{A}}(x_i) > | x_i \in X\}$ *and* $\widetilde{B} = \{< x_i, \mu_{\widetilde{B}}(x_i), \nu_{\widetilde{B}}(x_i) > | x_i \in X\}$*, if* $|\mu_{\widetilde{A}}(x_i) + \nu_{\widetilde{A}}(x_i)| = |\mu_{\widetilde{B}}(x_i) + \nu_{\widetilde{B}}(x_i)|$ *for all* $x_i \in X$*, then any entropy formula E above is adopted, and all of them have* $E(\widetilde{A}) = E(\widetilde{B})$.

(3) The entropy measure includes both the fuzziness and intuitionism of IF sets. However, some situations cannot be well distinguished.

For example, we show the IF entropy measure of Wang and Wang [32]:

$$E_W(A) = \frac{1}{n}\sum_{i=1}^{n} \cot\left(\frac{\pi}{4} + \frac{|\mu_A(x_i) - \nu_A(x_i)|}{4(1+\pi_A(x_i))}\pi\right)$$

The IF entropy measure of Wei et al. [33] is the following:

$$E_{WG}(A) = \frac{1}{n}\sum_{i=1}^{n} \cos\left(\frac{\mu_A(x_i)-\nu_A(x_i)}{2(1+\pi_A(x_i))}\pi\right)$$

Example 3. *Let* $A_5 = \{< x, 0.2, 0.5 > | x \in X\}$ *and* $A_6 = \{< x, 0.4, 0.04 > | x \in X\}$ *be two IF sets. Obviously, the fuzziness of* A_5 *is greater than that of* A_6*. Calculate the entropies of* A_5 *and* A_6 *with the entropy formulae* E_W *and* E_{WG}*.*

We can get the following results:

$$E_W(A_5) = E_W(A_6) = 0.6903, \; E_{WG}(A_5) = E_{WG}(A_6) = 0.9350$$

which are counterintuitive.

For example, the IF entropy measure of Liu and Ren [34] is

$$E_{LR}(A) = \frac{1}{n}\sum_{i=1}^{n} \cos\frac{{\mu_A}^2(x_i) - {\nu_A}^2(x_i)}{2}\pi$$

Example 4. *Let* $A_7 = \{< x, 0.2, 0.4 > | x \in X\}$ *and* $A_8 = \{< x, 0.4272, 0.25 > | x \in X\}$ *be two IF sets. Obviously, the fuzzinesses of* A_7 *and* A_8 *are not equal. However, calculating the entropy of* A_7 *and* A_8 *with the entropy formula* E_{LR}*, we have* $E_{LR}(A_7) = E_{LR}(A_8) = 0.9823$*.*

Motivation: we can see that some existing cosine and cotangent function-based entropy measures have no ability to discriminate some IF sets, and there are counterintuitive phenomena, such as the cases of Example 1 to 4. In this paper, we are also devoted to the development of IF entropy measures. We propose a new intuitionistic fuzzy entropy based on a cotangent function, which is an improvement of Wang's entropy [32], as follows:

$$E_{RZ}(A) = \frac{1}{n}\sum_{i=1}^{n} \cot\left(\frac{\pi}{4} + \frac{|\mu_A(x_i)-\nu_A(x_i)|}{4+\pi_A(x_i)}\pi\right) \tag{3}$$

which not only considers the deviation between membership and non-membership degrees $\mu_A(x_i) - \nu_A(x_i)$, but also considers the hesitancy degree $\pi_A(x_i)$ of the IF set.

Theorem 1. *The measure given by Equation (3) is an IF entropy.*

Proof. To prove the measure $E_{RZ}(A)$ given by Equation (3) is an IF entropy, we only need to prove it satisfies the properties in Definition 4. Obviously, for every x_i, we have:

$$0 \leq \frac{|\mu_A(x_i) - \nu_A(x_i)|}{4 + \pi_A(x_i)}\pi \leq \frac{\pi}{4},$$

then

$$0 \leq \cot\left(\frac{\pi}{4} + \frac{|\mu_A(x_i) - \nu_A(x_i)|}{4 + \pi_A(x_i)}\pi\right) \leq 1$$

Thus, we have $0 \leq E_{RZ}(A) \leq 1$.

(i) Let A be a crisp set, i.e., for $\forall x_i \in X$, we have $\mu_A(x_i) = 1, \nu_A(x_i) = 0$ or $\mu_A(x_i) = 0, \nu_A(x_i) = 1$. It is obvious that $E_{RZ}(A) = 0$.

If $E_{RZ}(A) = 0$, i.e., $E_{RZ}(A) = \frac{1}{n}\sum_{i=1}^{n} \cot(\frac{\pi}{4} + \frac{|\mu_A(x_i)-\nu_A(x_i)|}{4+\pi_A(x_i)}\pi) = 0$, then $\forall x_i \in X$, we have $\sum_{i=1}^{n} \cot(\frac{\pi}{4} + \frac{|\mu_A(x_i)-\nu_A(x_i)|}{4+\pi_A(x_i)}\pi) = 0$.

Thus $\frac{|\mu_A(x_i)-\nu_A(x_i)|}{4+\pi_A(x_i)} = \frac{1}{4}$, amd then we have $\mu_A(x_i) = 1$ $\nu_A(x_i) = 0$ or $\mu_A(x_i) = 0, \nu_A(x_i) = 1$. Therefore, A is a crisp set.

(ii) Let $\mu_A(x_i) = \nu_A(x_i), \forall x_i \in X$; according to Equation (3), we have $E_{RZ}(A) = \frac{1}{n}\sum_{i=1}^{n} \cot(\frac{\pi}{4}) = 1$.

Now we assume that $E_{RZ}(A) = 1$; then for all $x_i \in X$, we have: $\cot(\frac{\pi}{4} + \frac{|\mu_A(x_i)-\nu_A(x_i)|}{4+\pi_A(x_i)}\pi) = 1$, then $|\mu_A(x_i) - \nu_A(x_i)| = 0$, and we can obtain the conclusion $\mu_A(x_i) = \nu_A(x_i)$ for all $x_i \in X$.

(iii) By $A^C = \{< x_i, \nu_A(x_i), \mu_A(x_i) > | x_i \in X\}$ and Equation (3), we have:

$$E_{RZ}(A^C) = \frac{1}{n}\sum_{i=1}^{n} \cot(\frac{\pi}{4} + \frac{|\nu_A(x_i) - \mu_A(x_i)|}{4 + \pi_A(x_i)}\pi) = E_{RZ}(A).$$

(iv) Construct the function:

$$f(x,y) = \cot(\frac{\pi}{4} + \frac{|x-y|}{5-(x+y)}\pi), \text{ where } x, y \in [0,1].$$

Now, when $x \le y$, we have $f(x,y) = \cot(\frac{\pi}{4} + \frac{y-x}{5-(x+y)}\pi)$; we need to prove that the function $f(x,y)$ is increasing with x and decreasing with y.

We can easily derive the partial derivatives of $f(x,y)$ to x and to y, respectively:

$$\frac{\partial f}{\partial x} = -\csc^2(\frac{\pi}{4} + \frac{y-x}{5-(x+y)}\pi) \cdot \frac{(2y-5)\pi}{[5-(x+y)]^2}$$

$$\frac{\partial f}{\partial y} = -\csc^2(\frac{\pi}{4} + \frac{y-x}{5-(x+y)}\pi) \cdot \frac{(5-2x)\pi}{[5-(x+y)]^2}$$

When $x \le y$, we have $\frac{\partial f}{\partial x} \ge 0$, $\frac{\partial f}{\partial y} \le 0$; then, $f(x,y)$ is increasing with x and decreasing with y; thus, when $\mu_B(x_i) \le \nu_B(x_i)$ and $\mu_A(x_i) \le \mu_B(x_i), \nu_A(x_i) \ge \nu_B(x_i)$ are satisfied, we have $f(\mu_A(x_i), \nu_A(x_i)) \le f(\mu_B(x_i), \nu_B(x_i))$.

So $\cot(\frac{\pi}{4} + \frac{|\mu_A(x_i)-\nu_A(x_i)|}{4+\pi_A(x_i)}\pi) \le \cot(\frac{\pi}{4} + \frac{|\mu_B(x_i)-\nu_B(x_i)|}{4+\pi_B(x_i)}\pi)$, that is, $E_{RZ}(A) \prec E_{RZ}(B)$ holds.

Similarly, we can prove that when $x \ge y$, $\frac{\partial f}{\partial x} \le 0$, $\frac{\partial f}{\partial y} \ge 0$, then $f(x,y)$ is decreasing with x and increasing with y, thus when $\mu_B(x_i) \ge \nu_B(x_i)$ and $\mu_A(x_i) \ge \mu_B(x_i), \nu_A(x_i) \le \nu_B(x_i)$ is satisfied, so we have $f(\mu_A(x_i), \nu_A(x_i)) \le f(\mu_B(x_i), \nu_B(x_i))$.

Therefore, if $A \prec B$, we have $\frac{1}{n}\sum_{i=1}^{n} f(\mu_A(x_i), \nu_A(x_i)) \le \frac{1}{n}\sum_{i=1}^{n} f(\mu_B(x_i), \nu_B(x_i))$, i.e., $E_{RZ}(A) \prec E_{RZ}(B)$. □

From Equation (3), the entropies of $A_1, A_2, A_3, A_4, A_5, A_6, A_7$ and A_8 in Examples 1 to 4 can be obtained as follows:

$$\begin{aligned}
&E_{RZ}(A_1) = 0.8634, E_{RZ}(A_2) = 0.8694, E_{RZ}(A_1) \prec E_{RZ}(A_2).\\
&E_{RZ}(A_3) = 0.6298, E_{RZ}(A_4) = 0.8215, E_{RZ}(A_3) \prec E_{RZ}(A_4).\\
&E_{RZ}(A_5) = 0.6356, E_{RZ}(A_6) = 0.5959, E_{RZ}(A_5) \succ E_{RZ}(A_6).\\
&E_{RZ}(A_7) = 0.7486, E_{RZ}(A_5) = 0.7707, E_{RZ}(A_7) \prec E_{RZ}(A_8).
\end{aligned}$$

The calculation results are in agreement with our intuition.

According to the above examples, we see that the proposed entropy measure has a better performance than the entropy measures $E_Y, E_Z, E_{ZJ}, E_{VS}, E_{B_1}, E_W, E_{WG}, E_{LR}$. Furthermore, the new entropy measure considers the two aspects of the IF set (i.e., the uncertainty depicted by the derivation of membership and non-membership and the hesitancy degree reflected by the hesitation degree of the IF set), and thus the proposed entropy measure is a good entropy measure formula of the IF set.

4.2. Clustering Method of Group Decision Experts

For group decision-making problems, suppose that $X = \{x_1, x_2, \cdots, x_m\}$ is a set of m schemes, and $O = \{O_1, O_2, \cdots, O_n\}$ is a set of n decision makers. The evaluation values decision makers $O_j \in O$ to schemes $x_k \in X$ are expressed by IF number $< \mu_j(x_k), \nu_j(x_k) >$, where $\mu_j(x_k)$ and $\nu_j(x_k)$ are the membership (satisfaction) and non-membership (dissatisfaction) degrees of the decision maker $O_j \in O$ to the scheme $x_k \in X$ with respect to the fuzzy concept so that they satisfy the conditions $0 \le \mu_j(x_k) \le 1, 0 \le \nu_j(x_k) \le 1$ and $0 \le \mu_j(x_k) + \nu_j(x_k) \le 1$ $(j = 1, 2, \cdots, n; k = 1, 2, \cdots, m)$.

Thus, a group decision-making problem can be expressed by the decision matrix $O = [< \mu_{kj}, \nu_{kj} >]_{m \times n}$ as follows:

$$O = [< \mu_{kj}, \nu_{kj} >]_{m \times n} = \begin{array}{c} \\ x_1 \\ x_2 \\ \vdots \\ x_m \end{array} \begin{array}{c} \begin{array}{cccc} O_1 & O_2 & \cdots & O_n \end{array} \\ \left[\begin{array}{cccc} < \mu_{11}, \nu_{11} > & < \mu_{12}, \nu_{12} > & \cdots & < \mu_{1n}, \nu_{1n} > \\ < \mu_{21}, \nu_{21} > & < \mu_{22}, \nu_{22} > & \cdots & < \mu_{2n}, \nu_{2n} > \\ & \cdots & \cdots & \\ < \mu_{m1}, \nu_{m1} > & < \mu_{m2}, \nu_{m2} > & \cdots & < \mu_{mn}, \nu_{mn} > \end{array} \right]_{m \times n} \end{array}$$

4.2.1. A New IF Similarity Measure

To measure the similarities among any form of data is an important topic [44,45]. The measures used to find the resemblance between data is called a similarity measure. It has different applications in classification, medical diagnosis, pattern recognition, data mining, clustering [46], decision making and image processing. Khan et al. [47] proposed a newly similarity measure for a q-rung orthopair fuzzy set based on a cosine and cotangent function. Chen and Chang [48] proposed a new similarity measure between Atanassov's intuitionistic fuzzy sets (AIFSs) based on transformation techniques and applied the proposed similarity measure between AIFSs to deal with pattern recognition problems. Beliakov et al. [49] presented a new approach for defining similarity measures for AIFSs and applied it to image segmentation. Lohani et al. [50] presented a novel probabilistic similarity measure (PSM) for AIFSs and developed the novel probabilistic λ-cutting algorithm for clustering. Liu et al. [51] proposed a new intuitionistic fuzzy similarity measure, introduced it into intuitionistic fuzzy decision system and proposed an intuitionistic fuzzy three branch decision method based on intuitionistic fuzzy similarity. Mei [52] constructed a similarity model between intuitionistic fuzzy sets and applied it to dynamic intuitionistic fuzzy multi-attribute decision making.

At present, most of the existing similarity measures are expressed in real numbers, which is not in line with the characteristics of intuitionistic fuzzy sets. In this section, we define a new IF similarity measure whose value is an IF number.

For any two experts O_j and O_k, let

$$X_p(O_j, O_k) = \sqrt[p]{\sum_{i=1}^{m} w_i(\nu_{ij} - \nu_{ik})^p} \text{and} M_p(O_j, O_k) = \sqrt[p]{\sum_{i=1}^{m} w_i(\mu_{ij} - \mu_{ik})^p},$$

where w_i is the weight of scheme x_i for all $i \in \{1, 2, \cdots, m\}$ and $\sum_{i=1}^{m} w_i = 1$ and $p \ge 1$ is a parameter.

Let

$$\overline{\mu}_{jk} = 1 - \max\{X_p(O_j, O_k), M_p(O_j, O_k)\},$$

$$\overline{\nu}_{jk} = \min\{X_p(O_j, O_k), M_p(O_j, O_k)\}$$

Theorem 2. *Let O_j and O_k be two IF sets; then,*

$$\psi(O_j, O_k) = < \overline{\mu}_{jk}, \overline{\nu}_{jk} > \tag{4}$$

is the IF similarity measure of O_j and O_k.

Proof. To prove the measure given by Equation (4) is an IF similarity measure of O_j and O_k, we only need to prove that it satisfies the properties in Definition 6.

First, we prove that $\psi(O_j, O_k)$ is the form of an IFN.

Because $0 \leq X_p(O_j, O_k) = \sqrt[p]{\sum_{i=1}^{m} w_i(\nu_{ij} - \nu_{ik})^p} \leq 1$ and $0 \leq M_p(O_j, O_k) = \sqrt[p]{\sum_{i=1}^{m} w_i(\mu_{ij} - \mu_{ik})^p} \leq 1$, so $0 \leq 1 - \max\{X_p(O_j, O_k), M_p(O_j, O_k)\} \leq 1, 0 \leq \min\{X_p(O_j, O_k), M_p(O_j, O_k)\} \leq 1$ and $\overline{\mu}_{jk} + \overline{\nu}_{jk} \leq 1$. This proves that $\psi(O_j, O_k)$ is the form of an IFN.

Let $\psi(O_j, O_k) = < \overline{\mu}_{jk}, \overline{\nu}_{jk} > = < 1, 0 >$; we have

$$\overline{\mu}_{jk} = 1 - \max\{X_p(O_j, O_k), M_p(O_j, O_k)\} = 1$$

And $\overline{\nu}_{jk} = \min\{X_p(O_j, O_k), M_p(O_j, O_k)\} = 0$, so $X_p(O_j, O_k) = M_p(O_j, O_k)$. Because of the arbitrariness of w_i, we get $\mu_{ij} = \mu_{ik}$ and $\nu_{ij} = \nu_{ik}$ for all $i \in \{1, 2, \cdots, m\}$, that is, $O_j = O_k$.

Now we assume that $O_j = O_k$; then for all $i \in \{1, 2, \cdots, m\}$, we have $\mu_{ij} = \mu_{ik}, \nu_{ij} = \nu_{ik}$; we can obtain $X_p(O_j, O_k) = M_p(O_j, O_k) = 0$ and $\overline{\mu}_{jk} = 1 - \max\{X_p(O_j, O_k), M_p(O_j, O_k)\} = 1, \overline{\nu}_{jk} = \min\{X_p(O_j, O_k), M_p(O_j, O_k)\} = 0$, that is, $\psi(O_j, O_k) = < 1, 0 >$.

Property 3 clearly holds.

If $O_1 \subseteq O_2 \subseteq O_3$, i.e.,$\mu_{i1} \leq \mu_{i2} \leq \mu_{i3}, \nu_{i1} \geq \nu_{i2} \geq \nu_{i3}$ for all $i \in \{1, 2, \cdots, m\}$, then $(\mu_{i1} - \mu_{i2})^p \leq (\mu_{i1} - \mu_{i3})^p, (\nu_{i1} - \nu_{i2})^p \leq (\nu_{i1} - \nu_{i3})^p$ for all $i \in \{1, 2, \cdots, m\}$.

We have $X_p(O_1, O_2) \leq X_p(O_1, O_3)$ and $M_p(O_1, O_2) \leq M_p(O_1, O_3)$; therefore, $\overline{\mu}_{12} \geq \overline{\mu}_{13}$, and $\overline{\nu}_{12} \leq \overline{\nu}_{13}$, that is,$\psi(O_1, O_3) \subseteq \psi(O_1, O_2)$. Similarly, it can be proved that $\psi(O_1, O_3) \subseteq \psi(O_2, O_3)$.

This theorem is proved. □

For IF similarity measure Equation (4), since each scheme is equal, this paper takes $p = 2, w_i = \frac{1}{m}$ for all $i \in \{1, 2, \cdots, m\}$. Using this formula, the IF decision matrix $O = [< \mu_{kj}, \nu_{kj} >]_{m \times n}$ can be transformed into the IF similar matrix $Z = (z_{jk})_{n \times n}$, where $z_{jk} = \psi(O_j, O_k) = < \overline{\mu}_{jk}, \overline{\nu}_{jk} >$ is an IFN.

The IF decision matrix can be transformed into the IF similarity matrix $Z = (z_{jk})_{n \times n}$ by using the IF similarity formula proposed in this paper, where $z_{jk} = \psi(O_j, O_k) = < \overline{\mu}_{jk}, \overline{\nu}_{jk} >$ is an IFN.

People's pursuit of risk varies from person to person. Let $\beta \in [0, 1]$ be the risk factor; then the IF similarity matrix $Z = (z_{jk})_{n \times n}$ can be transformed into a real matrix $R = (r_{jk})_{n \times n}$ where $r_{jk} = \overline{\mu}_{jk} + \beta(1 - \overline{\mu}_{jk} - \overline{\nu}_{jk})$.

$$R = (r_{jk})_{n \times n} = \begin{bmatrix} r_{11} & r_{12} & \cdots & r_{1n} \\ r_{21} & r_{22} & \cdots & r_{2n} \\ & \cdots & \cdots & \\ r_{n1} & r_{n2} & \cdots & r_{nn} \end{bmatrix}$$

4.2.2. Threshold Change Rate Analysis Method

The method of Zhou et al. [40] is adopted in this section.

Let the clustering threshold $\theta = \theta_t$, where $\theta_t \in [0,1]$. If

$$r_{jk} \geq \theta_t, j \neq k \tag{5}$$

then elements O_k and O_j are considered to have the same properties. The closer the threshold is to 1, the finer the classification is.

In Zhou et al. [40], the selection of the optimal clustering threshold θ_i can be determined by analyzing the change rate C_i of θ_i. The rate of change C_i is given as follows:

$$C_i = \frac{\theta_{i-1} - \theta_i}{n_i - n_{i-1}} \tag{6}$$

where i is the clustering times of θ from large to small, n_i and n_{i-1} are the number of objects in the i-th and $(i-1)$-th clustering, respectively, and θ_i and θ_{i-1} are the thresholds for the i-th and $(i-1)$-th clustering, respectively. If

$$C_i = \max_j \{C_j\} \tag{7}$$

then the threshold value of i clustering is the best.

It can be seen from Equation (5) that the greater the change rate C_i of the clustering threshold θ is, the greater the difference between the corresponding two clusters and the more obvious the boundary between classes. When C_i is the maximum value, its corresponding θ is the optimal clustering threshold value, which can make the difference between the clusters obtained by the i-th clustering to be the largest, thus realizing the purpose and significance of classification.

4.3. Analysis of Group Decision Making Expert Group Weighting

In group decision-making problems, because each expert has a different specialty, experience and preference, their evaluation information should be treated differently. In order to reflect the status and importance of each expert in decision making, it is of great significance to determine the expert weight reasonably.

Two aspects need to be considered in expert weight, namely, the weight between categories and the weight within categories. The weight between categories mainly considers the number of experts in the category of experts. For the category with large capacity, the evaluation results given by experts represent the opinions of most experts, so the corresponding categories should be given a larger weight, which reflects the principle that the minority is subordinate to the majority, while the category with smaller capacity should be given a smaller weight.

Suppose that n experts are divided into t categories; the number of experts in the i category is $\varphi_i (\varphi_i \leq n)$; and the weights between the expert categories λ_i are as follows:

$$\lambda_i = \frac{{\varphi_i}^2}{\sum\limits_{k=1}^{t} {\varphi_t}^2}, k = 1, 2, \cdots, t. \tag{8}$$

The weight of experts within the category can be measured by the information contained in an IF evaluation value given by experts. Entropy is a measure of information uncertainty and information quantity. If the entropy of the evaluation information given by an expert is smaller, the uncertainty of the evaluation information is smaller, which means that the logic of the expert is clearer; the amount of information provided is greater; and the role of the expert in the comprehensive evaluation is greater, so the expert should be given more weight. Therefore, the weight of experts within the category can be measured by IF entropy.

The evaluation vector of expert k is $O_k = (< \mu_k(x_1), \nu_k(x_1) >, \cdots, < \mu_k(x_5), \nu_k(x_5) >)$. The IF entropy corresponding to Equation (1) is expressed as follows:

$$E(k) = \frac{1}{5}\sum_{i=1}^{5} \cot\left(\frac{\pi}{4} + \frac{|\mu_k(x_i) - \nu_k(x_i)|}{4 + \pi_k(x_i)}\pi\right) \tag{9}$$

The internal weight a_{ik} of the k expert in category i is as follows:

$$a_{ik} = \frac{1 - E(k)}{\sum_{i=1}^{\varphi(i)} [1 - E(i)]} \tag{10}$$

By linear weighting λ_i and a_{ik}, the total weight of experts ω_k is obtained:

$$\omega_k = \lambda_i \cdot a_{ik}, k = 1, 2, \cdots, n. \tag{11}$$

4.4. Intelligent Expert Combination Weighting Algorithm

A cluster analysis method is often used to realize the expert weighting in group decision making. The basic principle of expert cluster analysis is to measure the similarity degree of expert evaluation opinions according to certain standards and cluster experts based on the similarity degree. In short, Figure 4 shows the general scheme of the expert clustering method.

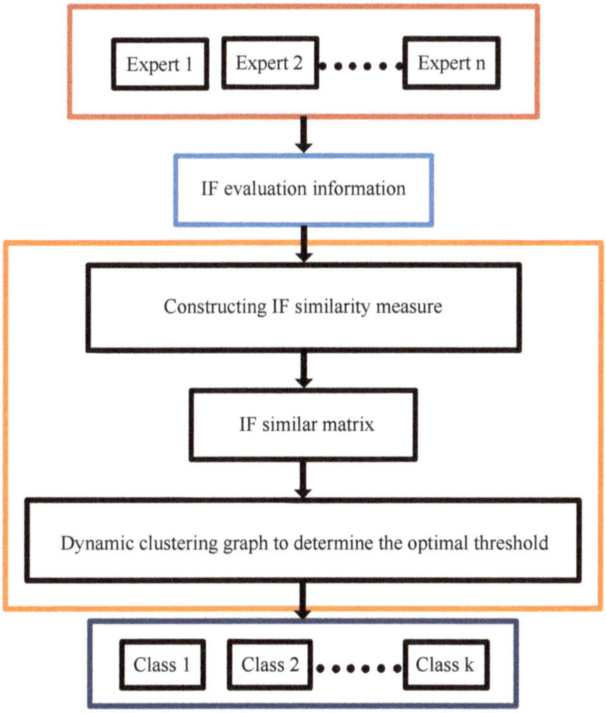

Figure 4. The general scheme of expert clustering method.

To sum up, this paper proposes an expert combination weighting scheme for group decision making, and obtains the following algorithm, which we call the intelligent expert combination weighting algorithm (see Algorithm 1).

Algorithm 1. Intelligent expert combination weighting algorithm

Input the IF decision matrix $O = [< \mu_{ij}, \nu_{ij} >]_{n\times m}$ given by experts where $I = \{1, 2, \cdots, n\}$ and $J = \{1, 2, \cdots, m\}$.
1: **For** $j \in I$ implement.
2: **For** $k \in I$ implement.
3: **For** $i \in J$ implement.
4: The IF similarity measure between experts $\psi(O_j, O_k) =< \overline{\mu}_{jk}, \overline{\nu}_{jk} >$ is calculated according to formula (4).
5: **End for**
6: Let $z_{jk} = \psi(O_j, O_k) =< \overline{\mu}_{jk}, \overline{\nu}_{jk} >$.
7: **End for**
8: **End for**
9: The IF decision matrix $O = [< \mu_{ij}, \nu_{ij} >]_{n\times m}$ is transformed into the similarity matrix $Z = (z_{jk})_{n\times n}$.
10: By selecting the risk factor β, the IF similarity matrix $Z = (z_{jk})_{n\times n}$ is transformed into the real matrix $R = (r_{jk})_{n\times n}$.
11: According to the real matrix $R = (r_{jk})_{n\times n}$, the dynamic clustering graph is drawn, and the optimal clustering threshold is determined by Formulae (6) and (7). According to this threshold, experts are classified into L categories.
12: **For** $l \in L$ implement.
13: Using Formula (8), the weight of experts between categories λ_l is determined.
14: **For** $k \in I$ implement.
15: Using Formula (8), the weight of experts between categories a_{lk} is determined.
16: Formula (11) is used to determine the total weight of experts. ω_k is calculated.
17: **End for**.
18: **End for**.
19: **For** $i \in J$ implement.
20: **For** $k \in I$ implement.
21: The weighted operator (2) of IF sets is used to aggregate expert IF group decision-making information.
22: **End for**.
23: According to definition 8, the scores and accuracy values of each scheme x_i are obtained.
24: **End for**.
25: **return** The results of the ranking of schemes x_i.

5. Performance Analysis

The railway is an important national infrastructure and livelihood project. It is a resource-saving and environment-friendly mode of transportation. In recent years, China's railway development has made remarkable achievements, but compared with the needs of economic and social development, other modes of transportation and advanced foreign railway technique, the railway in China is still a weak part of the whole transportation system [53,54]. In order to further accelerate railway construction, expand the scale of railway network and improve the layout structure and quality, the state promulgated the medium and long term railway network plan, which puts forward a series of railway plans, including the plan for railway reconstruction.

The railway reconstruction project is carried out under a series of communication, coordination and cooperation efforts, and the complex work is arranged in a limited work area, so it has encountered many unexpected challenges, such as carelessness or inadequate planning, which may lead to accidents and cause significant damage to life, assets, environment and society. According to literature [55], we can conclude that there are about seven types of risks in railway reconstruction projects, including financial and economic risks, contract and legal risks, subcontractor related risks, operation and safety risks, political and social risks, design risks and force majeure risks.

It is assumed that nine experts $O_i (i = 1, 2, \cdots, 9)$ form a decision-making group to rank five alternatives $x_j (j = 1, 2, 3, 4, 5)$ from the seven evaluation attributes above. Evaluation alternatives always contain ambiguity and diversity of meaning. In addition, in terms of qualitative attributes, human assessment is subjective and therefore inaccurate. In this case, an IF set is very advantageous; it can describe the decision process more accurately. IF sets are used in this study. After expert investigation and statistical analysis, we can get the satisfaction degree μ_{ij} and dissatisfaction ν_{ij} given by each expert $O_i(i = 1, 2, \cdots, 9)$ for each scheme $x_j(j = 1, 2, 3, 4, 5)$. The specific data are given in Table 1.

Table 1. Expert evaluation information on the program.

Expert	x_1	x_2	x_3	x_4	x_5
O_1	<0.43,0.45>	<0.24,0.70>	<0.57,0.40>	<0.29,0.55>	<0.25,0.60>
O_2	<0.58,0.30>	<0.37,0.52>	<0.30,0.50>	<0.55,0.35>	<0.35,0.50>
O_3	<0.31,0.61>	<0.74,0.22>	<0.70,0.25>	<0.50,0.40>	<0.70,0.20>
O_4	<0.44,0.45>	<0.31,0.60>	<0.56,0.40>	<0.31,0.52>	<0.24,0.60>
O_5	<0.31,0.60>	<0.70,0.20>	<0.75,0.20>	<0.60,0.30>	<0.68,0.20>
O_6	<0.70,0.20>	<0.58,0.32>	<0.52,0.40>	<0.20,0.70>	<0.60,0.30>
O_7	<0.38,0.52>	<0.72,0.21>	<0.68,0.22>	<0.61,0.30>	<0.70,0.22>
O_8	<0.41,0.40>	<0.28,0.60>	<0.55,0.35>	<0.30,0.55>	<0.26,0.60>
O_9	<0.56,0.34>	<0.40,0.50>	<0.30,0.40>	<0.71,0.10>	<0.38,0.45>

The calculation steps of the proposed method are given as follows:

Step 1. According to Equation (4), the IF similarity matrix Z is obtained as follows:

$$Z = \begin{bmatrix} <1,0> & <0.805,0.152> & <0.675,0.304> & <0.953,0.033> & <0.672,0.327> & <0.746,0.253> & <0.668,0.311> & <0.945,0.023> & <0.752,0.235> \\ & <1,0> & <0.685,0.261> & <0.821,0.125> & <0.685,0.274> & <0.758,0.211> & <0.706,0.246> & <0.816,0.133> & <0.876,0.075> \\ & & <1,0> & <0.694,0.271> & <0.946,0.051> & <0.751,0.245> & <0.938,0.060> & <0.691,0.276> & <0.689,0.255> \\ & & & <1,0> & <0.689,0.294> & <0.762,0.229> & <0.688,0.276> & <0.966,0.022> & <0.768,0.210> \\ & & & & <1,0> & <0.722,0.277> & <0.954,0.038> & <0.686,0.298> & <0.698,0.245> \\ & & & & & <1,0> & <0.745,0.250> & <0.755,0.216> & <0.705,0.286> \\ & & & & & & <1,0> & <0.683,0.279> & <0.720,0.220> \\ & & & & & & & <1,0> & <0.763,0.220> \\ & & & & & & & & <1,0> \end{bmatrix}$$

Step 2. By selecting the risk factor $\beta = 0.5$, i.e., moderate risk, the real matrix R is obtained.

$$R = \begin{bmatrix} 1 & 0.827 & 0.686 & 0.96 & 0.673 & 0.747 & 0.679 & 0.961 & 0.759 \\ & 1 & 0.712 & 0.848 & 0.706 & 0.774 & 0.73 & 0.842 & 0.901 \\ & & 1 & 0.712 & 0.948 & 0.753 & 0.939 & 0.708 & 0.717 \\ & & & 1 & 0.700 & 0.767 & 0.706 & 0.972 & 0.779 \\ & & & & 1 & 0.723 & 0.958 & 0.694 & 0.727 \\ & & & & & 1 & 0.748 & 0.770 & 0.710 \\ & & & & & & 1 & 0.702 & 0.75 \\ & & & & & & & 1 & 0.772 \\ & & & & & & & & 1 \end{bmatrix}$$

Step 3. According to Equation (5), let i take all the values in turn to get a series of classifications, and then draw a dynamic clustering graph according to Equations (5) and (6), as shown in Figure 5.

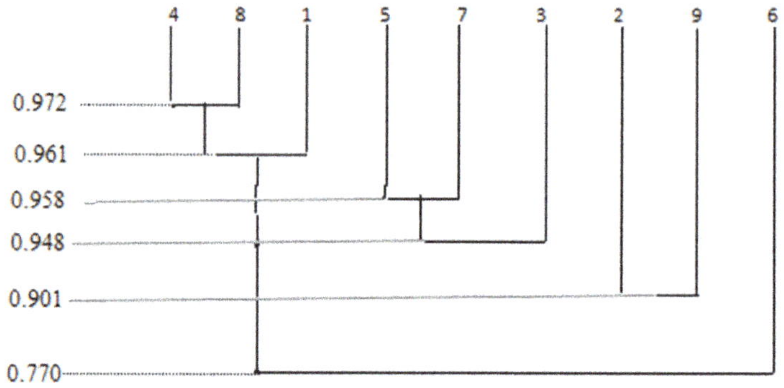

Figure 5. Dynamic clustering graph.

According to Equation (6), we have

$$C_1 = \frac{1-0.972}{2-0} = 0.014, C_2 = \frac{0.972-0.961}{3-2} = 0.011, C_3 = \frac{0.961-0.958}{5-3} = 0.0015,$$
$$C_4 = \frac{0.958-0.948}{6-5} = 0.01, C_5 = \frac{0.948-0.901}{8-6} = 0.0235, C_6 = \frac{0.901-0.770}{9-8} = 0.131.$$

Since it is meaningless for each expert to become a category or all experts to be classified into one category, we do not consider C_6; then, we have $C_5 = \max\{C_1, C_2, C_3, C_4, C_5\}$.

Therefore, taking $\theta = 0.891$ as the optimal clustering threshold, the clustering result is the most reasonable and consistent with the actual situation, and the clustering results are shown in Figure 6. We can see that the corresponding clustering results are as follows:

{(1 4 8), (3 5 7), (2 9), (6)}

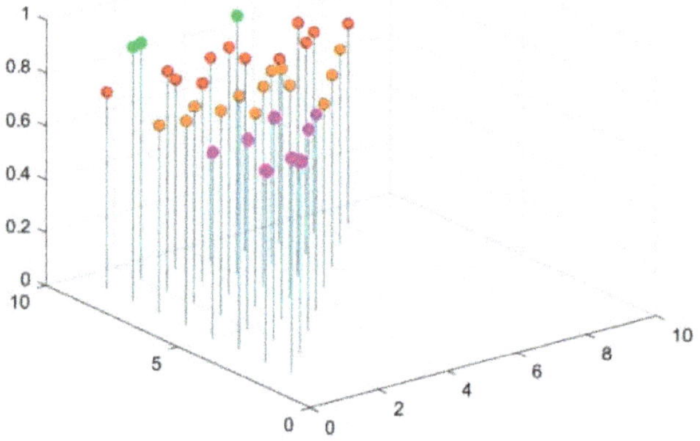

Figure 6. Clustering results.

Step 4. According to Equation (8), the weight of experts between categories is as follows:

$$\lambda_1 = 0.3913, \lambda_2 = 0.3913, \lambda_3 = 0.1739, \lambda_4 = 0.0435.$$

Step 5. According to Equation (9), the entropy vector of the expert group is obtained as follows:

$$(0.6868, 0.7405, 0.5538, 0.7364, 0.4995, 0.5935, 0.5507, 0.7159, 0.7339)$$

According to Equation (10), the weight of experts within the category is shown in Table 2.

Table 2. The weight of experts within the category.

Category	The Weight of Experts within the Category
Category 1	$a_{11} = 0.3638, a_{14} = 0.3062, a_{18} = 0.330$
Category 2	$a_{23} = 0.3196, a_{25} = 0.3585, a_{27} = 0.3219$
Category 3	$a_{32} = 0.4937, a_{39} = 0.4303$
Category 4	$a_{46} = 1$

Step 6. We weight λ_i and a_{ik} linearly to get the total weight vector ω_k of experts as follows:

$$(0.1424, 0.0859, 0.1251, 0.1198, 0.1403, 0.0435, 0.1260, 0.1291, 0.0748).$$

Step 7. According to the total weight of nine experts, the weighted aggregation operator given by Equation (2) is used to aggregate the expert information, and the comprehensive evaluation vector is obtained as follows:

$$(0.3616, 0.4504), (0.5226, 0.3878), (0.5932, 0.3218), (0.4749, 0.3853), (0.4972, 0.3718).$$

According to Equation (1), the scores and accuracy values of the comprehensive evaluation vector are calculated as follows:

$$M(x_1) = -0.089, M(x_2) = 0.1348, M(x_3) = 0.2714, M(x_4) = 0.0896, M(x_5) = 0.1254.$$
$$\Delta(x_1) = 0.812,\ \Delta(x_2) = 0.9104,\ \Delta(x_3) = 0.915,\ \Delta(x_4) = 0.8602, \Delta(x_5) = 0.869$$

Therefore, the priority of the five alternatives is $x_3 \succ x_2 \succ x_5 \succ x_4 \succ x_1$, and the optimal one is x_3.

6. Conclusions and Future Work

This article listed some counterintuitive phenomena of some existing intuitionistic fuzzy entropies. We defined an improved intuitionistic fuzzy entropy based on a cotangent function and a new IF similarity measure whose value is an IF number, applied them to the expert weight problem of group decision making and put forward the expert weight combination weighting scheme. Finally, this method was applied to a railway reconstruction case to illustrate the effectiveness of the method.

In the future, we will apply the expert weight combination weighting scheme proposed in this paper to situations in real life. We will also formulate this kind of entropy measure and similarity measures for an interval-valued IF set [56], Fermat fuzzy set, spherical fuzzy set, t-spherical fuzzy set, picture fuzzy set, single valued neutrosophic set [55,57], Plithogenic set [58] and linear fuzzy set.

While studying the theoretical method, this paper used numerical examples rather than the actual production data, which is the limitation of this paper. In the future research, we will apply the expert weight combination weighting scheme proposed in this paper to practical production problems.

Author Contributions: L.Z. and H.R. designed the method and wrote the paper; T.Y. and N.X. analyzed the data. All authors have read and agreed to the published version of the manuscript.

Funding: This work was mainly supported by the National Natural Science Foundation of China (No. 71661012) and scientific research project of the Jiangxi Provincial Department of Education (No. GJJ210827).

Institutional Review Board Statement: Not applicable.

Informed Consent Statement: Not applicable.

Data Availability Statement: The data used to support the findings of this study are included within the article.

Conflicts of Interest: The authors declared that they have no conflict of interest to this work.

References

1. Xie, Y.; Kou, Y.; Jiang, M.; Yu, W. Development and technical prospect of China railway. *High Speed Railway Technol.* **2020**, *11*, 11–16.
2. Li, Y.B. Research on the current situation and development direction of railway freight transportation in China. *Intell. City* **2019**, *5*, 133–134.
3. Fu, Z.; Zhong, M.; Li, Z. Development and innovation of Chinese railways over past century. *Chin. Rail.* **2021**, *7*, 1–7.
4. Li, Z.; Xie, R.; Sun, L.; Huang, T. A survey of mobile edge computing Telecommunications Science. *Chin. Rail.* **2014**, *2*, 9–13.
5. Lu, S.-T.; Yu, S.-H.; Chang, D.-S. Using fuzzy multiple criteria decision-making approach for assessing the risk of railway reconstruction project in Taiwan. *Sci. Word J.* **2014**, *2014*, 239793. [CrossRef]
6. Wang, M.; Xie, H. Present situation analysis and discussion on development of Chinese railway construction market. *J. Rail. Sci. Eng.* **2008**, *5*, 63–67.
7. Zhang, Z. Analysis on risks in construction of railway engineering projects and exploration for their prevention. *Rail. Stan. Desi.* **2010**, *9*, 51–52.
8. Zadeh, L. Fuzzy sets. *Inf. Control* **1965**, *8*, 338–353. [CrossRef]
9. Atanassov, K. Intuitionistic fuzzy sets. *Fuzzy Sets Syst.* **1986**, *20*, 87–96. [CrossRef]
10. Tao, P.; Liu, Z.; Cai, R.; Kang, H. A dynamic group MCDM model with intuitionistic fuzzy set: Perspective of alternative queuing method. *Inf. Sci.* **2021**, *555*, 85–103. [CrossRef]
11. Singh, M.; Rathi, R.; Antony, J.; Garza-Reyes, J. Lean six sigma project selection in a manufacturing environment using hybrid methodology based on intuitionistic fuzzy MADM approach. *IEEE Trans. Eng. Manag.* **2021**, *99*, 1–15. [CrossRef]
12. Chaira, T. An intuitionistic fuzzy clustering approach for detection of abnormal regions in mammogram images. *J. Digit. Imaging* **2021**, *34*, 428–439. [CrossRef] [PubMed]
13. Jiang, H.B.; Hu, B.Q. A novel three-way group investment decision model under intuitionistic fuzzy multi-attribute group decision-making environment. *Inf. Sci.* **2021**, *569*, 557–581. [CrossRef]
14. Wang, W.; Zhan, J.; Mi, J. A three-way decision approach with probabilistic dominance relations under intuitionistic fuzzy information. *Inf. Sci.* **2022**, *582*, 114–145. [CrossRef]
15. Wan, S.; Dong, J. A novel extension of best-worst method with intuitionistic fuzzy reference comparisons. *IEEE Trans. Fuzzy Syst.* **2021**, *99*, 1. [CrossRef]
16. Kumar, D.; Agrawal, R.; Kumar, P. Bias-corrected intuitionistic fuzzy c-means with spatial neighborhood information ap proach for human brain MRI image segmentation. *IEEE Trans. Fuzzy Syst.* **2020**. [CrossRef]
17. Senapati, T.; Yager, R. Fermatean fuzzy sets. *J. Ambient Intell. Humaniz. Comput.* **2020**, *11*, 663–674. [CrossRef]
18. Senapati, T.; Yager, R. Fermatean fuzzy weighted averaging/geometric operators and its application in multi-criteria decision-making methods—Science Direct. *Eng. Appl. Artif. Intell.* **2019**, *85*, 112–121. [CrossRef]
19. Senapati, T.; Yager, R. Some new operations over fermatean fuzzy numbers and application of fermatean fuzzy WPM in multiple criteria decision making. *Informatica* **2019**, *2*, 391–412. [CrossRef]
20. Ashraf, S.; Abdullah, S.; Mahmood, T.; Ghani, F. Spherical fuzzy sets and their applications in multi-attribute decision making problems. *J. Intell. Fuzzy Syst.* **2019**, *36*, 2829–2844. [CrossRef]
21. Khan, M.; Kumam, P.; Liu, P.; Kumam, W.; Rehman, H. An adjustable weighted soft discernibility matrix based on generalized picture fuzzy soft set and its applications in decision making. *J. Intell. Fuzzy Syst.* **2020**, *38*, 2103–2118. [CrossRef]
22. Riaz, M.; Hashmi, M. Linear Diophantine fuzzy set and its applications towards multi-attribute decision-making problems. *J. Intell. Fuzzy Syst.* **2019**, *37*, 5417–5439. [CrossRef]
23. Gao, K.; Han, F.; Dong, P.; Xiong, N.; Du, R. Connected vehicle as a mobile sensor for real time queue length at signalized intersections. *Sensors* **2019**, *19*, 2059. [CrossRef] [PubMed]
24. Zhang, Q.; Zhou, C.; Tian, Y.; Xiong, N.; Qin, Y.; Hu, B. A fuzzy probability Bayesian network approach for dynamic cybersecurity risk assessment in industrial control systems. *IEEE Trans. Ind. Inform.* **2017**, *14*, 2497–2506. [CrossRef]

25. Luca, A.; Termini, S. A definition of a nonprobabilistie entropy in the setting of fuzzy sets theory. *Inf. Control* **1972**, *3*, 301–312. [CrossRef]
26. Szmidt, E.; Kacprzyk, J. Entropy for intuitionistic fuzzy sets. *Fuzzy Sets Syst.* **2001**, *118*, 467–477. [CrossRef]
27. Ye, J. Two effective measures of intuitionistic fuzzy entropy. *Computing* **2010**, *87*, 55–62. [CrossRef]
28. Zeng, W.; Li, H. Relationship between similarity measure and entropy of interval valued fuzzy sets. *Fuzzy Sets Syst.* **2006**, *157*, 1477–1484. [CrossRef]
29. Zhang, Q.; Jiang, S. A note on information entropy measures for vague sets and its applications. *Inf. Sci.* **2008**, *178*, 4184–4191. [CrossRef]
30. Verma, R.; Sharma, B. Exponential entropy on intuitionistic fuzzy sets. *Kybernetika* **2013**, *49*, 114–127.
31. Burillo, P.; Bustince, H. Entropy on intuitionistic fuzzy sets and on interval-valued fuzzy sets. *Fuzzy Sets Syst.* **1996**, *78*, 305–316. [CrossRef]
32. Wang, J.; Wang, P. Intuitionistic linguistic fuzzy multi-criteria decision-making method based on intuitionistic fuzzy entropy. *Control Decis.* **2012**, *27*, 1694–1698.
33. Wei, C.; Gao, Z.; Guo, T. An intuitionistic fuzzy entropy measure based on trigonometric function. *Control Decis.* **2012**, *27*, 571–574.
34. Liu, M.; Ren, H. A new intuitionistic fuzzy entropy and application in multi-attribute decision making. *Informatica* **2014**, *5*, 587–601. [CrossRef]
35. Ai, C.; Feng, F.; Li, J.; Liu, K. AHP method of subjective group decision-making based on interval number judgment matrix and fuzzy clustering analysis. *Control Decis.* **2019**, *35*, 41–45.
36. He, Z.; Lei, Y. Research on intuitionistic fuzzy C-means clustering algorithm. *Control Decis.* **2011**, *26*, 847–850.
37. Zhang, H.; Xu, Z.; Chen, Q. On clustering approach to intuitionistic fuzzy sets. *Control Decis.* **2007**, *22*, 882–888.
38. He, Z.; Lei, Y.; Wang, G. Target recognition based on intuitionistic fuzzy clustering. *J. Syst. Eng. Electron.* **2011**, *6*, 1283–1286.
39. Wang, Z.; Xu, Z.; Liu, S.; Tang, J. A netting clustering analysis method under intuitionistic fuzzy environment. *Appl. Soft Comput.* **2011**, *11*, 5558–5564. [CrossRef]
40. Zhuo, X.; Zhang, F.; Hui, X.; Li, K. Method for determining experts' weights based on entropy and cluster analysis. *Control Decis.* **2011**, *26*, 153–156.
41. Xu, Z.; Chen, J. An overview of distance and similarity measures of intuitionistic fuzzy sets. *Int. J. Uncertain. Fuzz.* **2008**, *16*, 529–555. [CrossRef]
42. Zhang, Z.; Chen, S.; Wang, C. Group decision making with incomplete intuitionistic multiplicative preference relations. *Inf. Sci.* **2020**, *516*, 560–571. [CrossRef]
43. Xu, Z.S.; Yager, R.R. Some geometric aggregation operators based on intuitionistic fuzzy sets. *Int. J. Gen. Syst.* **2006**, *35*, 417–433. [CrossRef]
44. Huang, S.; Liu, A.; Zhang, S.; Wang, T.; Xiong, N. BD-VTE: A novel baseline data based verifiable trust evaluation scheme for smart network systems. *IEEE Trans. Netw. Sci. Eng.* **2020**, *8*, 2087–2105. [CrossRef]
45. Wu, M.; Tan, L.; Xiong, N. A structure fidelity approach for big data collection in wireless sensor networks. *Sensors* **2015**, *15*, 248–273. [CrossRef]
46. Li, H.; Liu, J.; Wu, K.; Yang, Z.; Liu, R.; Xiong, N. Spatio-temporal vessel trajectory clustering based on data mapping and density. *IEEE Access* **2018**, *6*, 58939–58954. [CrossRef]
47. Khan, M.; Kumam, P.; Alreshidi, N.; Kumam, W. Improved cosine and cotangent function-based similarity measures for q-rung orthopair fuzzy sets and TOPSIS method. *Complex Intell. Syst.* **2021**, *7*, 2679–2696. [CrossRef]
48. Chen, S.; Chang, C. A novel similarity measure between Atanassov's intuitionistic fuzzy sets based on transformation techniques with applications to pattern recognition. *Inf. Sci.* **2015**, *291*, 96–114. [CrossRef]
49. Beliakov, G.; Pagola, M.; Wilkin, T. Vector valued similarity measures for Atanassov's intuitionistic fuzzy sets. *Inf. Sci.* **2014**, *280*, 352–367. [CrossRef]
50. Lohani, Q.; Solanki, R.; Muhuri, P. Novel adaptive clustering algorithms based on a probabilistic similarity measure over Atanassov intuitionistic fuzzy set. *IEEE Trans. Fuzzy Syst.* **2018**, *6*, 3715–3729. [CrossRef]
51. Liu, J.; Zhou, X.; Li, H.; Huang, B.; Gu, P. An intuitionistic fuzzy three-way decision method based on intuitionistic fuzzy similarity degrees. *Syst. Eng. Theory Pract.* **2019**, *39*, 1550–1564.
52. Mei, X. Dynamic intuitionistic fuzzy multi-attribute decision making method based on similarity. *Stat. Decis.* **2016**, *15*, 22–24.
53. Tang, Y. Comparison and analysis of domestic and foreign railway energy consumption. *Rail. Tran. Econ.* **2018**, *40*, 97–103.
54. Gao, F. A study on the current situation and development strategies of China's railway restructuring. *Railw. Freight Transport.* **2020**, *38*, 15–19.
55. Chai, J.S.; Selvachandran, G.; Smarandache, F.; Gerogiannis, V.C.; Son, L.H.; Bui, Q.-T.; Vo, B. New similarity measures for single-valued neutrosophic sets with applications in pattern recognition and medical diagnosis problems. *Complex Intell. Syst.* **2021**, *7*, 703–723. [CrossRef]
56. Garg, H. Generalized intuitionistic fuzzy entropy-based approach for solving multi-attribute decision-making problems with unknown attribute weights. *Proc. Natl. Acad. Sci. USA* **2017**, *89*, 129–139. [CrossRef]
57. Majumdar, P. On new measures of uncertainty for neutrosophic sets. *Neutrosophic Sets Syst.* **2017**, *17*, 50–57.
58. Quek, S.G.; Selvachandran, G.; Smarandache, F.; Vimala, J.; Le, S.H.; Bui, Q.-T.; Gerogiannis, V.C. Entropy measures for Plithogenic sets and applications in multi-attribute decision making. *Mathematics* **2020**, *8*, 965. [CrossRef]

mathematics

MDPI

Article

Group Decision-Making Problems Based on Mixed Aggregation Operations of Interval-Valued Fuzzy and Entropy Elements in Single- and Interval-Valued Fuzzy Environments

Weiming Li [1] and Jun Ye [2,*]

[1] Yuanpei College, Shaoxing University, Shaoxing 312000, China; liweiming@usx.edu.cn
[2] School of Civil and Environmental Engineering, Ningbo University, Ningbo 315211, China
* Correspondence: yejun1@nbu.edu.cn; Tel.: +86-574-87602315

Abstract: Fuzzy sets and interval-valued fuzzy sets are two kinds of fuzzy information expression forms in real uncertain and vague environments. Their mixed multivalued information expression and operational problems are very challenging and indispensable issues in group decision-making (GDM) problems. To solve single- and interval-valued fuzzy multivalued hybrid information expression, operations, and GDM issues, this study first presents the notion of a single- and interval-valued fuzzy multivalued set/element (SIVFMS/SIVFME) with identical and/or different fuzzy values. To effectively solve operational problems for various SIVFME lengths, SIVFMS/SIVFME is converted into the interval-valued fuzzy and entropy set/element (IVFES/IVFEE) based on the mean and information entropy of SIVFME. Then, the operational relationships of IVFEEs and the expected value function and sorting rules of IVFEEs are defined. Next, the IVFEE weighted averaging and geometric operators and their mixed-weighted-averaging operation are proposed. In terms of the mixed-weighted-averaging operation and expected value function of IVFEEs, a GDM method is developed to solve multicriteria GDM problems in the environment of SIVFMSs. Finally, the proposed GDM method was utilized for a supplier selection problem in a supply chain as an actual sample to show the rationality and efficiency of SIVFMSs. Through the comparative analysis of relative decision-making methods, we found the superiority of this study in that the developed GDM method not only compensates for the defects of existing GDM methods, but also makes the GDM process more reasonable and flexible.

Keywords: single- and interval-valued fuzzy multivalued set; interval-valued fuzzy and entropy set; interval-valued fuzzy and entropy element weighted averaging operator; interval-valued fuzzy and entropy element weighted geometric operator; mixed-weighted-averaging operation; group decision making

MSC: 03E72; 91B06

Citation: Li, W.; Ye, J. Group Decision-Making Problems Based on Mixed Aggregation Operations of Interval-Valued Fuzzy and Entropy Elements in Single- and Interval-Valued Fuzzy Environments. *Mathematics* **2022**, *10*, 1077. https://doi.org/10.3390/math10071077

Academic Editor: Pasi Luukka

Received: 26 February 2022
Accepted: 25 March 2022
Published: 27 March 2022

1. Introduction

Fuzzy sets (FS) [1] and interval-valued fuzzy sets (IVFSs) [2] are two important tools of fuzzy information expressions in real uncertain and vague environments. A bag/fuzzy multiset [3,4] or an interval-valued fuzzy multiset (IVFM) [5] was proposed as the extension of FS or IVFS, where each element in a universe set can occur more times with different and/or identical fuzzy values or interval-valued fuzzy values. Therefore, they have been used in various areas [6–10]. In a hesitant situation, a hesitant fuzzy set (HFS) [11] can represent a set of a few of different fuzzy values of each element in the set. To express the hybrid information of HFS and IVFS, some researchers presented cubic HFSs and applied them to medical assessments of prostatic patients [12] and multicriteria decision-making problems [13]; then, other researchers introduced hesitant cubic fuzzy sets (HCFSs) and applied them to multicriteria (group) decision-making problems [14,15]. However, their hesitant information does not contain the same fuzzy values corresponding to the hesitant characteristics/concept [11], which is different from the fuzzy multiset concept.

Regarding the probability of an element belonging to a set, hesitant probabilistic fuzzy sets (HPFSs) [16,17] were introduced and applied to hesitant probabilistic fuzzy decision-making problems. However, an HPFS only contains the probabilistic values of a few of the same values, resulting in probabilistic distortion. Since the probabilistic method requires a lot of fuzzy data (more sample data) to maintain reasonable probabilistic values, the probabilistic values of small samples of data lead to irrationality/distortion. Therefore, it is difficult to apply the probabilistic method in actual group decision making (GDM) applications because the evaluation values of a lot of decision makers are required to ensure the rationality of the probabilistic values. Hence, it is obvious that the use of HPFSs may have some flaws from the perspective of probability.

Recently, Turkarslan et al. [18] introduced a consistency fuzzy set/element (CFS/CFE) based on the mean of a fuzzy sequence and the complement of the standard deviation of a fuzzy sequence in a fuzzy multiset to reasonably simplify the information expression and operation of different fuzzy sequence lengths, and then proposed a cosine similarity measure of CFSs for medical diagnosis in the case of fuzzy multisets. Furthermore, Du and Ye [19] presented cubic fuzzy multivalued sets (CFMSs) and converted them into cubic fuzzy consistency sets with the help of the mean of a fuzzy sequence and the complement of the standard deviation of a fuzzy sequence. Then, they developed a hybrid weighted arithmetic and geometric aggregation operator for GDM with CFMSs. In general, the concept of standard deviation is only applicable to the calculation of fuzzy sequences containing normal distributions, which exposes its limitations.

In real GDM problems, single- and interval-valued fuzzy hybrid multivalued information expression and operation problems are very challenging issues, due to the uncertainty and incompleteness of each decision-maker's judgement/cognition of the evaluated object. However, existing fuzzy multiset/HFS/HPFS/IVFM/CFMS cannot represent the single- and interval-valued fuzzy hybrid multivalued information with identical and/or different fuzzy values that are given by a group of decision makers in the GDM process. In the GDM problem, one of the experts/decision makers can assign his/her single-valued or interval-valued fuzzy evaluation value in terms of his cognition of the evaluated object in the assessment process. For example, five experts evaluate a car's "comfort" with a group of fuzzy values (0.5, 0.5, 0.6, [0.6, 0.7], [0.7, 0.8]). The fuzzy values 0.5, 0.5, and 0.6 are given by three of the five experts, and the interval-valued/uncertain fuzzy values [0.6, 0.7] and [0.7, 0.8] are given by the two of the five experts. In this issue, the existing fuzzy multiset/HFS/HPFS/IVFM/CFMS can only represent a fuzzy sequence or an interval-valued fuzzy sequence, but they cannot express such a group of single- and interval-valued fuzzy hybrid values (the hybrid set of two different fuzzy sequences) simultaneously. Meanwhile, there is no research on a single- and interval-valued fuzzy multivalued framework in the existing literature. Therefore, it is necessary to propose a new expression form to effectively express the single- and interval-valued fuzzy hybrid multivalued information to overcome the defect of existing various fuzzy expressions. Motivated by this new idea, this paper first puts forward the concept of a single- and interval-valued fuzzy multivalued set/element (SIVFMS/SIVFME). Then, a new information entropy measure of SIVFME is proposed to transform SIVFMS/SIVFME into an interval-valued fuzzy and entropy set/element (IVFES/IVFEE) based on the mean and information entropy of SIVFME, and then some operations of IVFEEs and the expected value function and sorting rules of IVFEEs are defined. Next, the IVFEE weighted averaging (IVFEEWA) and IVFEE weighted geometric (IVFEEWG) operators and their mixed-weighted-averaging operation are proposed to overcome the flaws of the IVFEEWA operator, which mainly attends to group arguments, and the IVFEEWG operator, which mainly attends to individual arguments [19], in the IVFEE aggregation process. According to the proposed mixed-weighted-averaging operation and the expected value function, a GDM method is developed to solve multicriteria GDM problems with SIVFMSs. Finally, the proposed GDM method is utilized for an actual supplier selection problem in a supply chain to show the rationality and effectiveness in the setting of SIVFMSs. The results indicate that the proposed GDM method makes the GDM process more reasonable and flexible.

This original study demonstrates the following main contributions and highlights:

(i). The proposed SIVFMS/SIVFME forms single- and interval-valued fuzzy multivalued framework to reasonably express the mixed information of the single-valued/certain fuzzy sequence and interval-valued/uncertain fuzzy sequence, which are given by different decision makers in the GDM process.

(ii). The IVFEE transformed based on the mean and information entropy of SIVFME can reasonably simplify the information expression and operation of different fuzzy sequence lengths in SIVFMEs; then, the proposed transformation method using the mean and information entropy of SIVFME can reveal the average level and consistency/consensus degree of the single- and interval-valued fuzzy sequence in SIVFME to keep much more useful information in the transformation process.

(iii). The mixed-weighted-averaging operation of the IVFEEWA and IVFEEWG operators can provide a useful modeling tool for their GDM method in the environment of SIVFMSs and overcome the flaw of having a single aggregation operator [19].

(iv). The developed GDM method can solve multicriteria GDM problems and make the decision results more flexible and more reasonable for SIVFMSs.

The remainder of this article is made up of the following structures. In Section 2, we present the concepts of SIVFMS, SIVFME, information entropy, and IVFEE. Then, we define the operational laws of IVFEEs, and the expected value function and sorting rules of IVFEEs. The IVFEEWA and IVFEEWG operators and their mixed-weighted-averaging operation are presented in Section 3. In Section 4, a GDM method is given by using the mixed-weighted-averaging operation and the expected value function. In Section 5, the proposed GDM method is applied to an actual supplier selection problem in a supply chain to show its rationality and effectiveness when dealing with SIVFMSs, and then the superiorities of the proposed method are indicated by comparative analysis. Section 6 depicts conclusions and future research.

2. SIVFMS and IVFES

Definition 1. *Let $U = \{u_1, u_2, \ldots, u_s\}$ be a finite universe set U. Then, a single- and interval-valued fuzzy multivalued set H in U is defined as follows:*

$$H = \{\langle u_k, F_H(u_k)\rangle | u_k \in U\} \tag{1}$$

where $F_H(u_k)$ for $u_k \in U$ ($k = 1, 2, \ldots, s$) is a single- and interval-valued fuzzy sequence of the element u_k in the set H, denoted as an increasing fuzzy sequence $F_H(u_k) = (\lambda_H^1(u_k), \lambda_H^2(u_k), \ldots, \lambda_H^{a_k}(u_k), [\lambda_H^{L1}(u_k), \lambda_H^{U1}(u_k)], [\lambda_H^{L2}(u_k), \lambda_H^{U2}(u_k)], \ldots, [\lambda_H^{Lb_k}(u_k), \lambda_H^{Ub_k}(u_k)])$ with identical and/or different fuzzy values, such that $0 \le \lambda_H^1(u_k) \le \lambda_H^2(u_k), \ldots, \le \lambda_H^{a_k}(u_k) \le 1$ with a_k single-valued fuzzy values and $[\lambda_H^{L1}(u_k), \lambda_H^{U1}(u_k)] \subseteq [\lambda_H^{L2}(u_k), \lambda_H^{U2}(u_k)] \subseteq, \ldots, \subseteq [\lambda_H^{Lb_k}(u_k), \lambda_H^{Ub_k}(u_k)] \subseteq [0, 1]$ with b_k interval-valued fuzzy values.

Especially when all $b_k = 0$ or $a_k = 0$ for $k = 1, 2, \ldots, s$, SIVFMS degenerates to a fuzzy multiset or an IVFM.

For simplicity, the kth element $F_H(u_k)$ in H is denoted as the kth SIVFME: $F_{Hk} = (\lambda_{Hk}^1, \lambda_{Hk}^2, \ldots, \lambda_{Hk}^{a_k}, [\lambda_{Hk}^{L1}, \lambda_{Hk}^{U1}], [\lambda_{Hk}^{L2}, \lambda_{Hk}^{U2}], \ldots, [\lambda_{Hk}^{Lb_k}, \lambda_{Hk}^{Ub_k}])$.

To solve the difficult conversions between different single- and interval-valued fuzzy sequence lengths, it is necessary to convert SIVFMS into IVFES in terms of the mean and information entropy of SIVFME.

First, the concept of the Shannon/probability entropy [20] is introduced below.

Set $R = \{r_1, r_2, \ldots, r_s\}$ as a probability distribution on a set of random variables. Thus, the Shannon entropy of the probability distribution R is denoted as [20]

$$E(R) = -\sum_{i=1}^{s} r_i \ln(r_i) \tag{2}$$

where $r_i \in [0, 1]$ and $\sum_{i=1}^{s} r_i = 1$.

If all probability values of r_i ($i = 1, 2, \ldots, s$) in R are the same, the probability entropy can reach the maximum value of $E(R)$, which reflects the perfect consistency (the same probabilities) of all r_i. Generally, the larger the probability entropy measure value, the better the consistency level of all probability values.

According to the probability entropy notion, the interval-valued entropy concept of SIVFME (an information entropy measure of SIVFME) is proposed, and SIVFMS is converted into IVFES based on the mean and information entropy of SIVFME, which is given by the following definition.

Definition 2. *An IVFES Z of a SIVFMS H in a finite universe set* $U = \{u_1, u_2, \ldots, u_s\}$ *is defined as*

$$Z = \{(u_k, m_Z(u_k), e_Z(u_k)) | u_k \in U\},$$

where $m_Z(u_k) \subseteq [0, 1]$ *and* $e_Z(u_k) \subseteq [0, 1]$ ($k = 1, 2, \ldots, s$) *are the interval-valued mean and interval-valued entropy of SIVFME, which are obtained by using the following formulae:*

$$m_Z(u_k) = [m_Z^L(u_k), m_Z^U(u_k)] = \begin{bmatrix} \frac{1}{a_k+b_k}\left(\sum_{i=1}^{a_k} \lambda_H^i(u_k) + \sum_{i=1}^{b_k} \lambda_H^{Li}(u_k)\right), \\ \frac{1}{a_k+b_k}\left(\sum_{i=1}^{a_k} \lambda_H^i(u_k) + \sum_{i=1}^{b_k} \lambda_H^{Ui}(u_k)\right) \end{bmatrix}, m_Z(u_k) \subseteq [0, 1], \tag{3}$$

$$e_Z(u_k) = [e_Z^L(u_k), e_Z^U(u_k)]$$
$$= \begin{bmatrix} \min\left(\begin{array}{l} -\frac{1}{\ln(a_k+b_k)}\left(\begin{array}{l} \sum_{i=1}^{a_k}\left(\frac{\lambda_H^i(u_k)}{\sum_{i=1}^{a_k}\lambda_H^i(u_k)+\sum_{i=1}^{b_k}\lambda_H^{Li}(u_k)}\ln\frac{\lambda_H^i(u_k)}{\sum_{i=1}^{a_k}\lambda_H^i(u_k)+\sum_{i=1}^{b_k}\lambda_H^{Li}(u_k)}\right)+ \\ \sum_{i=1}^{b_k}\left(\frac{\lambda_H^{Li}(u_k)}{\sum_{i=1}^{a_k}\lambda_H^i(u_k)+\sum_{i=1}^{b_k}\lambda_H^{Li}(u_k)}\ln\frac{\lambda_H^{Li}(u_k)}{\sum_{i=1}^{a_k}\lambda_H^i(u_k)+\sum_{i=1}^{b_k}\lambda_H^{Li}(u_k)}\right) \end{array}\right), \\ -\frac{1}{\ln(a_k+b_k)}\left(\begin{array}{l} \sum_{i=1}^{a_k}\left(\frac{\lambda_H^i(u_k)}{\sum_{i=1}^{a_k}\lambda_H^i(u_k)+\sum_{i=1}^{b_k}\lambda_H^{Ui}(u_k)}\ln\frac{\lambda_H^i(u_k)}{\sum_{i=1}^{a_k}\lambda_H^i(u_k)+\sum_{i=1}^{b_k}\lambda_H^{Ui}(u_k)}\right)+ \\ \sum_{i=1}^{b_k}\left(\frac{\lambda_H^{Ui}(u_k)}{\sum_{i=1}^{a_k}\lambda_H^i(u_k)+\sum_{i=1}^{b_k}\lambda_H^{Ui}(u_k)}\ln\frac{\lambda_H^{Ui}(u_k)}{\sum_{i=1}^{a_k}\lambda_H^i(u_k)+\sum_{i=1}^{b_k}\lambda_H^{Ui}(u_k)}\right) \end{array}\right) \end{array}\right), \\ \max\left(\begin{array}{l} -\frac{1}{\ln(a_k+b_k)}\left(\begin{array}{l} \sum_{i=1}^{a_k}\left(\frac{\lambda_H^i(u_k)}{\sum_{i=1}^{a_k}\lambda_H^i(u_k)+\sum_{i=1}^{b_k}\lambda_H^{Li}(u_k)}\ln\frac{\lambda_H^i(u_k)}{\sum_{i=1}^{a_k}\lambda_H^i(u_k)+\sum_{i=1}^{b_k}\lambda_H^{Li}(u_k)}\right)+ \\ \sum_{i=1}^{b_k}\left(\frac{\lambda_H^{Li}(u_k)}{\sum_{i=1}^{a_k}\lambda_H^i(u_k)+\sum_{i=1}^{b_k}\lambda_H^{Li}(u_k)}\ln\frac{\lambda_H^{Li}(u_k)}{\sum_{i=1}^{a_k}\lambda_H^i(u_k)+\sum_{i=1}^{b_k}\lambda_H^{Li}(u_k)}\right) \end{array}\right), \\ -\frac{1}{\ln(a_k+b_k)}\left(\begin{array}{l} \sum_{i=1}^{a_k}\left(\frac{\lambda_H^i(u_k)}{\sum_{i=1}^{a_k}\lambda_H^i(u_k)+\sum_{i=1}^{b_k}\lambda_H^{Ui}(u_k)}\ln\frac{\lambda_H^i(u_k)}{\sum_{i=1}^{a_k}\lambda_H^i(u_k)+\sum_{i=1}^{b_k}\lambda_H^{Ui}(u_k)}\right)+ \\ \sum_{i=1}^{b_k}\left(\frac{\lambda_H^{Ui}(u_k)}{\sum_{i=1}^{a_k}\lambda_H^i(u_k)+\sum_{i=1}^{b_k}\lambda_H^{Ui}(u_k)}\ln\frac{\lambda_H^{Ui}(u_k)}{\sum_{i=1}^{a_k}\lambda_H^i(u_k)+\sum_{i=1}^{b_k}\lambda_H^{Ui}(u_k)}\right) \end{array}\right) \end{array}\right) \end{bmatrix}, e_Z(u_k) \subseteq [0, 1] \tag{4}$$

It is obvious that the IVFES Z consists of interval-valued fuzzy average values and entropy values to reasonably solve the expression and operation problems of different sequence lengths in SIVFMEs.

Remark 1.

(1) *The entropy value indicates a degree of difference among various fuzzy values in the SIVFME $F_H(u_k)$. The larger the entropy value, the better the consistency of various fuzzy values in the SIVFME $F_H(u_k)$.*
(2) *All fuzzy values in $F_H(u_k)$ are identical when $e_Z(u_k) = \left[e_Z^L(u_k), e_Z^U(u_k)\right] = [1,1]$, which can indicate the complete consistency of the multiple fuzzy values.*
(3) *In GDM problems, the larger the average value and entropy value of the group evaluation, the better the group evaluation values and their consistency/consensus. When the entropy value of the group evaluation values is equal to one, this reflects complete consistency/consensus of the group evaluation values.*

Example 1. *Let us consider a GDM problem. When a group of four decision makers/experts is asked to assess product quality (u_1) and service quality (u_2) in $U = \{u_1, u_2\}$ regarding a supplier A, they can give two groups of fuzzy assessment values, (u_1, 0.7, 0.8, [0.6, 0.8], [0.7, 0.9]) and (u_2, [0.6, 0.7], [0.6, 0.7], [0.6, 0.7], [0.6, 0.7], [0.6, 0.7]). Therefore, using Equations (2) and (3), their interval-valued fuzzy average values and entropy values are [0.7, 0.8] and [0.9963, 0.9972] for u_1 and [0.6, 0.7] and [1, 1] for u_2, respectively, which are expressed as the IVFES Z = {(u_1, [0.7, 0.8], [0.9963, 0.9972]), (u_2, [0.6, 0.7], [1, 1])} in the GDM example.*

In this example, it can be seen that the average values and entropy values can reflect the magnitude and consistency/consensus degree of the group evaluation values. The larger the entropy value, the better the consistency/consensus of the group evaluation values.

Then, the simplified expression form of a basic element $z(u_k) = \left(u_k, m_Z(u_k), e_{Zk}(u_k)\right)$ for $\left[m_Z^L(u_k), m_Z^U(u_k)\right] \subseteq [0,1]$ and $\left[e_Z^L(u_k), e_Z^U(u_k)\right] \subseteq [0,1]$ in the IVFES Z can be denoted as $z_k = (m_{Zk}, e_{Zk})$ for $\left[m_{Zk}^L, m_{Zk}^U\right] \subseteq [0,1]$ and $\left[e_{Zk}^L, e_{Zk}^U\right] \subseteq [0,1]$, which is named IVFEE.

Definition 3. *Set two IVFEEs as $z_1 = ([m_{Z1}^L, m_{Z1}^U], [e_{Z1}^L, e_{Z1}^U])$ and $z_2 = ([m_{Z2}^L, m_{Z2}^U], [e_{Z2}^L, e_{Z2}^U])$. Thus, their operational relationships are defined as follows:*

(1) *$z_1 \supseteq z_2$ if and if then $m_{Z1}^L \geq m_{Z2}^L$, $m_{Z1}^U \geq m_{Z2}^U$, $e_{Z1}^L \geq e_{Z2}^L$, and $e_{Z1}^U \geq e_{Z2}^U$;*
(2) *$z_1 = z_2$ if and if then $z_1 \supseteq z_2$ and $z_2 \supseteq z_1$;*
(3) *$z_1 \cup z_2 = \left([m_{Z1}^L \vee m_{Z2}^L, m_{Z1}^U \vee m_{Z2}^U], [e_{Z1}^L \vee e_{Z2}^L, e_{Z1}^U \vee e_{Z2}^U]\right)$;*
(4) *$z_1 \cap z_2 = \left([m_{Z1}^L \wedge m_{Z2}^L, m_{Z1}^U \wedge m_{Z2}^U], [e_{Z1}^L \wedge e_{Z2}^L, e_{Z1}^U \wedge e_{Z2}^U]\right)$.*

Definition 4. *Set two IVFEEs as $z_1 = ([m_{Z1}^L, m_{Z1}^U], [e_{Z1}^L, e_{Z1}^U])$ and $z_2 = ([m_{Z2}^L, m_{Z2}^U], [e_{Z2}^L, e_{Z2}^U])$. Thus, their operational laws are defined as follows:*

(1) $z_1 \oplus z_2 = \left(\begin{array}{c} [m_{Z1}^L + m_{Z2}^L - m_{Z1}^L m_{Z2}^L, m_{Z1}^U + m_{Z2}^U - m_{Z1}^U m_{Z2}^U], \\ [e_{Z1}^L + e_{Z2}^L - e_{Z1}^L e_{Z2}^L, e_{Z1}^U + e_{Z2}^U - e_{Z1}^U e_{Z2}^U] \end{array} \right)$;
(2) $z_1 \otimes z_2 = \left([m_{Z1}^L m_{Z2}^L, m_{Z1}^U m_{Z2}^U], [e_{Z1}^L e_{Z2}^L, e_{Z1}^U e_{Z2}^U]\right)$;
(3) $z_1^\lambda = \left([(m_{Z1}^L)^\lambda, (m_{Z1}^U)^\lambda], [(e_{Z1}^L)^\lambda, (e_{Z1}^U)^\lambda]\right)$ *for* $\lambda > 0$;
(4) $\lambda z_1 = \left([1 - (1 - m_{Z1}^L)^\lambda, 1 - (1 - m_{Z1}^U)^\lambda], [1 - (1 - e_{Z1}^L)^\lambda, 1 - (1 - e_{Z1}^U)^\lambda]\right)$ *for* $\lambda > 0$.

However, it is obvious that the above operational results are still IVFEEs.

To compare two IVFEEs $z_k = ([m_{Zk}^L, m_{Zk}^U], [e_{Zk}^L, e_{Zk}^U])$ for $k = 1, 2$, the expected value function is defined as

$$Q(z_k) = (m_{Zk}^L e_{Zk}^L + m_{Zk}^U e_{Zk}^U)/2 \text{ for } Q(z_k) \in [0,1] \quad (5)$$

Then, the sorting rules of the two IVFEEs are given as follows:

(1) If $Q(z_1) > Q(z_2)$, then $z_1 > z_2$;
(2) If $Q(z_1) = Q(z_2)$, then $z_1 \cong z_2$.

Example 2. *Assume that two IVFEEs are z_1 = ([0.7, 0.8], [0.8, 0.9]) and z_2 = ([0.6, 0.7], [0.7, 0.8]). Then, their sorting is yielded below:*

Using Equation (5), there are $Q(z_1) = (0.7 \times 0.8 + 0.8 \times 0.9)/2 = 0.56$ and $Q(z_2) = (0.6 \times 0.7 + 0.7 \times 0.8)/2 = 0.54$. Since $Q(z_1) > Q(z_2)$, their sorting is $z_1 > z_2$.

3. Two Weighted Aggregation Operators of IVFEEs and Their Mixed-Weighted-Averaging Operation

In this section, we propose the IVFEEWA and IVFEEWG operators according to the operational laws in Definition 4, and then define their mixed-weighted-averaging operation to make up for their flaws in aggregating IVFEEs; that is, the weighted averaging aggregation operator mainly tends to group arguments, and the weighted geometric aggregation operator tends to group personal arguments.

3.1. Weighted Averaging Aggregation Operator of IVFEEs

Based on the operational laws in Definition 4, the IVFEEWA operator is defined to aggregate IVFEE information.

Definition 5. *Let $z_k = ([m_{Zk}^L, m_{Zk}^U], [e_{Zk}^L, e_{Zk}^U])$ (k = 1, 2, ..., s) be a group of IVFEEs and IVFEEWA: $\Omega^s \to \Omega$. Then, the IVFEEWA operator is defined as*

$$IVFEEWA(z_1, z_2, \ldots, z_s) = \overset{s}{\underset{k=1}{\oplus}} \lambda_k z_k \tag{6}$$

where λ_k is the weight of z_k with $0 \le \lambda_k \le 1$ and $\sum_{k=1}^{s} \lambda_k = 1$.

Theorem 1. *Let $z_k = ([m_{Zk}^L, m_{Zk}^U], [e_{Zk}^L, e_{Zk}^U])$ (k = 1, 2, ..., s) be a group of IVFEEs with the weight vector = ($\lambda_1, \lambda_2, \ldots, \lambda_n$) for $0 \le \lambda_k \le 1$ and $\sum_{k=1}^{s} \lambda_k = 1$. Then, the aggregated result of the IVFEEWA operator is still IVFEE, which is obtained by the equation:*

$$\begin{aligned} &IVFEEWA(z_1, z_2, \ldots, z_s) = \overset{s}{\underset{k=1}{\oplus}} \lambda_k z_k \\ &= \left(\left[1 - \prod_{k=1}^{s} (1 - m_{Zk}^L)^{\lambda_k}, 1 - \prod_{k=1}^{s} (1 - m_{Zk}^U)^{\lambda_k} \right], \left[1 - \prod_{k=1}^{s} (1 - e_k^L)^{\lambda_k}, 1 - \prod_{k=1}^{s} (1 - e_k^U)^{\lambda_k} \right] \right) \end{aligned} \tag{7}$$

Proof. Regarding mathematical induction, Equation (7) can be proved.

(1) When s = 2, by the operational laws in Definition 4, the aggregation result is yielded as follows:

$$\begin{aligned} &IVFEEWA(z_1, z_2) = \lambda_1 z_1 \oplus \lambda_1 z_2 \\ &= \left(\begin{array}{c} \left[\begin{array}{c} 1 - (1 - m_{Z1}^L)^{\lambda_1} + 1 - (1 - m_{Z2}^L)^{\lambda_2} - \left(1 - (1 - m_{Z1}^L)^{\lambda_1}\right)\left(1 - (1 - m_{Z2}^L)^{\lambda_2}\right), \\ 1 - (1 - m_{Z1}^U)^{\lambda_1} + 1 - (1 - m_{Z2}^U)^{\lambda_2} - \left(1 - (1 - m_{Z1}^U)^{\lambda_1}\right)\left(1 - (1 - m_{Z2}^U)^{\lambda_2}\right) \end{array} \right], \\ \left[\begin{array}{c} 1 - (1 - e_{Z1}^L)^{\lambda_1} + 1 - (1 - e_{Z2}^L)^{\lambda_2} - \left(1 - (1 - e_{Z1}^L)^{\lambda_1}\right)\left(1 - (1 - e_{Z2}^L)^{\lambda_2}\right), \\ 1 - (1 - e_{Z1}^U)^{\lambda_1} + 1 - (1 - e_{Z2}^U)^{\lambda_2} - \left(1 - (1 - e_{Z1}^U)^{\lambda_1}\right)\left(1 - (1 - e_{Z2}^U)^{\lambda_2}\right) \end{array} \right] \end{array} \right) \\ &= \left(\left[1 - \prod_{k=1}^{2} (1 - m_{Zk}^L)^{\lambda_k}, 1 - \prod_{k=1}^{2} (1 - m_{Zk}^U)^{\lambda_k} \right], \left[1 - \prod_{k=1}^{2} (1 - e_{Zk}^L)^{\lambda_k}, 1 - \prod_{k=1}^{2} (1 - e_{Zk}^U)^{\lambda_k} \right] \right). \end{aligned} \tag{8}$$

(2) When s = n, Equation (7) can keep the following result:

$$IVFEEWA(z_1, z_2, \ldots, z_n) = \overset{n}{\underset{k=1}{\oplus}} \lambda_k z_k = \left(\begin{array}{c} \left[1 - \prod_{k=1}^{n} (1 - m_{Zk}^L)^{\lambda_k}, 1 - \prod_{k=1}^{n} (1 - m_{Zk}^U)^{\lambda_k} \right], \\ \left[1 - \prod_{k=1}^{n} (1 - e_{Zk}^L)^{\lambda_k}, 1 - \prod_{k=1}^{n} (1 - e_{Zk}^U)^{\lambda_k} \right] \end{array} \right) \tag{9}$$

(3) When $s = n + 1$, by the operational laws in Definition 4 and Equations (8) and (9), the aggregated result is given as follows:

$$\begin{aligned}
&IVFEEWA(z_1, z_2, \ldots, z_n, z_{n+1}) = \mathop{\oplus}_{k=1}^{n} \lambda_k z_k \oplus \lambda_{n+1} z_{n+1} \\
&= \left(\left[1 - \prod_{k=1}^{n} (1 - m_{Zk}^L)^{\lambda_k}, 1 - \prod_{k=1}^{n} (1 - m_{Zk}^U)^{\lambda_k} \right], \left[1 - \prod_{k=1}^{n} (1 - e_{Zk}^L)^{\lambda_k}, 1 - \prod_{k=1}^{n} (1 - e_{Zk}^U)^{\lambda_k} \right] \right) \oplus \lambda_{n+1} z_{n+1} \\
&= \left(\begin{array}{c} \left[1 - \prod_{k=1}^{n} (1 - m_{Zk}^L)^{\lambda_k} (1 - m_{Zn+1}^L)^{\lambda_{n+1}}, 1 - \prod_{k=1}^{n} (1 - m_{Zk}^U)^{\lambda_k} (1 - m_{Zn+1}^U)^{\lambda_{n+1}} \right], \\ \left[1 - \prod_{k=1}^{n} (1 - e_{Zk}^L)^{\lambda_k} (1 - e_{Zs+1}^L)^{\lambda_{n+1}}, 1 - \prod_{k=1}^{n} (1 - e_{Zk}^U)^{\lambda_k} (1 - e_{Zs+1}^U)^{\lambda_{n+1}} \right] \end{array} \right) \\
&= \left(\left[1 - \prod_{k=1}^{n+1} (1 - m_{Zk}^L)^{\lambda_k}, 1 - \prod_{k=1}^{n+1} (1 - m_{Zk}^U)^{\lambda_k} \right], \left[1 - \prod_{k=1}^{n+1} (1 - e_{Zk}^L)^{\lambda_k}, 1 - \prod_{k=1}^{n+1} (1 - e_{Zk}^U)^{\lambda_k} \right] \right).
\end{aligned} \tag{10}$$

Obviously, Equation (7) exists for any s. □

Theorem 2. *The IVFEEWA operator implies these properties:*

(1) *Idempotency: Set $z_k = ([m_{Zk}^L, m_{Zk}^U], [e_{Zk}^L, e_{Zk}^U])$ ($k = 1, 2, \ldots, s$) as a group of IVFEEs. There is $IVFEEWA(z_1, z_2, \ldots, z_s) = z$ if $z_k = z = ([m_Z^L, m_Z^U], [e_Z^L, e_Z^U])$ ($k = 1, 2, \ldots, s$).*

(2) *Boundedness: Set $z_k = ([m_{Zk}^L, m_{Zk}^U], [e_{Zk}^L, e_{Zk}^U])$ ($k = 1, 2, \ldots, s$) as a group of IVFEEs and let $z_{\min} = \left(\left[\min_k (m_{Zk}^L), \min_k (m_{Zk}^U) \right], \left[\min_k (e_{Zk}^L), \min_k (e_{Zk}^U) \right] \right)$ and $z_{\max} = \left(\left[\max_k (m_{Zk}^L), \max_k (m_{Zk}^U) \right], \left[\max_k (e_{Zk}^L), \max_k (e_{Zk}^U) \right] \right)$ be the minimum IVFEE and the maximum IVFEE, respectively. Then, $z_{\min} \le IVFEEWA(z_1, z_2, \ldots, z_s) \le z_{\max}$ exists.*

(3) *Monotonicity: Set $z_k = ([m_{Zk}^L, m_{Zk}^U], [e_{Zk}^L, e_{Zk}^U])$ and $z_k^* = ([m_{Zk}^{L*}, m_{Zk}^{U*}], [e_{Zk}^{L*}, e_{Zk}^{U*}])$ ($k = 1, 2, \ldots, s$) as two groups of IVFEEs. Then, there exists $IVFEEWA(z_1, z_2, \ldots, z_s) \le IVFEEWA(z_1^*, z_2^*, \ldots, z_s^*)$ if $z_k \le z_k^*$.*

Proof. (1) For $z_k = z = ([m_Z^L, m_Z^U], [e_Z^L, e_Z^U])$ ($k = 1, 2, \ldots, s$), by Equation (7) the result is yielded below:

$$\begin{aligned}
&IVFEEWA(z_1, z_2, \ldots, z_s) = \mathop{\oplus}_{k=1}^{s} \lambda_k z_k = \left(\begin{array}{c} \left[1 - \prod_{k=1}^{s} (1 - m_{Zk}^L)^{\lambda_k}, 1 - \prod_{k=1}^{s} (1 - m_{Zk}^U)^{\lambda_k} \right], \\ \left[1 - \prod_{k=1}^{s} (1 - e_{Zk}^L)^{\lambda_k}, 1 - \prod_{k=1}^{s} (1 - e_{Zk}^U)^{\lambda_k} \right] \end{array} \right) \\
&= \left(\left[1 - (1 - m_Z^L)^{\sum_{k=1}^{s} \lambda_k}, 1 - (1 - m_Z^U)^{\sum_{k=1}^{s} \lambda_k} \right], \left[1 - (1 - e_Z^L)^{\sum_{k=1}^{s} \lambda_k}, 1 - (1 - e_Z^U)^{\sum_{k=1}^{s} \lambda_k} \right] \right) \\
&= \left([m_Z^L, m_Z^U], [e_Z^L, e_Z^U] \right) = z.
\end{aligned} \tag{11}$$

(2) There exists the inequality $z_{min} \le z_k \le z_{max}$ when z_{min} and z_{max} are the minimum and maximum IVFEEs. Thus, there also exists $\mathop{\oplus}_{k=1}^{s} \lambda_k z_{\min} \le \mathop{\oplus}_{k=1}^{s} \lambda_k z_k \le \mathop{\oplus}_{k=1}^{s} \lambda_k z_{\max}$. Then, the inequality $z_{min} \le \mathop{\oplus}_{k=1}^{s} \lambda_k z_k \le z_{max}$ can be kept regarding the above property (1); i.e., there is $z_{\min} \le IVFEEWA(z_1, z_2, \ldots, z_s) \le z_{\max}$.

(3) For $z_k \le z_k^*$, there is the inequality $\mathop{\oplus}_{k=1}^{s} \lambda_k z_k \le \mathop{\oplus}_{k=1}^{s} \lambda_k z_k^*$; i.e., $IVFEEWA(z_1, z_2, \ldots, z_s) \le IVFEEWA(z_1^*, z_2^*, \ldots, z_s^*)$ exists.

Therefore, all the above properties are true. □

3.2. Weighted Geometric Aggregation Operator of IVFEEs

Definition 6. *Let* $z_k = ([m_{Zk}^L, m_{Zk}^U], [e_{Zk}^L, e_{Zk}^U])$ *(k = 1, 2, ..., s) be a group of IVFEEs and IVFEEWG:* $\Omega^s \to \Omega$*. Then, the IVFEEWG operator is defined as*

$$IVFEEWG(z_1, z_2, \ldots, z_s) = \overset{s}{\underset{k=1}{\otimes}} z_k^{\lambda_k} \quad (12)$$

where λ_k *is the weight of* z_k *with* $0 \le \lambda_k \le 1$ *and* $\sum_{k=1}^{s} \lambda_k = 1$.

Theorem 3. *Let* $z_k = ([m_{Zk}^L, m_{Zk}^U], [e_{Zk}^L, e_{Zk}^U])$ *(k = 1, 2, ..., s) be a group of IVFEEs along with the weight vector* $\lambda = (\lambda_1, \lambda_2, \ldots, \lambda_s)$ *for* $0 \le \lambda_k \le 1$ *and* $\sum_{k=1}^{s} \lambda_k = 1$*. Then, the aggregated result of the IVFEEWG operator is still IVFEE, which is yielded by the equation:*

$$IVFEEWG(z_1, z_2, \ldots, z_s) = \overset{s}{\underset{k=1}{\otimes}} z_k^{\lambda_k} = \left(\left[\prod_{k=1}^{s} \left(m_{Zk}^L\right)^{\lambda_k}, \prod_{k=1}^{s} \left(m_{Zk}^U\right)^{\lambda_k} \right], \left[\prod_{k=1}^{s} \left(e_{Zk}^L\right)^{\lambda_k}, \prod_{k=1}^{s} \left(e_{Zk}^U\right)^{\lambda_k} \right] \right) \quad (13)$$

Similarly to Theorem 1, Theorem 3 can easily be proved, which is omitted here.

Theorem 4. *The IVFEEWG operator implies these properties:*

(1) *Idempotency: Let* $z_k = ([m_{Zk}^L, m_{Zk}^U], [e_{Zk}^L, e_{Zk}^U])$ *(k = 1, 2, ..., s) be a group of IVFEEs. If* $z_k = z = ([m_Z^L, m_Z^U], [e_Z^L, e_Z^U])$ *(k= 1, 2, ..., s), then* $IVFEEWG(z_1, z_2, \ldots, z_s) = z$.

(2) *Boundedness: Let* $z_k = ([m_{Zk}^L, m_{Zk}^U], [e_{Zk}^L, e_{Zk}^U])$ *(k = 1, 2, ..., s) be a group of IVFEEs, and let* $z_{\min} = \left(\left[\min_k \left(m_{Zk}^L\right), \min_k \left(m_{Zk}^U\right) \right], \left[\min_k \left(e_{Zk}^L\right), \min_k \left(e_{Zk}^U\right) \right] \right)$ *and* $z_{\max} = \left(\left[\max_k \left(m_{Zk}^L\right), \max_k \left(m_{Zk}^U\right) \right], \left[\max_k \left(e_{Zk}^L\right), \max_k \left(e_{Zk}^U\right) \right] \right)$ *be the minimum and maximum IVFEEs. Then,* $z_{\min} \le IVFEEWG(z_1, z_2, \ldots, z_s) \le z_{\max}$ *exists.*

(3) *Monotonicity: Let* $z_k = ([m_{Zk}^L, m_{Zk}^U], [e_{Zk}^L, e_{Zk}^U])$ *and* $z_k^* = ([m_{Zk}^{L*}, m_{Zk}^{U*}], [e_{Zk}^{L*}, e_{Zk}^{U*}])$ *(k = 1, 2, ..., s) be two groups of IVFEEs. Then, there exists* $IVFEEWG(z_1, z_2, \ldots, z_s) \le IVFEEWG(z_1^*, z_2^*, \ldots, z_s^*)$ *for* $z_k \le z_k^*$.

Theorem 4 can be proved similarly to Theorem 2 (omitted).

3.3. Mixed-Weighted-Averaging Operation for the IVFEEWA and IVFEEWG Operators

Since the IVFEEWA operator and the IVFEEWG operator mainly tend to group arguments and individual arguments, respectively, here we propose a mixed-weighted-averaging operation for the IVFEEWA and IVFEEWG operators.

Definition 7. *Set* $\eta \in [0, 1]$ *as a weight parameter. Then, a mixed-weighted-averaging operation of the IVFEEWA and IVFEEWG operators with a weight parameter* η *is defined below:*

$$z(\eta) = \eta \times IVFEEWA(z_1, z_2, \ldots, z_s) \oplus (1 - \eta) \times IVFEEWG(z_1, z_2, \ldots, z_s) \quad (14)$$

Theorem 5. *Let $\eta \in [0, 1]$ be a weight parameter. Then, the operational result of Equation (14) with a weight parameter η is still IVFEE, which is obtained by the following equation:*

$$z(\eta) = \eta \times IVFEEWA(z_1, z_2, \ldots, z_s) \oplus (1-\eta) \times IVFEEWG(z_1, z_2, \ldots, z_s)$$
$$= \left(\begin{array}{c} \left[\begin{array}{c} 1 - \left(\prod_{k=1}^{s} (1 - m_{Zk}^{L})^{\lambda_k} \right)^{\eta} \left(1 - \prod_{k=1}^{s} (m_{Zk}^{L})^{\lambda_k} \right)^{(1-\eta)}, \\ 1 - \left(\prod_{k=1}^{s} (1 - m_{Zk}^{U})^{\lambda_k} \right)^{\eta} \left(1 - \prod_{k=1}^{s} (m_{Zk}^{U})^{\lambda_k} \right)^{(1-\eta)} \end{array} \right], \\ \left[\begin{array}{c} 1 - \left(\prod_{k=1}^{s} (1 - e_{Zk}^{L})^{\lambda_k} \right)^{\eta} \left(1 - \prod_{k=1}^{s} (e_{Zk}^{L})^{\lambda_k} \right)^{(1-\eta)}, \\ 1 - \left(\prod_{k=1}^{s} (1 - e_{Zk}^{U})^{\lambda_k} \right)^{\eta} \left(1 - \prod_{k=1}^{s} (e_{Zk}^{U})^{\lambda_k} \right)^{(1-\eta)} \end{array} \right] \end{array} \right) \quad (15)$$

Proof. Based on Equations (7), (13), and (14), along with the operational laws in Definition 4, the following result is obtained below:

$$z(\eta) = \eta \times IVFEEWA(z_1, z_2, \ldots, z_s) \oplus (1-\eta) \times IVFEEWG(z_1, z_2, \ldots, z_s)$$
$$= \left(\left[1 - \left(\prod_{k=1}^{s} (1 - m_{Zk}^{L})^{\lambda_k} \right)^{\eta}, 1 - \left(\prod_{k=1}^{s} (1 - m_{Zk}^{U})^{\lambda_k} \right)^{\eta} \right], \left[1 - \left(\prod_{k=1}^{s} (1 - e_{Zk}^{L})^{\lambda_k} \right)^{\eta}, 1 - \left(\prod_{k=1}^{s} (1 - e_{Zk}^{U})^{\lambda_k} \right)^{\eta} \right] \right) \oplus$$
$$\left(\left[1 - \left(1 - \prod_{k=1}^{s} (m_{Zk}^{L})^{\lambda_k} \right)^{(1-\eta)}, 1 - \left(1 - \prod_{k=1}^{s} (m_{Zk}^{U})^{\lambda_k} \right)^{(1-\eta)} \right], \left[1 - \left(1 - \prod_{k=1}^{s} (e_{Zk}^{L})^{\lambda_k} \right)^{(1-\eta)}, 1 - \left(1 - \prod_{k=1}^{s} (e_{Zk}^{U})^{\lambda_k} \right)^{(1-\eta)} \right] \right)$$
$$= \left(\begin{array}{c} \left[\begin{array}{c} 1 - \left(\prod_{k=1}^{s} (1 - m_{Zk}^{L})^{\lambda_k} \right)^{\eta} + 1 - \left(1 - \prod_{k=1}^{s} (m_{Zk}^{L})^{\lambda_k} \right)^{(1-\eta)} - \left(1 - \left(\prod_{k=1}^{s} (1 - m_{Zk}^{L})^{\lambda_k} \right)^{\eta} \right) \left(1 - \left(1 - \prod_{k=1}^{s} (m_{Zk}^{L})^{\lambda_k} \right)^{(1-\eta)} \right), \\ 1 - \left(\prod_{k=1}^{s} (1 - m_{Zk}^{U})^{\lambda_k} \right)^{\eta} + 1 - \left(1 - \prod_{k=1}^{s} (m_{Zk}^{U})^{\lambda_k} \right)^{(1-\eta)} - \left(1 - \left(\prod_{k=1}^{s} (1 - m_{Zk}^{U})^{\lambda_k} \right)^{\eta} \right) \left(1 - \left(1 - \prod_{k=1}^{s} (m_{Zk}^{U})^{\lambda_k} \right)^{(1-\eta)} \right) \end{array} \right], \\ \left[\begin{array}{c} 1 - \left(\prod_{k=1}^{s} (1 - e_{Zk}^{L})^{\lambda_k} \right)^{\eta} + 1 - \left(1 - \prod_{k=1}^{s} (e_{Zk}^{L})^{\lambda_k} \right)^{(1-\eta)} - \left(1 - \left(\prod_{k=1}^{s} (1 - e_{Zk}^{L})^{\lambda_k} \right)^{\eta} \right) \left(1 - \left(1 - \prod_{k=1}^{s} (e_{Zk}^{L})^{\lambda_k} \right)^{(1-\eta)} \right), \\ 1 - \left(\prod_{k=1}^{s} (1 - e_{Zk}^{U})^{\lambda_k} \right)^{\eta} + 1 - \left(1 - \prod_{k=1}^{s} (e_{Zk}^{U})^{\lambda_k} \right)^{(1-\eta)} - \left(1 - \left(\prod_{k=1}^{s} (1 - e_{Zk}^{U})^{\lambda_k} \right)^{\eta} \right) \left(1 - \left(1 - \prod_{k=1}^{s} (e_{Zk}^{U})^{\lambda_k} \right)^{(1-\eta)} \right) \end{array} \right] \end{array} \right) \quad (16)$$
$$= \left(\begin{array}{c} \left[1 - \left(\prod_{k=1}^{s} (1 - m_{Zk}^{L})^{\lambda_k} \right)^{\eta} \left(1 - \prod_{k=1}^{s} (m_{Zk}^{L})^{\lambda_k} \right)^{(1-\eta)}, 1 - \left(\prod_{k=1}^{s} (1 - m_{Zk}^{U})^{\lambda_k} \right)^{\eta} \left(1 - \prod_{k=1}^{s} (m_{Zk}^{U})^{\lambda_k} \right)^{(1-\eta)} \right], \\ \left[1 - \left(\prod_{k=1}^{s} (1 - e_{Zk}^{L})^{\lambda_k} \right)^{\eta} \left(1 - \prod_{k=1}^{s} (e_{Zk}^{L})^{\lambda_k} \right)^{(1-\eta)}, 1 - \left(\prod_{k=1}^{s} (1 - e_{Zk}^{U})^{\lambda_k} \right)^{\eta} \left(1 - \prod_{k=1}^{s} (e_{Zk}^{U})^{\lambda_k} \right)^{(1-\eta)} \right] \end{array} \right).$$

When η = 1, 0, $z(\eta)$ degenerates into the IVFEEWA operator of Equation (7) and the IVFEEWG operator of Equation (13), respectively. □

4. GDM Method Using the Mixed-Weighted-Averaging Operation and Expected Value Function

Here we propose a multicriteria GDM method using the mixed-weighted-averaging operation and expected value function for SIVFMSs.

A multicriteria GDM problem usually contains a set of alternatives $Y = \{Y_1, Y_2, \ldots, Y_m\}$, which is assessed by a set of criteria $U = \{u_1, u_2, \ldots, u_s\}$. To consider the importance of different criteria u_k (k = 1, 2, …, s) in U, decision makers specify a weigh vector $\boldsymbol{\lambda} = (\lambda_1, \lambda_2, \ldots, \lambda_s)$ for the set of criteria. Regarding the uncertainty and certainty of decision makers' cognitions/judgments for the suitability assessment of alternatives over the criteria, the single- and interval-valued fuzzy values of the alternatives Y_j (j = 1, 2, …, m) over the criteria u_k (k = 1, 2, …, s) will be specified by various decision makers. Thus, the multicriteria GDM method is depicted by the following decision steps.

Step 1. A group of decision makers/experts is invited to give their single- and interval-valued fuzzy values of the alternatives Y_j (j = 1, 2, …, m) over the criteria u_k (k = 1, 2, …, s) and to set up the SIVFME decision matrix $D = (F_{Hjk})_{m \times s}$, where $F_{Hjk} = (\lambda_{Hjk}^{1}, \lambda_{Hjk}^{2}, \ldots, \lambda_{Hjk}^{a_{jk}}, [\lambda_{Hjk}^{L1}, \lambda_{Hjk}^{U1}], [\lambda_{Hjk}^{L2}, \lambda_{Hjk}^{U2}], \ldots, [\lambda_{Hjk}^{Lb_{jk}}, \lambda_{Hjk}^{Ub_{jk}}])$ composed of a_{jk} single-valued fuzzy values and b_{jk} interval-valued fuzzy values (j = 1, 2, …, m; k = 1, 2, …, s) are

SIVFMEs, such that $0 \le \lambda_{Hjk}^{1} \le \lambda_{Hjk}^{2}, \ldots, \le \lambda_{Hjk}^{a_{jk}} \le 1$ and $[\lambda_{Hjk}^{L1}, \lambda_{Hjk}^{U1}] \subseteq [\lambda_{Hjk}^{L2}, \lambda_{Hjk}^{U2}] \subseteq, \ldots, \subseteq [\lambda_{Hjk}^{Lb_{jk}}(u_k), \lambda_{Hjk}^{Ub_{jk}}] \subseteq [0,1]$ with identical and/or different fuzzy values.

Step 2. Using Equations (3) and (4) for the decision matrix $D = (F_{Hjk})_{m \times s}$, the interval-valued fuzzy average values m_{Zjk} and entropy values e_{Zjk} are obtained and IVFEEs are assembled by $z_{jk} = (m_{Zjk}, e_{Zjk})$ for $m_{Zjk} = \left[m_{Zjk}^{L}, m_{Zjk}^{U}\right] \subseteq [0,1]$ and $e_{Zjk} = \left[e_{Zjk}^{L}, e_{Zjk}^{U}\right] \subseteq [0,1]$ ($k = 1, 2, \ldots, s; j = 1, 2, \ldots, m$), which are constructed as the IVFEE decision matrix $M = (z_{jk})_{m \times s}$.

Step 3. Using Equation (15) with some values of η, the operational values of $z_j(\eta)$ for Y_j ($j = 1, 2, \ldots, m$) are obtained by the following equation:

$$z_j(\eta) = \eta \times IVFEEWA(z_{j1}, z_{j2}, \ldots, z_{js}) \oplus (1-\eta) \times IVFEEWG(z_{j1}, z_{j2}, \ldots, z_{js})$$
$$= \left(\begin{bmatrix} 1 - \left(\prod_{k=1}^{s} (1 - m_{Zjk}^{L})^{\lambda_k}\right)^{\eta} \left(1 - \prod_{k=1}^{s} (m_{Zjk}^{L})^{\lambda_k}\right)^{(1-\eta)}, \\ 1 - \left(\prod_{k=1}^{s} (1 - m_{Zjk}^{U})^{\lambda_k}\right)^{\eta} \left(1 - \prod_{k=1}^{s} (m_{Zjk}^{U})^{\lambda_k}\right)^{(1-\eta)} \end{bmatrix}, \begin{bmatrix} 1 - \left(\prod_{k=1}^{s} (1 - e_{Zjk}^{L})^{\lambda_k}\right)^{\eta} \left(1 - \prod_{k=1}^{s} (e_{Zjk}^{L})^{\lambda_k}\right)^{(1-\eta)}, \\ 1 - \left(\prod_{k=1}^{s} (1 - e_{Zjk}^{U})^{\lambda_k}\right)^{\eta} \left(1 - \prod_{k=1}^{s} (e_{Zjk}^{U})^{\lambda_k}\right)^{(1-\eta)} \end{bmatrix} \right) \quad (17)$$

Step 4. The expected values of $Q(z_j(\eta))$ ($j = 1, 2, \ldots, m$) are given by Equation (5).

Step 5. Alternatives are sorted in descending order of the expected values, and the optimal one is selected depending on some specified value of η.

Step 6. End.

5. GDM Example of a Supplier Selection Problem and Comparative Analysis

5.1. Actual GDM Example

This section reports the application of the proposed GDM method to an actual example of a supplier selection problem in a supply chain to show the rationality and effectiveness of SIVFMSs.

Any enterprise tries to reduce the supply chain risks and uncertainty to improve customer service, inventory levels, and cycle times, which will increasing its competitiveness and profitability. Assume that a group of five suppliers is provided as a set of preliminary alternatives $Y = \{Y_1, Y_2, Y_3, Y_4, Y_5\}$. Then, a group of decision makers is invited to evaluate the five suppliers with three criteria: performance (e.g., quality, delivery, and price) (u_1), technology (e.g., design capability, manufacturing capability, and ability to deal with technology changes) (u_2), and organizational culture and strategy (e.g., external and internal integration of suppliers, feeling of trust, compatibility across levels, and functions of the supplier and buyer) (u_3). The weight vector of the three criteria is specified as λ = (0.3, 0.33, 0.37). Thus, the proposed GDM method can be applied to this GDM problem, which is depicted below.

Step 1. Suppose that three decision makers are invited to evaluate a set of five suppliers $Y = \{Y_1, Y_2, Y_3, Y_4, Y_5\}$ with a set of three criteria $U = \{u_1, u_2, u_3\}$. For instance, the three decision makers can declare the degree that an alternative Y_1 should satisfy a criterion u_1, and these values could be a group of three single- and interval-valued fuzzy values (0.7, 0.8, [0.7, 0.9]). In this manner, all their evaluation values of SIVFMEs are indicated in Table 1.

Table 1. Evaluation values of SIVFMEs provided by the three decision makers.

	u_1	u_2	u_3
Y_1	(0.7, 0.8, [0.7, 0.9])	(0.6, [0.6, 0.7], [0.7, 0.8])	(0.6, 0.7, [0.7, 0.8])
Y_2	(0.7, 0.8, [0.6, 0.7])	(0.6, 0.7, [0.7, 0.8])	(0.7, 0.7, [0.6, 0.9])
Y_3	(0.8, [0.8, 0.9], [0.8, 0.9])	(0.8, [0.7, 0.9], [0.8, 0.9])	(0.6, 0.7, [0.7, 0.9])
Y_4	(0.6, 0.6, [0.7, 0.8])	(0.6, 0.8, [0.7, 0.9])	(0.8, 0.8, [0.7, 0.9])
Y_5	(0.8, 0.9, [0.7, 0.8])	(0.8, 0.9, [0.7, 0.8])	(0.7, [0.6, 0.8], [0.7, 0.8])

Step 2. Using Equations (3) and (4) on Table 1, IVFEEs can be obtained based on the average values and entropy values of various SIVFMVEs, and the IVFEE decision matrix $M = (z_{jk})_{5\times 3}$ is established as follows:

$$M = \begin{bmatrix} ([0.7333, 0.8000], [0.9952, 0.9981]) & ([0.6333, 0.7000], [0.9938, 0.9975]) & ([0.6667, 0.7000], [0.9938, 0.9977]) \\ ([0.7000, 0.7333], [0.9938, 0.9981]) & ([0.6667, 0.7000], [0.9938, 0.9977]) & ([0.6667, 0.7667], [0.9933, 0.9977]) \\ ([0.8000, 0.8667], [0.9986, 1.0000]) & ([0.7667, 0.8667], [0.9983, 0.9986]) & ([0.6667, 0.7333], [0.9870, 0.9977]) \\ ([0.6333, 0.6667], [0.9912, 0.9975]) & ([0.7000, 0.7667], [0.9876, 0.9938]) & ([0.7667, 0.8333], [0.9983, 0.9986]) \\ ([0.8000, 0.8333], [0.9952, 0.9986]) & ([0.8000, 0.8333], [0.9952, 0.9986]) & ([0.6667, 0.7667], [0.9977, 0.9983]) \end{bmatrix}$$

Step 3. Using Equation (17) with η = 0, 0.3, 0.5, 0.7, and 1, the operational values of $z_j(\eta)$ for Y_j (j = 1, 2, 3, 4, 5) and the decision results are indicated in Table 2.

Table 2. Decision results of the proposed GDM method with various weight values of η.

η	$z_1(\eta)$, $z_2(\eta)$, $z_3(\eta)$, $z_4(\eta)$, $z_5(\eta)$	$E(z_1(\eta))$, $E(z_2(\eta))$, $E(z_3(\eta))$, $E(z_4(\eta))$, $E(z_5(\eta))$	Sorting	Optimal One
0	([0.6745, 0.7286], [0.9942, 0.9978]), ([0.6765, 0.7341], [0.9936, 0.9978]), ([0.7374, 0.8147], [0.9942, 0.9987]), ([0.7025, 0.7582], [0.9926, 0.9967]), ([0.7478, 0.8080], [0.9961, 0.9984])	0.6988, 0.7024, 0.7734, 0.7265, 0.7758	$Y_5 > Y_3 > Y_4 > Y_2 > Y_1$	Y_5
0.3	([0.6756, 0.7303], [0.9942, 0.9978]), ([0.6767, 0.7347], [0.9936, 0.9978]), ([0.7399, 0.8187], [0.9951, 1.0000]), ([0.7047, 0.7620], [0.9933, 0.9969]), ([0.7510, 0.8090], [0.9962, 0.9984])	0.7002, 0.7027, 0.7775, 0.7298, 0.7780	$Y_5 > Y_3 > Y_4 > Y_2 > Y_1$	Y_5
0.5	([0.6764, 0.7315], [0.9942, 0.9978]), ([0.6768, 0.7351], [0.9936, 0.9978]), ([0.7416, 0.8213], [0.9956, 1.0000]), ([0.7061, 0.7645], [0.9937, 0.9970]), ([0.7532, 0.8096], [0.9963, 0.9985])	0.7012, 0.7030, 0.7798, 0.7320, 0.7794	$Y_3 > Y_5 > Y_4 > Y_2 > Y_1$	Y_3

Table 2. *Cont.*

η	$z_1(\eta)$, $z_2(\eta)$, $z_3(\eta)$, $z_4(\eta)$, $z_5(\eta)$	$E(z_1(\eta))$, $E(z_2(\eta))$, $E(z_3(\eta))$, $E(z_4(\eta))$, $E(z_5(\eta))$	Sorting	Optimal One
0.7	([0.6772, 0.7326], [0.9943, 0.9978]), ([0.6769, 0.7355], [0.9936, 0.9978]), ([0.7433, 0.8239], [0.9960, 1.0000]), ([0.7076, 0.7670], [0.9940, 0.9971]), ([0.7553, 0.8103], [0.9963, 0.9985])	0.7021, 0.7032, 0.7821, 0.7341, 0.7807	$Y_3 > Y_5 > Y_4 > Y_2 > Y_1$	Y_3
1	([0.6783, 0.7344], [0.9943, 0.9978]), ([0.6770, 0.7361], [0.9936, 0.9978]), ([0.7458, 0.8277], [0.9966, 1.0000]), ([0.7097, 0.7707], [0.9946, 0.9973]), ([0.7584, 0.8112], [0.9964, 0.9985])	0.7036, 0.7036, 0.7855, 0.7372, 0.7828	$Y_3 > Y_5 > Y_4 > Y_1 = Y_2$	Y_3

Step 4. By Equation (5), the expected values of $E(z_j(\eta))$ (j = 1, 2, 3, 4, 5) are given in Table 2.

Step 5. The sorting orders of the alternatives are $Y_5 > Y_3 > Y_4 > Y_2 > Y_1$, $Y_3 > Y_5 > Y_4 > Y_2 > Y_1$, and $Y_3 > Y_5 > Y_4 > Y_1 = Y_2$. The optimal one is Y_5 or Y_3, depending on some specified value of η.

Regarding the decision results in Table 2, there are different sorting orders for the IVFEEWA operator and the IVFEEWG operator when η = 0, 1 (two special cases), since the IVFEEWA operator tends to group arguments and the IVFEEWG operator tends to group personal arguments. The mixed-weighted-averaging operation of the IVFEEWA and IVFEEWG operators can compensate for the different tendencies of both when $\eta \neq 0, 1$.

5.2. Comparative Analysis

To verify the efficiency of the proposed GDM method, the proposed GDM method is compared with the existing consistency fuzzy decision-making method and various fuzzy decision-making methods.

First, the proposed GDM method is compared with the existing consistency fuzzy decision-making method [19]. For a convenient comparison with the existing consistency fuzzy decision-making method [19], assume that all interval-valued fuzzy values and entropy values in the IVFEE decision matrix M are fuzzy average values and consistency degrees as a special case of the actual example mentioned above. Thus, the IVFEE decision matrix M is reduced to the decision matrix of CFEs:

$$M' = \begin{bmatrix} (0.7667, 0.9881) & (0.6667, 0.9957) & (0.6834, 0.9958) \\ (0.7167, 0.9960) & (0.6834, 0.9958) & (0.7167, 0.9955) \\ (0.8334, 0.9993) & (0.8167, 0.9985) & (0.7000, 0.9924) \\ (0.6500, 0.9944) & (0.7334, 0.9907) & (0.8000, 0.9985) \\ (0.8167, 0.9969) & (0.8167, 0.9969) & (0.7167, 0.9980) \end{bmatrix}$$

Thus, the existing decision-making method [19] can be applied to the special case of the above actual example by the following CFE weighted averaging (CFEWA) and CFE weighted geometric (CFEWG) operators and score function [19]:

$$z'_j = CFEWA(z'_{j1}, z'_{j2}, \ldots, z'_{js}) = \overset{s}{\underset{k=1}{\oplus}} \lambda_k z'_{jk} = \left(1 - \prod_{k=1}^{s} (1 - m'_{Zjk})^{\lambda_k}, 1 - \prod_{k=1}^{s} (1 - e'_{Zjk})^{\lambda_k}\right) \tag{18}$$

$$z'_j = CFEWG(z'_{j1}, z'_{j2}, \ldots, z'_{js}) = \overset{s}{\underset{k=1}{\otimes}} (z'_{jk})^{\lambda_k} = \left(\prod_{k=1}^{s} (m'_{Zjk})^{\lambda_k}, \prod_{k=1}^{s} (e'_{Zjk})^{\lambda_k}\right) \tag{19}$$

$$F(z'_j) = (m'_{Zj}e'_{Zj} + (m'_{Zj} + e'_{Zj})/2)/2 \text{ for } F(z'_j) \in [0, 1] \tag{20}$$

Using Equations (18)–(20), the aggregated values of the CFEWA and CFEWG operators, the score values of $F(z'_j)$ for Y_i (i = 1, 2, 3, 4, 5), and the decision results were achieved. They are shown in Table 3.

Table 3. Decision results of the existing decision-making method in the case of CFEs [19].

Aggregation Operator	$z'_1, z'_2, z'_3, z'_4, z'_5$	$F(z'_1), F(z'_2), F(z'_3), F(z'_4), F(z'_5)$	Sorting	Optimal One
CFEWA	(0.7061, 0.9960), (0.7061, 0.9957), (0.7862, 0.9978), (0.7399, 0.9958), (0.7846, 0.9974)	0.7772, 0.7770, 0.8383, 0.8023, 0.8368	$Y_3 > Y_5 > Y_4 > Y_1 > Y_2$	Y_3
CFEWG	(0.7016, 0.9960), (0.7055, 0.9957), (0.7761, 0.9964), (0.7304, 0.9946), (0.7781, 0.9973)	0.7738, 0.7765, 0.8298, 0.7945, 0.8319	$Y_5 > Y_3 > Y_4 > Y_2 > Y_1$	Y_5

In the decision results in Table 3, there exists their sorting difference, since there are the different tendencies for the CFEWA and CFEWG operators. The optimal alternatives are Y_3 and Y_5 according to the existing decision-making method with CFE information. Although the optimal ones, Y_3 and Y_5, are the same according to the proposed GDM method and the existing decision-making method [19] in the example, the superiorities of the proposed GDM method over the existing decision-making method [19] are as follows:

(1) SIVFMSs can effectively express group evaluation values using identical and/or different single- and interval-valued fuzzy values, whereas CFMS introduced in [19] cannot.
(2) IVFEEs can reasonably reflect the mean and consistency/consensus degrees of the group evaluation values with the help of quantitative calculations corresponding to the mean and information entropy of a SIVFME in a SIVFMS. The transformation method introduced in [19] is only suitable for the normal distribution of fuzzy data, and there is no distribution limitation for the new transformation method proposed in this paper.
(3) The proposed GDM method not only demonstrated its decision flexibility, but also overcomes the flaws of the existing decision-making method using the single CFEWA operator or the CFEWG operator.

In comparison with the PFDM methods [16,17], the PFDM methods need a lot of fuzzy data to maintain the rationality (no distortion) of probabilistic fuzzy values from the probabilistic viewpoint; otherwise, the probabilistic fuzzy values are infeasible and irrational, since a lot of fuzzy data are created with difficultly by several decision makers and obviously unrealized in the GDM application. Hence, the PFDM methods cannot represent this decision example involving three decision makers and also cannot express single- and interval-valued fuzzy data. In the case of SIVFMSs, the proposed GDM method with the mean and information entropy only needs a few of decision makers to perform GDM problems with several single- and interval-valued fuzzy data, which are easily handled in actual applications. In this case, the proposed GDM method showed its rationality and efficiency and is superior to the existing PFDM methods regarding SIVFMSs.

Furthermore, with respect to the above GDM example in the SIVFMS setting, existing fuzzy multiset/IVFM/HFS/CHFS [9–15,18] cannot express SIVFMS, and then they also cannot be applied to this GDM problem with SIVFMS information.

However, our method not only solves the expression and operation problems of SIVFMEs, but also enhances the flexibility and rationality of GDM, which serve to highlight its advantages in the setting of SIVFMSs.

6. Conclusions

In this study, the presented SIVFMSs could effectively express single- and interval-valued fuzzy sequences in hybrid fuzzy multivalued situations to solve the difficult problems of various existing fuzzy expressions. The proposed information entropy of SIVFME provides a reasonable mathematical tool for converting SIVFMEs into IVFEEs when dealing with SIVFMSs. IVFEEs converted by the mean and information entropy of SIVFMEs in SIVFMS can reasonably reflect the average and consistency level of group evaluation values and effectively solve the operational problems of different fuzzy sequence lengths in SIVFMSs. In addition, the proposed mixed-weighted-averaging operation of the IVFEEWA and IVFEEWG operators can reasonably and flexibly aggregate IVFEE information with a changeable weight parameter and compensate for the flaws of the IVFEEWA and IVFEEWG operators. Next, the multicriteria GDM method developed based on the proposed mixed-weighted-averaging operation solved flexible decision-making problems involving SIVFMSs. Furthermore, the proposed GDM method was utilized for an actual example of a supplier selection problem to indicate its application. Through the comparative analysis with existing relative decision-making methods, the proposed GDM method demonstrated its rationality and effectiveness. However, this study not only effectively solved the expression and operation problems of the mixed information of single- and interval-valued fuzzy sequences with identical and/or different fuzzy values, but also strengthened the GDM rationality and flexibility with the help of the presented information entropy and the proposed mixed-weighted-averaging operation, which highlighted its merits when dealing with SIVFMSs.

This original study demonstrated new contributions in mixed fuzzy information expression, presented a transformation method based on the mean and information entropy of SIVFME, and presented mixed aggregation operations of IVFEEs and their GDM method in the environment of SIVFMSs. However, the new techniques proposed in this paper can only handle GDM problems with SIVFMSs, but cannot solve GDM problems with the fuzzy information of truth and falsity membership degrees. Regarding future research, this study will be further extended to image processing, pattern recognition, clustering analysis, and their applications in the setting of SIVFMSs. Then, the Aczel–Alsina operations and aggregation operators [21,22], and their applications, will be further developed in the intuitionistic and interval-valued intuitionistic fuzzy multivalued context.

Author Contributions: Conceptualization, W.L. and J.Y.; methodology, W.L. and J.Y.; software, J.Y.; validation, W.L. and J.Y.; formal analysis, W.L.; investigation, W.L.; resources, W.L.; data curation, W.L.; writing—original draft preparation, W.L.; writing—review and editing, W.L.; visualization, J.Y.; supervision, J.Y.; project administration, J.Y. All authors have read and agreed to the published version of the manuscript.

Funding: This research received no external funding.

Institutional Review Board Statement: Not applicable.

Informed Consent Statement: Not applicable.

Data Availability Statement: We did not use any data for this research work.

Conflicts of Interest: The authors declare no conflict of interest.

References

1. Zadeh, L.A. Fuzzy sets. *Inf. Control* **1965**, *8*, 338–353. [CrossRef]
2. Gorzałczany, M.B. A method of inference in approximate reasoning based on interval valued fuzzy sets. *Fuzzy Sets Syst.* **1987**, *21*, 1–17. [CrossRef]
3. Yager, R.R. On the theory of bags. *Int. J. Gen. Syst.* **1986**, *13*, 23–37. [CrossRef]
4. Miyamoto, S. *Fuzzy Multisets and Their Generalizations*; Springer: Berlin, Germany, 2000; pp. 225–235.
5. Kreinovich, V.; Sriboonchitta, S. For multi-interval-valued fuzzy sets, centroid defuzzification is equivalent to defuzzifying its interval hull: A theorem. In *Advances in Computational Intelligence, Lecture Notes in Computer Science*; Sidorov, G., Herrera-Alcántara, O., Eds.; Springer: Cham, Switzerland, 2016; Volume 100612017.
6. Li, B. Fuzzy bags and application. *Fuzzy Sets Syst.* **1999**, *34*, 67–71. [CrossRef]
7. Miyamoto, S. Fuzzy multisets and fuzzy clustering of documents. In Proceedings of the 10th TEEE International Conference on Fuzzy Systems, Melbourne, Australia, 2–5 December 2001; pp. 1539–1542.
8. Banatre, J.P.; Le Metayer, D. Programming by multiset transformation. *Commun. ACM* **1993**, *36*, 98–111. [CrossRef]
9. Miyamoto, S. Generalized bags, bag relations, and applications to data analysis and decision making. *Modeling Decis. Artif. Intell.* **2009**, *5861*, 37–54.
10. El-Azab, M.S.; Shokry, M.; Abo khadra, R.A. Correlation measure for fuzzy multisets. *J. Egypt. Math. Soc.* **2017**, *25*, 263–267. [CrossRef]
11. Torra, V. Hesitant fuzzy sets. *Int. J. Intell. Syst.* **2010**, *25*, 529–539. [CrossRef]
12. Fu, J.; Ye, J.; Cui, W.H. An evaluation method of risk grades for prostate cancer using similarity measure of cubic hesitant fuzzy sets. *J. Biomed. Inform.* **2018**, *87*, 131–137. [CrossRef] [PubMed]
13. Yong, R.; Zhu, A.; Ye, J. Multiple attribute decision method using similarity measure of cubic hesitant fuzzy sets. *J. Intell. Fuzzy Syst.* **2019**, *37*, 1075–1083. [CrossRef]
14. Khan, Q.; Mahmood, T.; Mehmood, F. Cubic hesitant fuzzy sets and their applications to multi criteria decision making. *Int. J. Algebra Stat.* **2016**, *5*, 19–51.
15. Fahmi, A.; Amin, F. Precursor selection for Sol–Gel synthesis of titanium carbide nanopowders by a new hesitant cubic fuzzy multi-attribute group decision-making model. *New Math. Nat. Comput.* **2019**, *15*, 145–167. [CrossRef]
16. Xu, Z.S.; Zhou, W. Consensus building with a group of decision makers under the hesitant probabilistic fuzzy environment. *Fuzzy Optim. Decis. Mak.* **2017**, *16*, 481–503. [CrossRef]
17. Park, J.H.; Park, Y.K.; Son, M.J. Hesitant probabilistic fuzzy information aggregation using Einstein operations. *Information* **2018**, *9*, 226. [CrossRef]
18. Turkarslan, E.; Ye, J.; Unver, M.; Olgun, M. Consistency fuzzy sets and a cosine similarity measure in fuzzy multiset setting and application to medical diagnosis. *Math. Probl. Eng.* **2021**, *2021*, 9975983. [CrossRef]
19. Du, C.; Ye, J. Hybrid weighted aggregation operator of cubic fuzzy-consistency elements and their group decision-making model in cubic fuzzy multi-valued setting. *J. Intell. Fuzzy Syst.* **2021**, *41*, 7373–7386. [CrossRef]
20. Shannon, C.E. A mathematical theory of communication. *Bell Syst. Tech. J.* **1948**, *27*, 379–423. [CrossRef]
21. Senapati, T.; Chen, G.; Mesiar, R.; Yager, R.R. Novel Aczel–Alsina operations-based interval-valued intuitionistic fuzzy aggregation operators and their applications in multiple attribute decision-making process. *Int. J. Intell. Syst.* **2021**. [CrossRef]
22. Mesiar, R.; Kolesárová, A.; Senapati, T. Aggregation on lattices isomorphic to the lattice of closed subintervals of the real unit interval. *Fuzzy Sets Syst.* **2022**. [CrossRef]

mathematics MDPI

Article

A Hybrid Intuitionistic Fuzzy Group Decision Framework and Its Application in Urban Rail Transit System Selection

Bing Yan [1], Yuan Rong [2], Liying Yu [2] and Yuting Huang [2,*]

1 School of Urban Railway Transportation, Shanghai University of Engineering Science, Shanghai 201620, China; bingyan@sues.edu.cn
2 School of Management, Shanghai University, Shanghai 200444, China; ry1995@shu.edu.cn (Y.R.); yuliying@shu.edu.cn (L.Y.)
* Correspondence: huangyuting1018@shu.edu.cn

Abstract: The selection of an urban rail transit system from the perspective of green and low carbon can not only promote the construction of an urban rail transit system but also have a positive impact on urban green development. Considering the uncertainty caused by different conflict criteria and the fuzziness of decision-making experts' cognition in the selection process of a rail transit system, this paper proposes a hybrid intuitionistic fuzzy MCGDM framework to determine the priority of a rail transit system. To begin with, the weights of experts are determined based on the improved similarity method. Secondly, the subjective weight and objective weight of the criterion are calculated, respectively, according to the DEMATEL and CRITIC methods, and the comprehensive weight is calculated by the linear integration method. Thirdly, considering the regret degree and risk preference of experts, the COPRAS method based on regret theory is propounded to determine the prioritization of urban rail transit system ranking. Finally, urban rail transit system selection of City N is selected for the case study to illustrate the feasibility and effectiveness of the developed method. The results show that a metro system (P_1) is the most suitable urban rail transit system for the construction of city N, followed by a municipal railway system (P_7). Sensitivity analysis is conducted to illustrate the stability and robustness of the designed decision framework. Comparative analysis is also utilized to validate the efficacy, feasibility and practicability of the propounded methodology.

Keywords: urban rail transit; intuitionistic fuzzy set; regret theory; DEMATEL; CRITIC; COPRAS

MSC: 90B50; 94D05

Citation: Yan, B.; Rong, Y.; Yu, L.; Huang, Y. A Hybrid Intuitionistic Fuzzy Group Decision Framework and Its Application in Urban Rail Transit System Selection. *Mathematics* **2022**, *10*, 2133. https://doi.org/10.3390/math10122133

Academic Editors: Jun Ye and Yanhui Guo

Received: 22 May 2022
Accepted: 16 June 2022
Published: 19 June 2022

1. Introduction

At present, environmental problems, such as acid rain, air pollution and global warming, are prominent. One of the important reasons for this series of environmental problems is the emission of a large number of greenhouse gases caused by urban traffic operation. Severe environmental problems affect the ecological balance and human health [1]. The large-scale increase in the number of cars stems from the deepening degree of urbanization. The process of urbanization is accelerating, the construction of urban infrastructure is gradually improving and many cities have successfully entered the automotive era with the progress of society and economic development. However, although the popularity of cars has greatly facilitated people's lives, a series of problems, such as vehicle exhaust pollution and traffic congestion, need to be paid attention to. Urban environmental problems caused by automobile operation restrict the green development of the city. As the center of population, economy and transportation, it is particularly important to realize urban sustainable development.

Green travel can save energy, alleviate traffic congestion, reduce environmental pollution and promote sustainable urban development. An urban public transport system plays an important role in promoting urban sustainable development [2]. As one of the most effective green and low-carbon transportation modes, urban public transport is an important part of green travel. It is mainly composed of buses and urban rail transit. Buses can meet the daily travel of the public in small cities, but buses are far from meeting the daily travel of the public in medium and large cities with a large population density, wide range of activities and large passenger flow. Therefore, in order to alleviate the traffic pressure in urban areas, the construction of an urban rail transit system has become the focus of attention. As the backbone of urban public transport, urban rail transit has the characteristics of being fast, convenient, efficient, safe and comfortable [3]. With the development of the economy and the progress of science and technology, urban rail transit has developed rapidly, but, in this process, its green standard has been formulated relatively late and a perfect development system has not been formed, resulting in a series of problems regarding that the existing urban rail transit does not adapt to the green development in terms of environment, resources and equipment allocation. Hence, it is particularly important to select the urban rail system from the perspective of green and low-carbon transportation.

Since the problem of urban rail transit system selection involves multiple criteria and different types of urban rail transit systems, it requires the joint discussion of experts in various fields to make decisions. Therefore, the problem of urban rail transit system selection can be regarded as an MCGDM problem. In addition, limited by the complexity of the decision-making environment and the inherent uncertainty of practical problems, a traditional deterministic decision is difficult to solve such complex and uncertain decision problems. As an effective tool to describe uncertainty, IFS [4] are proposed to use membership degree, non-membership degree and hesitation degree to express uncertain information more comprehensively by expanding fuzzy set theory. In terms of information measurement, Das et al. [5] studied the relationship between intuitionistic fuzzy information measurement and its similarity measurement, distance measurement and knowledge measurement based on the intuitionistic fuzzy framework. Mishra et al. [6] proposed a series of similarity measures and entropy measures based on the cosine function and logarithmic function under an intuitionistic fuzzy environment. In terms of decision methods, Ecer and Pamucar [7] proposed a method to rank insurance companies according to Marcos under an intuitionistic fuzzy environment. Schitea et al. [8] proposed a MCDM method based on IFS to select the best location for the summary location of hydrogen mobility in Romania. Mishra et al. [9] developed a fuzzy decision method for ranking and evaluating low-carbon sustainable suppliers by combining IFS and distance-based combined evaluation. As for the intuitionistic fuzzy preference relationship, Zhang et al. [10] studied the distance-based consistency measure in group-decision-making with an intuitionistic multiplication preference relationship and proposed some new distance measures between intuitionistic multiplication sets. Meng et al. [11] studied group-decision-making with heterogeneous intuitionistic fuzzy preference relations, including intuitionistic fuzzy preference relations, multiplicative intuitionistic fuzzy preference relations, etc.

Considering that the dimensions of different criteria are different, and there are differences, conflicts and mutual influences between criteria, the DEMATEL [12] method developed by the Geneva center of Battelle Geneva Research Centre can represent the causal logical relationship between criteria, which can visualize the structure of a complex causal relationship with the help of a matrix or graph. In the DEMATEL method, by calculating the cause degree and centrality of each criterion according to the relative importance of each criterion provided by experts, that is, the influence degree and influence degree of each criterion on other criteria, the subjective weight of each standard can then be determined according to the cause degree and centrality. This structured approach helps to analyze the interdependencies between criteria. The DEMATEL method is widely used. Many researchers use the DEMATEL method for criterion evaluation or factor analysis. For example, Topgul et al. [13] used the IF-DEMATEL method to evaluate the green degree of four stages of incoming logistics in plant logistics, outgoing logistics and reverse logistics in the supply chain. Roostaie et al. [14] used the DEMATEL method to analyze the factors affecting the sustainability of buildings. Tseng et al. [15] and Liu et al. [16], respectively, analyzed the obstacles to the adoption of renewable energy and China's sustainable food consumption and production by using the DEMATEL method under the triangular fuzzy environment. In addition, DEMATEL can also be used to determine the subjective weight of criteria in MCDM problems, then evaluate the alternatives in combination with different evaluation methods and, finally, select the optimal alternative. For example, Hosseini et al. [17] and Li et al. [18], respectively, used the DEMATEL and VIKOR methods to evaluate solutions for ecotourism centers during the COVID-19 pandemic and select for a machine tool under the triangular fuzzy environment. Fang et al. [19] used the DEMATEL and TOPSIS methods to evaluate the energy investment risk and safety management system.

Experts have bounded rationality in the reality decision analysis procedure [20], and the psychological preference of experts will affect the decision-making results, so it is necessary to consider the psychological behavior of experts. As an important branch of behavioral decision-making theory, the regret theory proposed by Lomes and Suggen [21] and Bell [22] describes the regret avoidance behavior of decision-makers in the decision process through the regret–rejoice function and the risk preference coefficient of decision-makers. For the application of regret theory, many researchers combine regret theory with decision methods to put forward a group decision framework [23,24]. In other respects, Zhang et al. [25] developed a case retrieval method based on regret theory. Liu and Cheng [26] combined the likelihood-based MABAC method with regret theory to establish a new MCGDM method. Liang and Wang [20] developed an extended scoring method of gain and loss of advantage based on regret theory and the interval evidence reasoning method. Huang and Zhan [27] proposed a three-way decision-making method based on regret theory. Liu et al. [28] proposed a new method combining regret theory and the evaluation method based on average solution distance.

In the past few decades, researchers have proposed many new methods to deal with MCDM problems in real life, such as TOPSIS, VIKOR, MABAC, COPRAS and so on. COPRAS is an MCDM method proposed by Zavadskas et al. [29] in 1994. This method can effectively evaluate the scheme step by step in combination with the importance and effectiveness of the evaluation criteria to obtain the best scheme. It has the characteristics of wide application range and good evaluation effect [30]. The COPRAS method is also widely used. For example, Büyüközkan and Göçer [31] combine AHP and COPRAS to select the best digital supply chain partner. Balali et al. [32] used ANP and COPRAS to rank the effective risks of human resource threats in natural gas supply projects. Mishra et al. [33] and Alipour et al. [34] proposed the combination of SWARA and COPRAS for the sustainability evaluation of the bioenergy production process and the selection of fuel cell and hydrogen component suppliers, respectively. Yuan et al. [35] and Narayanamoorty et al. [36], respectively, used DEMATEL and COPRAS to evaluate and select the third-party logistics suppliers and the best alternative fuel, but both of them only used the subjective weight determination method to determine the attribute weight. In addition, although the methodological framework proposed by many scholars takes into account the regret theory, there are, however, no studies combining regret theory with the COPRAS method to provide decision support for the selection of an urban rail transit system.

Based on the above analysis, the motivations of this study are as follows:

(1) The selection of an urban rail transit system plays an important role in the sustainable development of the city, but now there is no unified standard for the selection of an urban rail transit system, and the construction of urban rail transit involves many aspects. Therefore, it is necessary to determine the corresponding evaluation criteria to select the appropriate type of urban rail transit system.
(2) In the MCGDM problem, the weight of the criterion is a very important part. In the existing decision-making models, most studies only consider the subjective or objective weight model, and the criterion weight determination method is single, which is difficult to comprehensively consider the subjective and objective importance of the criterion so as to affect the final decision-making results. Therefore, it is necessary to establish the comprehensive weight of a criterion determination model

considering the subjective and objective influence to obtain more reasonable and credible decision-making results.

(3) Through literature analysis, it is found that the intuitionistic fuzzy group decision methods in the existing research rarely consider the interaction between criteria in the decision-making process, and most decision-making methods determine the optimal alternatives based on the traditional utility theory, ignoring the psychological behavior of experts in the decision process.

According to the above research motivation, the main contributions of this study are outlined as follows:

(1) Determine evaluation criteria of an urban rail transit system. In order to solve the problem that the existing urban rail transit system selection lacks unified standards, this study establishes the urban rail transit system selection evaluation criteria from four aspects: characteristics, technology, economy and environment.
(2) Build a comprehensive weight determination model of criteria. In order to determine the criterion weight more reasonably, based on the intuitionistic fuzzy environment, the objective weight and subjective weight of the criterion are calculated, respectively, according to DEMATEL and CRITIC, and then the comprehensive weight of the criterion is calculated by the linear integration method and a new comprehensive weight determination model of the criterion is built.
(3) Develop a hybrid intuitionistic fuzzy group decision framework. Based on the proposed intuitionistic fuzzy distance measurement method, the comprehensive weight of the criterion determination model and COPRAS method combined with regret theory, a hybrid group-decision-making framework for urban rail transit system selection is established. Meanwhile, taking city N as an example, the effectiveness and rationality of the method framework proposed in this study are verified.

The rest of this study is organized as follows: the Section 2 is the introduction of preliminaries, including IFS and regret theory. The Section 3 first introduces the proposed intuitionistic fuzzy distance measurement model, and then introduces the detailed steps of the hybrid intuitionistic fuzzy group decision framework proposed in this study. The Section 4 is the application of practical cases and the corresponding sensitivity analysis and comparative analysis and the Section 5 provides the conclusions of this study.

2. Preliminaries

This section briefly introduces the background knowledge needed in this paper, including IFS theory and regret theory.

2.1. Intuitionistic Fuzzy Sets

The following introduces the basic concepts and related theories of IFS.

Definition 1 ([4]). *Let X be a non-empty set, and then*

$$\widetilde{A} = \{(x, \mu_{\widetilde{A}}(x), \gamma_{\widetilde{A}}(x)) | x \in X\} \tag{1}$$

is called intuitionistic fuzzy set on X. Where $\mu_{\widetilde{A}}(x) : X \to [0,1]$ *and* $\gamma_{\widetilde{A}}(x) : X \to [0,1]$ *represent the membership degree and non-membership degree of the subset* $\widetilde{A}$ *of element x in X, respectively, and hold true for all* $x \in X, 0 \le \mu_{\widetilde{A}}(x) + \gamma_{\widetilde{A}}(x) \le 1$ *on* $\widetilde{A}$. $\pi_{\widetilde{A}}(x) = 1 - \mu_{\widetilde{A}}(x) - \gamma_{\widetilde{A}}(x)$, $0 \le \pi_{\widetilde{A}}(x) \le 1$ *represents the hesitation degree or uncertainty degree that element x in X belongs to* $\widetilde{A}$. *The ordinal number pair* $(\mu_{\widetilde{A}}(x), \gamma_{\widetilde{A}}(x))$ *composed of membership degree* $\mu_{\widetilde{A}}(x)$ *and non-membership degree* $\gamma_{\widetilde{A}}(x)$ *are IFNs.*

Definition 2 ([4]). *Let* $\widetilde{\alpha} = (\mu_{\widetilde{\alpha}}, \gamma_{\widetilde{\alpha}})$ *and* $\widetilde{\beta} = \left(\mu_{\widetilde{\beta}}, \gamma_{\widetilde{\beta}}\right)$ *be two IFN, the operational laws of IFNs are:*

(1) $\widetilde{\alpha} \oplus \widetilde{\beta} = \left(x; \mu_{\widetilde{\alpha}} + \mu_{\widetilde{\beta}} - \mu_{\widetilde{\alpha}}\mu_{\widetilde{\beta}}, \gamma_{\widetilde{\alpha}}\gamma_{\widetilde{\beta}}\right)$;

(2) $\widetilde{\alpha} \otimes \widetilde{\beta} = \left(x; \mu_{\widetilde{\alpha}}\mu_{\widetilde{\beta}}, \gamma_{\widetilde{\alpha}} + \gamma_{\widetilde{\beta}} - \gamma_{\widetilde{\alpha}}\gamma_{\widetilde{\beta}}\right)$;

(3) $\widetilde{\alpha} \wedge \widetilde{\beta} = \left(x; \min\left(\mu_{\widetilde{\alpha}}, \mu_{\widetilde{\beta}}\right), \max\left(\gamma_{\widetilde{\alpha}}, \gamma_{\widetilde{\beta}}\right)\right)$;

(4) $\widetilde{\alpha} \vee \widetilde{\beta} = \left(x; \max\left(\mu_{\widetilde{\alpha}}, \mu_{\widetilde{\beta}}\right), \min\left(\gamma_{\widetilde{\alpha}}, \gamma_{\widetilde{\beta}}\right)\right)$;

(5) $\lambda\widetilde{\alpha} = \left(x; 1 - (1 - \mu_{\widetilde{\alpha}})^{\lambda}, (\gamma_{\widetilde{\alpha}})^{\lambda}\right), \lambda > 0$;

(6) $\widetilde{\alpha}^{\lambda} = \left(x; (\mu_{\widetilde{\alpha}})^{\lambda}, 1 - (1 - \gamma_{\widetilde{\alpha}})^{\lambda}\right), \lambda > 0$.

Definition 3. *The score function S and accuracy function H of IFN* $\widetilde{\alpha} = (\mu_{\widetilde{\alpha}}, \gamma_{\widetilde{\alpha}})$ *are defined as* $S(\widetilde{\alpha}) = \mu_{\alpha} - \gamma_{\alpha}$ *and* $H(\widetilde{\alpha}) = \mu_{\alpha} + \gamma_{\alpha}$*; however, when the membership degree is equal to the non-membership degree, the score function cannot be directly used to compare intuitionistic fuzzy numbers. So, Zeng et al.* [37] *proposed a novel score function as below:*

$$S(\widetilde{\alpha}) = \mu_{\widetilde{\alpha}} - \gamma_{\widetilde{\alpha}} - \pi_{\widetilde{\alpha}} \times \frac{\log_2(1 + \pi_{\widetilde{\alpha}})}{100}, S(\widetilde{\alpha}) \in [-1, 1]. \quad (2)$$

Definition 4. *Let* $\widetilde{\alpha} = (\mu_{\widetilde{\alpha}}, \gamma_{\widetilde{\alpha}})$ *and* $\widetilde{\beta} = (\mu_{\widetilde{\beta}}, \gamma_{\widetilde{\beta}})$ *be two IFNs; the order relations between them are defined as follows:*

(1) *If* $S(\widetilde{\alpha}) > S\left(\widetilde{\beta}\right)$*, then* $\widetilde{\alpha}$ *is better than* $\widetilde{\beta}$*, written as* $\widetilde{\alpha} \succ \widetilde{\beta}$.

(2) *If* $S(\widetilde{\alpha}) = S\left(\widetilde{\beta}\right)$, then

  (i) *If* $H(\widetilde{\alpha}) > H\left(\widetilde{\beta}\right)$*, then* $\widetilde{\alpha}$ *is better than* $\widetilde{\beta}$*, written as* $\widetilde{\alpha} \succ \widetilde{\beta}$;

  (ii) *If* $H(\widetilde{\alpha}) = H\left(\widetilde{\beta}\right)$*, then* $\widetilde{\alpha}$ *is equal to* $\widetilde{\beta}$*, written as* $\widetilde{\alpha} = \widetilde{\beta}$.

Definition 5 ([38])**.** *Let* $\widetilde{\alpha_j} = \left(\mu_{\widetilde{\alpha}_j}, \gamma_{\widetilde{\alpha}_j}\right) (j = 1, 2, \cdots, n)$ *be a set of IFNs; the intuitionistic fuzzy weighted aggregation operator is defined as:*

$$IFWA_{\omega}(\widetilde{\alpha_1}, \widetilde{\alpha_2}, \cdots, \widetilde{\alpha_n}) = \left(1 - \prod_{j=1}^{n}\left(1 - \mu_{\widetilde{\alpha}_j}\right)^{\omega_j}, \prod_{j=1}^{n}\left(\gamma_{\widetilde{\alpha}_j}\right)^{\omega_j}\right). \quad (3)$$

where ω_j *is the weight of* $\widetilde{\alpha_j} = \left(\mu_{\widetilde{\alpha}_j}, \gamma_{\widetilde{\alpha}_j}\right), j = 1, 2, \cdots, n, \omega_j \in [0, 1]$ *and* $\sum_{j=1}^{n} \omega_j = 1$.

2.2. Regret Theory

The main idea of regret theory is to compare the results obtained by the selected alternative with the possible results obtained by other alternatives and then characterize the degree of rejoice and regret of decision experts and select the optimal alternative that they will not regret.

Definition 6 ([39])**.** *Let* y_1 *and* y_2 *be the evaluation values of alternatives* P_1 *and* P_2*, and then the perceived utility value of experts on alternative* P_1 *is*

$$u(y_1, y_2) = v(y_1) + R(v(y_1) - v(y_2)). \quad (4)$$

where $v(\cdot)$ *is a monotonically increasing concave utility function satisfying* $v'(\cdot) > 0$ *and* $v''(\cdot) < 0$. $R(\cdot)$ *is a monotonically increasing concave regret–rejoice function satisfying* $R(0) = 0$, $R'(\cdot) > 0$ *and* $R''(\cdot) < 0$. $\Delta v = v(x_1) - v(x_2)$ *represents the utility increment of alternatives* P_1 *and* P_1. $R(\Delta v) > 0$ *means that the decision-maker is willing to choose option* P_1 *and abandon option* P_2*; otherwise, he will regret.*

3. A Hybrid Intuitionistic Fuzzy Group Decision Framework

This part introduces the proposed hybrid intuitionistic fuzzy group decision framework. Firstly, a new intuitionistic fuzzy distance measure is proposed, then the MCGDM problem studied in this paper is described and, finally, the detailed steps of the decision framework are given.

3.1. A Novel Intuitionistic Fuzzy Distance Measure

In this paper, IFS are used to deal with the fuzziness and uncertainty of decision information. In the process of decision-making, intuitionistic fuzzy distance needs to be used many times. In order to better measure intuitionistic fuzzy distance and reduce the lack of information, a novel intuitionistic fuzzy distance measurement method needs to be proposed.

Definition 7. *Let* $\widetilde{\alpha} = \{\widetilde{\alpha}_j | j = 1,2,\cdots,n\}$ *and* $\widetilde{\beta} = \left\{\widetilde{\beta}_j | j = 1,2,\cdots,n\right\}$ *be two intuitionistic fuzzy number vectors, where* $\widetilde{\alpha}_j = (\mu_{\widetilde{\alpha}_j}, \gamma_{\widetilde{\alpha}_j})$, $\widetilde{\beta}_j = (\mu_{\widetilde{\beta}_j}, \gamma_{\widetilde{\beta}_j})$. *The new generalized intuitionistic fuzzy distance measure is defined as follows:*

$$D^\sigma\left(\widetilde{\alpha},\widetilde{\beta}\right) = \left(\frac{1}{3n}\sum_{j=1}^{n}\left(\left|\widetilde{\mu}_j^{\widetilde{\alpha}} - \widetilde{\mu}_j^{\widetilde{\beta}}\right|^\sigma + \left|\widetilde{\gamma}_j^{\widetilde{\alpha}} - \widetilde{\gamma}_j^{\widetilde{\beta}}\right|^\sigma + \left|\widetilde{\pi}_j^{\widetilde{\alpha}} - \widetilde{\pi}_j^{\widetilde{\beta}}\right|^\sigma + \left|\frac{1}{2}\left(s(\widetilde{\alpha}_j) - s\left(\widetilde{\beta}_j\right)\right)\right|^\sigma\right)\right)^{\frac{1}{\sigma}}. \tag{5}$$

Theorem 1. *Let* $\widetilde{\alpha} = \{\widetilde{\alpha}_j | j = 1,2,\cdots,n\}$, $\widetilde{\beta} = \left\{\widetilde{\beta}_j | j = 1,2,\cdots,n\right\}$ *and* $\widetilde{\chi} = \{\widetilde{\chi}_j | j = 1,2,\cdots,n\}$ *be three intuitionistic fuzzy number vectors, and then* $D^\sigma\left(\widetilde{\alpha},\widetilde{\beta}\right)$ *is the intuitionistic fuzzy distance measure.*

(1) $0 \le D^\sigma\left(\widetilde{\alpha},\widetilde{\beta}\right) \le 1$;

(2) $D^\sigma\left(\widetilde{\alpha},\widetilde{\beta}\right) = 0$ if and only if $\widetilde{\alpha} = \widetilde{\beta}$;

(3) $D^\sigma\left(\widetilde{\alpha},\widetilde{\beta}\right) = D^\sigma\left(\widetilde{\beta},\widetilde{\alpha}\right)$,

(4) If $\widetilde{\alpha} \subseteq \widetilde{\beta} \subseteq \widetilde{\chi}$, then $D^\sigma\left(\widetilde{\alpha},\widetilde{\beta}\right) \le D^\sigma(\widetilde{\alpha},\widetilde{\chi})$, $D^\sigma\left(\widetilde{\beta},\widetilde{\chi}\right) \le D^\sigma(\widetilde{\alpha},\widetilde{\chi})$.

(2) and (3) can be proved directly; only (1) and (4) are proved here.

(1) Since $0 \le \widetilde{\mu}_j^{\widetilde{\alpha}}, \widetilde{\mu}_j^{\widetilde{\beta}} \le 1$, $0 \le \widetilde{\gamma}_j^{\widetilde{\alpha}}, \widetilde{\gamma}_j^{\widetilde{\beta}} \le 1$, then

$$0 \le \left|\widetilde{\mu}_j^{\widetilde{\alpha}} - \widetilde{\mu}_j^{\widetilde{\beta}}\right| \le 1,\ 0 \le \left|\widetilde{\gamma}_j^{\widetilde{\alpha}} - \widetilde{\gamma}_j^{\widetilde{\beta}}\right| \le 1,\ \left|\widetilde{\pi}_j^{\widetilde{\alpha}} - \widetilde{\pi}_j^{\widetilde{\beta}}\right| \to 0,$$

$$0 \le \left|\frac{1}{2}\left(s(\widetilde{\alpha}_j) - s\left(\widetilde{\beta}_j\right)\right)\right| \le 1.$$

Hence, $0 \le \left(\left|\widetilde{\mu}_j^{\widetilde{\alpha}} - \widetilde{\mu}_j^{\widetilde{\beta}}\right|^\sigma + \left|\widetilde{\gamma}_j^{\widetilde{\alpha}} - \widetilde{\gamma}_j^{\widetilde{\beta}}\right|^\sigma + \left|\widetilde{\pi}_j^{\widetilde{\alpha}} - \widetilde{\pi}_j^{\widetilde{\beta}}\right|^\sigma + \left|\frac{1}{2}\left(s(\widetilde{\alpha}_j) - s\left(\widetilde{\beta}_j\right)\right)\right|^\sigma\right) \le 3$, for $\sigma \ge 1$, i.e.,

$$0 \le \frac{1}{3n}\sum_{j=1}^{n}\left(\left|\widetilde{\mu}_j^{\widetilde{\alpha}} - \widetilde{\mu}_j^{\widetilde{\beta}}\right|^\sigma + \left|\widetilde{\gamma}_j^{\widetilde{\alpha}} - \widetilde{\gamma}_j^{\widetilde{\beta}}\right|^\sigma + \left|\widetilde{\pi}_j^{\widetilde{\alpha}} - \widetilde{\pi}_j^{\widetilde{\beta}}\right|^\sigma + \left|\frac{1}{2}\left(s(\widetilde{\alpha}_j) - s\left(\widetilde{\beta}_j\right)\right)\right|^\sigma\right) \le 1,$$

$$0 \le \left(\frac{1}{3n}\sum_{j=1}^{n}\left(\left|\widetilde{\mu}_j^{\widetilde{\alpha}} - \widetilde{\mu}_j^{\widetilde{\beta}}\right|^\sigma + \left|\widetilde{\gamma}_j^{\widetilde{\alpha}} - \widetilde{\gamma}_j^{\widetilde{\beta}}\right|^\sigma + \left|\widetilde{\pi}_j^{\widetilde{\alpha}} - \widetilde{\pi}_j^{\widetilde{\beta}}\right|^\sigma + \left|\frac{1}{2}\left(s(\widetilde{\alpha}_j) - s\left(\widetilde{\beta}_j\right)\right)\right|^\sigma\right)\right)^{\frac{1}{\sigma}} \le 1.$$

(4) Since $\widetilde{\alpha} \subseteq \widetilde{\beta} \subseteq \widetilde{\chi}$, then $\widetilde{\mu}_j^{\widetilde{\alpha}} \leq \widetilde{\mu}_j^{\widetilde{\beta}} \leq \widetilde{\mu}_j^{\widetilde{\chi}}$, $\widetilde{\gamma}_j^{\widetilde{\alpha}} \leq \widetilde{\gamma}_j^{\widetilde{\beta}} \leq \widetilde{\gamma}_j^{\widetilde{\chi}}$, $S(\widetilde{\alpha}_j) \geq S\left(\widetilde{\beta}_j\right) \geq S(\widetilde{\chi}_j)$ for all $x_j \in X$. Then, we have

$$\left|\widetilde{\mu}_j^{\widetilde{\alpha}} - \widetilde{\mu}_j^{\widetilde{\beta}}\right|^\sigma \leq \left|\widetilde{\mu}_j^{\widetilde{\alpha}} - \widetilde{\mu}_j^{\widetilde{\chi}}\right|^\sigma, \left|\widetilde{\mu}_j^{\widetilde{\beta}} - \widetilde{\mu}_j^{\widetilde{\chi}}\right|^\sigma \leq \left|\widetilde{\mu}_j^{\widetilde{\alpha}} - \widetilde{\mu}_j^{\widetilde{\chi}}\right|^\sigma;$$

$$\left|\widetilde{\gamma}_j^{\widetilde{\alpha}} - \widetilde{\gamma}_j^{\widetilde{\beta}}\right|^\sigma \leq \left|\widetilde{\gamma}_j^{\widetilde{\alpha}} - \widetilde{\gamma}_j^{\widetilde{\chi}}\right|^\sigma, \left|\widetilde{\gamma}_j^{\widetilde{\beta}} - \widetilde{\gamma}_j^{\widetilde{\chi}}\right|^\sigma \leq \left|\widetilde{\gamma}_j^{\widetilde{\alpha}} - \widetilde{\gamma}_j^{\widetilde{\chi}}\right|^\sigma;$$

$$\left|\frac{1}{2}\left(S(\widetilde{\alpha}_j) - S\left(\widetilde{\beta}_j\right)\right)\right|^\sigma \leq \left|\frac{1}{2}(S(\widetilde{\alpha}_j) - S(\widetilde{\chi}_j))\right|^\sigma \leq, \left|\frac{1}{2}\left(S\left(\widetilde{\beta}_j\right) - S\chi\right)\right|^\sigma \leq \left|\frac{1}{2}(S(\widetilde{\alpha}_j) - S(\widetilde{\chi}_j))\right|^\sigma.$$

Thus,

$$\left(\left|\widetilde{\mu}_j^{\widetilde{\alpha}} - \widetilde{\mu}_j^{\widetilde{\beta}}\right|^\sigma + \left|\widetilde{\gamma}_j^{\widetilde{\alpha}} - \widetilde{\gamma}_j^{\widetilde{\beta}}\right|^\sigma + \left|\widetilde{\pi}_j^{\widetilde{\alpha}} - \widetilde{\pi}_j^{\widetilde{\beta}}\right|^\sigma + \left|\tfrac{1}{2}\left(S(\widetilde{\alpha}_j) - S\left(\widetilde{\beta}_j\right)\right)\right|^\sigma\right) \leq \left(\left|\widetilde{\mu}_j^{\widetilde{\alpha}} - \widetilde{\mu}_j^{\widetilde{\chi}}\right|^\sigma + \left|\widetilde{\gamma}_j^{\widetilde{\alpha}} - \widetilde{\gamma}_j^{\widetilde{\chi}}\right|^\sigma + \left|\widetilde{\pi}_j^{\widetilde{\alpha}} - \widetilde{\pi}_j^{\widetilde{\chi}}\right|^\sigma + \left|\tfrac{1}{2}(S(\widetilde{\alpha}_j) - S(\widetilde{\chi}_j))\right|^\sigma\right)$$

$$\left(\left|\widetilde{\mu}_j^{\widetilde{\beta}} - \widetilde{\mu}_j^{\widetilde{\chi}}\right|^\sigma + \left|\widetilde{\gamma}_j^{\widetilde{\beta}} - \widetilde{\gamma}_j^{\widetilde{\chi}}\right|^\sigma + \left|\widetilde{\pi}_j^{\widetilde{\beta}} - \widetilde{\pi}_j^{\widetilde{\chi}}\right|^\sigma + \left|\tfrac{1}{2}\left(S\left(\widetilde{\beta}_j\right) - S(\widetilde{\chi}_j)\right)\right|^\sigma\right) \leq \left(\left|\widetilde{\mu}_j^{\widetilde{\alpha}} - \widetilde{\mu}_j^{\widetilde{\chi}}\right|^\sigma + \left|\widetilde{\gamma}_j^{\widetilde{\alpha}} - \widetilde{\gamma}_j^{\widetilde{\chi}}\right|^\sigma + \left|\widetilde{\pi}_j^{\widetilde{\alpha}} - \widetilde{\pi}_j^{\widetilde{\chi}}\right|^\sigma + \left|\tfrac{1}{2}(S(\widetilde{\alpha}_j) - S(\widetilde{\chi}_j))\right|^\sigma\right)$$

Furthermore,

$$\begin{aligned} D^\sigma\left(\widetilde{\alpha}, \widetilde{\beta}\right) &= \left(\tfrac{1}{3n}\sum_{j=1}^{n}\left(\left|\widetilde{\mu}_j^{\widetilde{\alpha}} - \widetilde{\mu}_j^{\widetilde{\beta}}\right|^\sigma + \left|\widetilde{\gamma}_j^{\widetilde{\alpha}} - \widetilde{\gamma}_j^{\widetilde{\beta}}\right|^\sigma + \left|\widetilde{\pi}_j^{\widetilde{\alpha}} - \widetilde{\pi}_j^{\widetilde{\beta}}\right|^\sigma + \left|\tfrac{1}{2}\left(S(\widetilde{\alpha}_j) - S\left(\widetilde{\beta}_j\right)\right)\right|^\sigma\right)\right)^{\frac{1}{\sigma}} \\ &\leq \left(\tfrac{1}{3n}\sum_{j=1}^{n}\left(\left|\widetilde{\mu}_j^{\widetilde{\alpha}} - \widetilde{\mu}_j^{\widetilde{\chi}}\right|^\sigma + \left|\widetilde{\gamma}_j^{\widetilde{\alpha}} - \widetilde{\gamma}_j^{\widetilde{\chi}}\right|^\sigma + \left|\widetilde{\pi}_j^{\widetilde{\alpha}} - \widetilde{\pi}_j^{\widetilde{\chi}}\right|^\sigma + \left|\tfrac{1}{2}(S(\widetilde{\alpha}_j) - S(\widetilde{\chi}_j))\right|^\sigma\right)\right)^{\frac{1}{\sigma}} \\ &= D^\sigma(\widetilde{\alpha}, \widetilde{\chi}) \end{aligned}$$

$$\begin{aligned} D^\sigma\left(\widetilde{\beta}, \widetilde{\chi}\right) &= \left(\tfrac{1}{3n}\sum_{j=1}^{n}\left(\left|\widetilde{\mu}_j^{\widetilde{\beta}} - \widetilde{\mu}_j^{\widetilde{\chi}}\right|^\sigma + \left|\widetilde{\gamma}_j^{\widetilde{\beta}} - \widetilde{\gamma}_j^{\widetilde{\chi}}\right|^\sigma + \left|\widetilde{\pi}_j^{\widetilde{\beta}} - \widetilde{\pi}_j^{\widetilde{\chi}}\right|^\sigma + \left|\tfrac{1}{2}\left(S\left(\widetilde{\beta}_j\right) - S(\widetilde{\chi}_j)\right)\right|^\sigma\right)\right)^{\frac{1}{\sigma}} \\ &\leq \left(\tfrac{1}{3n}\sum_{j=1}^{n}\left(\left|\widetilde{\mu}_j^{\widetilde{\alpha}} - \widetilde{\mu}_j^{\widetilde{\chi}}\right|^\sigma + \left|\widetilde{\gamma}_j^{\widetilde{\alpha}} - \widetilde{\gamma}_j^{\widetilde{\chi}}\right|^\sigma + \left|\widetilde{\pi}_j^{\widetilde{\alpha}} - \widetilde{\pi}_j^{\widetilde{\chi}}\right|^\sigma + \left|\tfrac{1}{2}(S(\widetilde{\alpha}_j) - S(\widetilde{\chi}_j))\right|^\sigma\right)\right)^{\frac{1}{\sigma}} \\ &= D^\sigma(\widetilde{\alpha}, \widetilde{\chi}) \end{aligned}$$

Accordingly, $D^\sigma\left(\widetilde{\alpha}, \widetilde{\beta}\right) \leq D^\sigma(\widetilde{\alpha}, \widetilde{\chi})$ and $D^\sigma\left(\widetilde{\beta}, \widetilde{\chi}\right) \leq D^\sigma(\widetilde{\alpha}, \widetilde{\chi})$.

Definition 8. *Let $\widetilde{A} = (\widetilde{\alpha_{ij}})_{m \times n}$ and $\widetilde{B} = \left(\widetilde{\beta_{ij}}\right)_{m \times n}$ be two intuitionistic fuzzy matrices, where $\widetilde{\alpha_{ij}} = \left(\mu_{\widetilde{\alpha_{ij}}}, \gamma_{\widetilde{\alpha_{ij}}}\right)$ and $\widetilde{\beta_{ij}} = \left(\mu_{\widetilde{\beta_{ij}}}, \gamma_{\widetilde{\beta_{ij}}}\right)$ are IFNs. Then, the distance between intuitionistic fuzzy matrices $\widetilde{A}$ and $\widetilde{B}$ is defined as follows:*

$$\hat{D}^\sigma\left(\widetilde{A}, \widetilde{B}\right) = \left(\frac{1}{3mn}\sum_{i=1}^{m}\sum_{j=1}^{n}\left(\left|\widetilde{\mu}_{ij}^{\widetilde{A}} - \widetilde{\mu}_{ij}^{\widetilde{B}}\right|^\sigma + \left|\widetilde{\gamma}_{ij}^{\widetilde{A}} - \widetilde{\gamma}_{ij}^{\widetilde{B}}\right|^\sigma + \left|\widetilde{\pi}_{ij}^{\widetilde{A}} - \widetilde{\pi}_{ij}^{\widetilde{B}}\right|^\sigma + \left|\frac{1}{2}\left(S\left(\widetilde{A}_{ij}\right) - S\left(\widetilde{B}_{ij}\right)\right)\right|^\sigma\right)\right)^{\frac{1}{\sigma}}. \tag{6}$$

when $\sigma = 1$, $\sigma = 2$ and $\sigma = +\infty$, $\hat{D}^\sigma\left(\widetilde{A}, \widetilde{B}\right)$ are degenerated to the corresponding intuitionistic fuzzy Hamming distance $\hat{D}^1\left(\widetilde{A}, \widetilde{B}\right)$, Euclidean distance $\hat{D}^2\left(\widetilde{A}, \widetilde{B}\right)$ and Chebyshev distance $\hat{D}^{+\infty}\left(\widetilde{A}, \widetilde{B}\right)$.

$$\hat{D}^1\left(\widetilde{A},\widetilde{B}\right)=\frac{1}{3mn}\sum_{i=1}^{m}\sum_{j=1}^{n}\left(\left|\widetilde{\mu}_{ij}^{\widetilde{A}}-\widetilde{\mu}_{ij}^{\widetilde{B}}\right|+\left|\widetilde{\gamma}_{ij}^{\widetilde{A}}-\widetilde{\gamma}_{ij}^{\widetilde{B}}\right|+\left|\widetilde{\pi}_{ij}^{\widetilde{A}}-\widetilde{\pi}_{ij}^{\widetilde{B}}\right|+\left|\frac{1}{2}\left(s\left(\widetilde{A}_{ij}\right)-s\left(\widetilde{B}_{ij}\right)\right)\right|\right). \quad (7)$$

$$\hat{D}^2\left(\widetilde{A},\widetilde{B}\right)=\sqrt{\frac{1}{3mn}\sum_{i=1}^{m}\sum_{j=1}^{n}\left(\left|\widetilde{\mu}_{ij}^{\widetilde{A}}-\widetilde{\mu}_{ij}^{\widetilde{B}}\right|^2+\left|\widetilde{\gamma}_{ij}^{\widetilde{A}}-\widetilde{\gamma}_{ij}^{\widetilde{B}}\right|^2+\left|\widetilde{\pi}_{ij}^{\widetilde{A}}-\widetilde{\pi}_{ij}^{\widetilde{B}}\right|^2+\left|\frac{1}{2}\left(s\left(\widetilde{A}_{ij}\right)-s\left(\widetilde{B}_{ij}\right)\right)\right|^2\right)}. \quad (8)$$

$$\hat{D}^{+\infty}\left(\widetilde{A},\widetilde{B}\right)=\max_{\substack{1\le i\le m\\ 1\le j\le m}}\left(\left|\widetilde{\mu}_{ij}^{\widetilde{A}}-\widetilde{\mu}_{ij}^{\widetilde{B}}\right|,\left|\widetilde{\gamma}_{ij}^{\widetilde{A}}-\widetilde{\gamma}_{ij}^{\widetilde{B}}\right|,\left|\widetilde{\pi}_{ij}^{\widetilde{A}}-\widetilde{\pi}_{ij}^{\widetilde{B}}\right|,\left|\frac{1}{2}\left(s\left(\widetilde{A}_{ij}\right)-s\left(\widetilde{B}_{ij}\right)\right)\right|\right). \quad (9)$$

3.2. Problem Statement

For the MCGDM problem under the intuitionistic fuzzy environment, let $P_i(i=1,2,\cdots,m)$ be the set of urban rail transit system types, $Q_j(j=1,2,\cdots,n)$ be the set of criteria. $\omega_j(j=1,2,\cdots,n)$ is the weight of the criterion $Q_j(j=1,2,\cdots,n)$ and satisfying $0\le\omega_j\le 1$, $\sum_{j=1}^{n}\omega_j=1$. $D_k(k=1,2,\cdots,K)$ is the set of experts. The corresponding weight of expert is expressed as $\lambda_k(k=1,2,\cdots,K)$ and satisfying $0\le\lambda_k\le 1$, $\sum_{k=1}^{K}\lambda_k=1$. $\widetilde{E}^k=(\widetilde{e}_{ij}^k)_{m\times n}$ represents the evaluation value of the urban rail transit system P_i under criterion Q_j given by the kth expert.

3.3. Detailed Steps of the Hybrid Intuitionistic Fuzzy Group Decision Framework

This paper developed a hybrid group decision framework considering the psychological behavior of experts under the intuitionistic fuzzy environment. Firstly, experts express their qualitative evaluation through linguistic variables and then obtain the intuitionistic fuzzy decision matrix of experts. Secondly, the weight information of experts is determined by similarity method based on the proposed intuitionistic fuzzy distance measure, and then the aggregation decision matrix is obtained. Thirdly, the subjective weight and objective weight of attributes are obtained by DEMATEL and CRITIC methods, respectively, and the comprehensive weights of criteria are obtained by linear integration method. DEMATEL method can fully consider the relationship between criteria, making the final subjective weight results more accurate. CRITIC method is based on the contrast strength of criteria and the conflict between criteria to comprehensively measure the objective weight of criteria. The objective attribute of the data itself is fully used for scientific evaluation. In the stage of ranking, the COPRAS method based on regret theory is used to calculate the comprehensive evaluation value of the scheme and finally determine the ranking of the urban rail transit systems. COPRAS method is simple to operate, does not need standardization process and can reduce the lack of evaluation information. The detailed steps are as follows and the method framework is shown in Figure 1.

(1) Stage 1 Collect the evaluation information

Step 1.1: Obtain the linguistic decision matrix.

The evaluation value of $P_i(i=1,2,\cdots,m)$ in criterion $Q_j(j=1,2,\cdots,n)$ is given by expert $D_k(k=1,2,\cdots,K)$ in the form of linguistic variables.

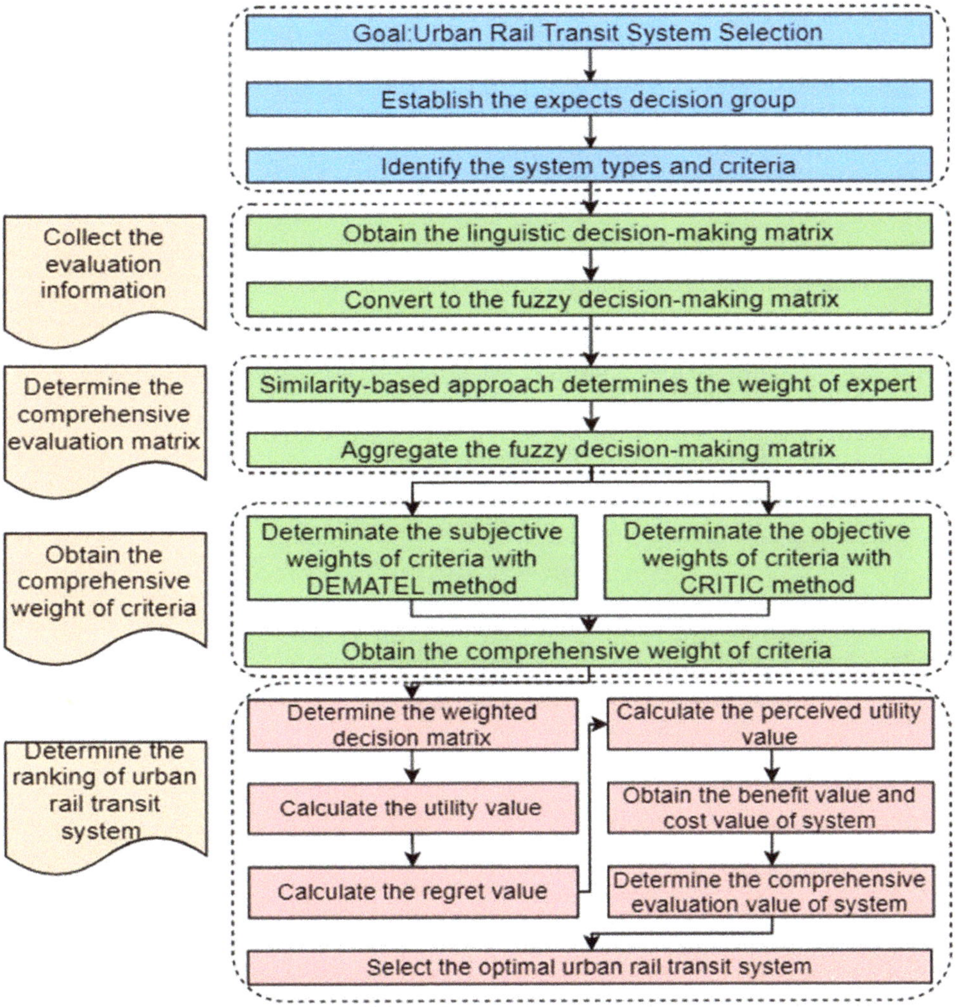

Figure 1. The framework of the proposed method.

Step 1.2: Convert to the fuzzy decision matrix.

The linguistic evaluation value is transformed into intuitionistic fuzzy number, and then obtain the intuitionistic fuzzy evaluation matrix. Table 1 lists the linguistic variables, which reflect the transformation relationship between linguistic variables of decision matrix and IFNs.

$$\widetilde{E}^k = \begin{pmatrix} \widetilde{e}^k_{11} & \widetilde{e}^k_{12} & \cdots & \widetilde{e}^k_{1n} \\ \widetilde{e}^k_{21} & \widetilde{e}^k_{22} & \cdots & \widetilde{e}^k_{2n} \\ \vdots & \vdots & \vdots & \vdots \\ \widetilde{e}^k_{m1} & \widetilde{e}^k_{m2} & \cdots & \widetilde{e}^k_{mn} \end{pmatrix}, \widetilde{e}^k_{ij} = (\widetilde{\mu}^k_{ij}, \widetilde{\gamma}^k_{ij}).$$

Table 1. The transformation relationship of decision-making matrix linguistic variables [40].

Linguistic Variables	IFNs
Extremely Low (EL)	(0.10, 0.90, 0.00)
Very Low (VL)	(0.10, 0.75, 0.15)
Low (L)	(0.25, 0.60, 0.15)
Medium Low (ML)	(0.40, 0.50, 0.10)
Medium (M)	(0.50, 0.40, 0.10)
Medium High (MH)	(0.60, 0.30, 0.10)
High (H)	(0.70, 0.20, 0.10)
Very High (VH)	(0.80, 0.10, 0.10)
Extremely High (EH)	(0.90, 0.10, 0.00)

(2) Stage 2 Determine the comprehensive evaluation matrix

Step 2.1: Similarity-based approach determines the weight of expert.

The determination of weights of experts is a key to MCGDM problem. In this study, the weights of experts are determined by similarity method. Generally speaking, the closer the expert's evaluation is to the evaluation of the whole expert group, the greater the expert's weight is.

Step 2.1.1 Obtain the average evaluation matrix of the expert group from Equation (10)

$$\overline{e_{ij}} = (\overline{\mu_{ij}}, \overline{\gamma_{ij}}) = IFWA_{\omega}\left(e_{ij}^{1}, e_{ij}^{2}, \cdots, e_{ij}^{K}\right) = \left(1 - \prod_{k=1}^{K}\left(1-\mu_{ij}^{k}\right)^{\frac{1}{K}}, \prod_{k=1}^{K}\left(\gamma_{ij}^{k}\right)^{\frac{1}{K}}\right). \quad (10)$$

where *IFWA* is intuitionistic fuzzy weighted average operator.

Step 2.1.2 According to Definition 8, the distance between the kth expert's evaluation matrix $E^k = \left(e_{ij}^k\right)_{m\times n}$ and the average evaluation matrix $\overline{E} = (\overline{e_{ij}})_{m\times n}$ of the expert group is expressed as:

$$\begin{aligned} \hat{D}^{1}\left(E^{k},\overline{E}\right) &= \frac{1}{3mn}\sum_{i=1}^{m}\sum_{j=1}^{n}\left(\left|\mu_{ij}^{k}-\overline{\mu}_{ij}\right| + \left|\gamma_{ij}^{k}-\overline{\gamma}_{ij}\right| + \left|\pi_{ij}^{k}-\overline{\pi}_{ij}\right| + \left|\tfrac{1}{2}\left(S(e_{ij}) - S\left(\overline{e}_{ij}\right)\right)\right|\right). \\ \hat{D}^{+\infty}\left(E^{k},\overline{E}\right) &= \max_{\substack{1\le i\le m \\ 1\le j\le n}}\left\{\left|\mu_{ij}^{k}-\overline{\mu}_{ij}\right|, \left|\gamma_{ij}^{k}-\overline{\gamma}_{ij}\right|, \left|\pi_{ij}^{k}-\overline{\pi}_{ij}\right|, \left|\tfrac{1}{2}\left(S(e_{ij}) - S\left(\overline{e}_{ij}\right)\right)\right|\right\}. \end{aligned} \quad (11)$$

Step 2.1.3 Through the control parameters, the comprehensive distance calculated from Equation (12) is:

$$D^{*}\left(E^{k},\overline{E}\right) = \theta\hat{D}^{1}\left(E^{k},\overline{E}\right) + (1-\theta)\hat{D}^{+\infty}\left(E^{k},\overline{E}\right). \quad (12)$$

where $D^{*}\left(E^{k},\overline{E}\right)$ represents comprehensive distance, θ represents balance coefficient, $0 \le \theta \le 1$.

Step 2.1.4 The smaller the distance $d\left(E^{k},\overline{E}\right)$, the greater the weight of the expert. The corresponding weight λ_k is obtained from Equation (13):

$$\lambda_k = \frac{1 - D^{*}\left(E^{k},\overline{E}\right)}{\sum_{k=1}^{K}\left(1 - D^{*}\left(E^{k},\overline{E}\right)\right)}, k = 1, 2, \cdots, K. \quad (13)$$

Step 2.2: Aggregate the fuzzy decision-making matrix.

Using Equation (14), expert decision matrices are aggregated to obtain the comprehensive evaluation decision matrix:

$$e_{ij} = IFWA_{\omega}\left(e_{ij}^1, e_{ij}^2, \cdots, e_{ij}^K\right) = \left(1 - \prod_{k=1}^{K}\left(1 - \tilde{\mu}_{ij}^k\right)^{\lambda_k}, \prod_{k=1}^{K}\left(\tilde{\gamma}_{ij}^k\right)^{\lambda_k}\right). \tag{14}$$

(3) Stage 3 Obtain the comprehensive weight of criteria

Firstly, the subjective weights of criteria are calculated by DEMATEL method, and then the objective weights of criteria are calculated by CRITIC. Finally, the comprehensive weights of criteria are obtained by combining the weight preference coefficient with the subjective and objective weight.

Step 3.1: Determine the subjective weights of criteria with DEMATEL method.

Step 3.1.1 Construct the fuzzy direct-influence matrix

The direct influence relation matrix of criterion Q_j to Q_l is given by expert $D_k(k = 1, 2, \cdots, K)$ in the form of linguistic variables and then transformed into intuitionistic fuzzy numbers to obtain the intuitionistic fuzzy direct-influence matrix $T^k = \left(t_{jl}^k\right)_{n\times n}$.

Step 3.1.2 Aggregate the direct-influence matrices with Equation (15) to determine the group direct-influence matrix $\widetilde{T} = \left(\widetilde{t_{jl}}\right)_{n\times n}$:

$$\widetilde{t_{jl}} = IFWA_{\omega}(t_{jl}^1, t_{jl}^2, \cdots, t_{jl}^K) = \left(1 - \prod_{k=1}^{K}\left(1 - \mu_{jl}^k\right)^{\lambda_k}, \prod_{k=1}^{K}\left(\gamma_{jl}^k\right)^{\lambda_k}\right). \tag{15}$$

where $\mu_{jl} = 1 - \prod\limits_{k=1}^{K}\left(1 - \mu_{jl}^k\right)^{\lambda_k}$, $\gamma_{jl} = \prod\limits_{k=1}^{K}\left(\gamma_{jl}^k\right)^{\lambda_k}$, λ_k is the weight of kth expert, $\lambda_k = \frac{1}{K}$.

Step 3.1.3 Use Equation (16) to standardize the direct-influence matrix to obtain the standardized direct-influence matrix $T' = \left(t'_{jl}\right)_{n\times n}$:

$$t'_{jl} = \frac{t_{jl}}{\max\limits_{1\le j\le n}\left(\sum\limits_{l=1}^{n} t_{jl}\right)}, j, l = 1, 2, \cdots, n. \tag{16}$$

where $t_{jl} = \mu_{jl} - \gamma_{jl} - \pi_{jl} \times \frac{\log_2(1+\pi_{jl})}{100}$, $\pi_{jl} = 1 - \mu_{jl} - \gamma_{jl}$.

Step 3.1.4 Utilize Equation (17) to calculate the total impact matrix $T^* = \left(t_{jl}^*\right)_{n\times n}$:

$$T^* = T' \times \left(I - T'\right)^{-1}. \tag{17}$$

where I is the identity matrix.

Step 3.1.5 Employ Equations (18) and (19) to calculate importance ξ and influence ζ:

$$\xi_j = R_j + C_j, j = 1, 2, \cdots, n. \tag{18}$$

$$\zeta_j = R_j - C_j, j = 1, 2, \cdots, n. \tag{19}$$

where $R_j = \sum\limits_{l=1}^{n} t_{jl}$, $C_j = \sum\limits_{j=1}^{n} t_{jl}$.

Step 3.1.6 Use Equation (20) to obtain the subjective weight ω_j^s of criterion Q_j:

$$\omega_j^s = \frac{\sqrt{\xi_j^2 + \zeta_j^2}}{\sum\limits_{j=1}^{n}\sqrt{\xi_j^2 + \zeta_j^2}}. \tag{20}$$

Step 3.2: Determinate the objective weights of criteria with CRITIC method.

Step 3.2.1 Use Equation (21) to normalize the fuzzy decision-making matrix $\widetilde{E}^k = \left(\widetilde{e}_{ij}^k\right)_{m\times n}$:

$$\begin{aligned} e_{ij}^k &= \left(\mu_{ij}^k, \gamma_{ij}^k\right) = \left(\widetilde{\mu}_{ij}^k, \widetilde{\gamma}_{ij}^k\right), \text{ for benefit criterion} \\ e_{ij}^k &= \left(\mu_{ij}^k, \gamma_{ij}^k\right) = \left(\widetilde{\gamma}_{ij}^k, \widetilde{\mu}_{ij}^k\right), \text{ for cost criterion} \end{aligned} \quad (21)$$

Step 3.2.2 Use Equation (22) to aggregate the fuzzy decision-making matrix:

$$e_{ij}^* = IFWA_\omega\left(e_{ij}^1, e_{ij}^2, \cdots, e_{ij}^K\right) = \left(1 - \prod_{k=1}^{K}\left(1-\mu_{ij}^k\right)^{\lambda_k}, \prod_{k=1}^{K}\left(\gamma_{ij}^k\right)^{\lambda_k}\right). \quad (22)$$

Step 3.2.3 Use Equation (23) to obtain the standard deviation τ_j of the criterion:

$$\tau_j = \sqrt{\frac{1}{n-1}\sum_{i=1}^{m}\left(D^\sigma\left(e_{ij}^*, \overline{e_j}\right)\right)^2}, j = 1, 2, \cdots, n. \quad (23)$$

where $\overline{e_j} = \frac{1}{m}\sum_{i=1}^{m} e_{ij}^* = IFWA_\omega\left(e_{1j}^*, e_{2j}^*, \cdots, e_{mj}^*\right) = \left(1 - \prod_{i=1}^{m}\left(1-\mu_{ij}^*\right)^{\frac{1}{m}}, \prod_{i=1}^{m}\left(\gamma_{ij}^*\right)^{\frac{1}{m}}\right)$.

Use Equation (24) to evaluate correlation coefficient ρ_{jl} between criteria:

$$\rho_{jl} = \frac{\sum_{i=1}^{m}\left[D^\sigma\left(e_{ij}^*, \overline{e_j}\right)\cdot D^\sigma\left(e_{ij}^*, \overline{e_j}\right)\right]}{\sqrt{\sum_{i=1}^{m}\left(D^\sigma\left(e_{ij}^*, \overline{e_j}\right)\right)^2}\sqrt{\sum_{i=1}^{m}\left(D^\sigma\left(e_{ij}^*, \overline{e_j}\right)\right)^2}}, j, l = 1, 2, \cdots, n. \quad (24)$$

where $\overline{e_j} = \frac{1}{m}\sum_{i=1}^{m} e_{ij}^*, \overline{e_l} = \frac{1}{m}\sum_{i=1}^{m} e_{ij}^*, j, l = 1, 2, \cdots, n$.

Step 3.2.4 Use Equation (25) to obtain the objective weight ω_j^o of criterion Q_j:

$$\omega_j^o = \frac{\tau_j\sum_{l=1}^{n}\left(1-\rho_{jl}\right)}{\sum_{j=1}^{n}\left[\tau_j\sum_{l=1}^{n}\left(1-\rho_{jl}\right)\right]}, j = 1, 2, \cdots, n. \quad (25)$$

Step 3.3: Obtain the comprehensive weights ω_j of criteria.

$$\omega_j = \varphi\omega_j^s + (1-\varphi)\omega_j^o. \quad (26)$$

where $\varphi(0 \le \varphi \le 1)$ indicates the relative importance of subjective weight and objective weight severally. Here, it is assumed that the subjective and objective weights are of equal importance, so $\varphi = 0.5$.

(4) Stage 4 Determine the ranking of urban rail transit systems

In this paper, the power function $u(x) = x^\varepsilon$ is used as the utility function of attribute value, where $\varepsilon(0 \le \varepsilon \le 1)$ is risk aversion coefficient, to describe the risk attitude of experts in decision-making, and, the smaller it is, the higher the risk aversion degree of experts is. $R(x) = 1 - \exp(-\vartheta \cdot x)$ is used as the regret and joy function, where it is the regret avoidance coefficient of experts, and the greater the $\vartheta(\vartheta \in [0, +\infty])$ is, the higher the expert's regret avoidance degree is [41].

Let the evaluation value of $P_i(i = 1, 2, \cdots, m)$ be $y_i(i = 1, 2, \cdots, m)$, and then the perceived utility value of experts on P_i is $u_i = v(y_i) + R(v(y_i) - v(y^*))$. Where $y^* = \max_{1\le i\le m}\{y_i\}$ is the utility value of the ideal urban rail transit system type. $R(v(y_i) - v(y^*)) \le 0$ indicates the regret value when the decision-maker chooses P_i and abandons the ideal urban rail transit system type. Therefore, the perceived utility value of experts on the urban rail transit system type includes the utility value of the P_i and the regret value of P_i compared with the ideal urban rail transit system type.

Step 4.1: Determinate comprehensive evaluation value of urban rail transit systems based on COPRAS method considering regret theory.

Step 4.1.1 Determinate the weighted decision matrix:

$$\widehat{U} = \begin{pmatrix} \widehat{u_{ij}} & \widehat{u_{ij}} & \cdots & \widehat{u_{ij}} \\ \widehat{u_{ij}} & \widehat{u_{ij}} & \cdots & \widehat{u_{ij}} \\ \vdots & \vdots & \ddots & \vdots \\ \widehat{u_{ij}} & \widehat{u_{ij}} & \cdots & \widehat{u_{ij}} \end{pmatrix}$$

where $\widehat{u_{ij}} = w_j \cdot u_{ij}$.

Step 4.1.2 Use Equation (27) to calculate the utility value of the P_i under the criterion:

$$\kappa_{ij} = \left(D^{\sigma}\left(e_{ij}, e_j^*\right)\right)^{\varepsilon}. \tag{27}$$

where ε is the risk aversion coefficient of decision-making experts. Based on the previous studies [39,42], $\varepsilon = 0.88$, e_j^* is ideal point. For benefit criteria, $e_j^* = \left(\max\limits_{1\le i\le m} \mu_{ij}, \min\limits_{1\le i\le m} \gamma_{ij}\right)$; for cost criteria, $e_j^* = \left(\min\limits_{1\le i\le m} \mu_{ij}, \max\limits_{1\le i\le m} \gamma_{ij}\right)$.

Step 4.1.3 Use Equation (28) to calculate the regret value of P_i:

$$\xi_{ij} = 1 - \exp(-\vartheta\cdot(\Delta u)). \tag{28}$$

where $\Delta u = \kappa_j^* - \kappa_{ij}$, $\kappa_j^* = \min\limits_{1\le i\le m}\{\kappa_{ij}\}$ is the utility value of ideal point. ϑ is the regret avoidance coefficient of expert.

Step 4.1.4 Utilize Equation (29) to calculate the perceived utility value of P_i:

$$u_{ij} = \kappa_{ij} + \xi_{ij}. \tag{29}$$

Step 4.1.5 Obtain the benefit value and cost value of P_i:

For benefit criteria, use Equation (30) to calculate comprehensive benefit value G_i^+ of P_i:

$$G_i^+ = \sum_{j=1}^{r} {u_{ij}}^+, i = 1, 2, \cdots, m. \tag{30}$$

For cost criteria, use Equation (31) to calculate comprehensive cost value G_i^+ of P_i:

$$G_i^- = \sum_{j=r+1}^{n} {u_{ij}}^-, i = 1, 2, \cdots, m. \tag{31}$$

where "+" and "-" represent "benefit" and "cost", respectively, r is the number of benefit criteria.

Step 4.1.6 Use Equation (32) to determine the comprehensive evaluation value of P_i:

$$H_i = G_i^+ + \frac{\min\limits_i G_i^- \sum\limits_{i=1}^{m} G_i^-}{G_i^- \sum\limits_{i=1}^{m} \frac{\min\limits_i G_i^-}{G_i^-}} = G_i^+ + \frac{\sum\limits_{i=1}^{m} G_i^-}{G_i^- \sum\limits_{i=1}^{m} \frac{1}{G_i^-}}, i = 1, 2, \cdots, m; \min_i G_i^- = \min_{1\le i\le m}\{G_i^-\} \tag{32}$$

Step 4.2: Select the optimal urban rail transit system.

During the process of urban rail transit system selection, the optimal urban rail transit system shall be determined according to the comprehensive utility value H_i calculated by Equation (32). That is, sort H_i from small to large. The larger H_i is, the better the scheme is.

4. Case Study

In this part, firstly, seven types of urban rail transit systems and eight criteria are listed. Secondly, the proposed hybrid decision model is used for the selection of the urban rail transit system of City N, and the optimal urban rail transit system is selected to prove the applicability and effectiveness of the proposed method. Finally, the stability and robustness of the model are verified through sensitivity analysis and comparative analysis.

Therefore, the types and related evaluation criteria of urban rail transit are systematically studied. Based on the existing research and discussion with four experts (Table 2 for experts' background), seven types of urban rail transit and eight criteria were determined to evaluate the types of urban rail transit (Table 3). After the preliminary analysis, an expert group composed of four experts was responsible for the evaluation of urban rail transit types. These decision-makers have played a role in rail transit, universities and government agencies. Next, the steps of the developed method in evaluating the type selection of urban rail transit will be introduced.

Table 2. The background of experts.

Experts	Major	Occupation	Working Experience
D_1	Transportation	Professor	26 years
D_2	Transportation	Professor	22 years
D_3	Transportation	Associate professor	15 years
D_4	Transportation	Researcher	8 years

Table 3. The evaluation criteria of urban rail transit system.

Primary Index	Secondary Index	Type	Description
Characteristic	Transportation capacity (Q_1)	Benefit	It refers to the average number of passengers transported by the rail transit system per hour.
	Transportation speed (Q_2)	Benefit	It refers to the average operating distance of the rail transit system per hour.
Technology	Technology maturity (Q_3)	Benefit	It refers to the maturity of the technology used in the construction of the rail transit system.
	Application degree of green technology (Q_4)	Benefit	It refers to the degree of application of green technology in the design and construction stage of the rail transit system, such as land saving, energy saving, environmental protection technology, etc.
	Construction difficulty (Q_5)	Cost	It refers to the environmental conditions required for the construction of the rail transit system, such as underground, ground, soil requirements, etc.
Economy	Construction cost (Q_6)	Cost	It refers to the average construction cost per kilometer of the rail transit system.
	Operation and maintenance cost (Q_7)	Cost	It refers to the cost required for the operation and maintenance of the rail transit system after the completion of construction.
Environment	Environmental harmony (Q_8)	Benefit	It refers to the influence degree of the noise generated during the operation of the rail transit system on the environment and the environmental quality and aesthetics of the internal environment (vehicles and stations).

4.1. The Types of Urban Rail Transit System

As the backbone of urban public transport, urban rail transit has the characteristics of being fast, convenient, efficient, safe and comfortable. Under the current green and sustainable development policy, this type of system caters to the needs of the new era. According to the research on the classification of various forms of urban rail transit systems, this paper divides the urban rail transit system into seven forms: metro system, light rail system, monorail system, modern tram system, mid–low-speed maglev system, automatic guided track system and municipal railway system.

(1) Metro System (P_1). A metro system is a kind of urban rail transit. It adopts a steel wheel and rail system and mainly operates in tunnels built in underground space of big cities. When conditions permit, it can also pass through the ground and operate on the ground or viaduct.
(2) Light Rail System (P_2). A light rail system refers to the tram or train running on all streets or viaducts. It is a kind of urban rail transit system.
(3) Monorail System (P_3). A monorail system is a medium-volume rail transportation system in which vehicles and special track beams are combined into one. Its track beam is not only the load-bearing structure of vehicles but also the guide track for vehicle operation.
(4) Modern Tram System (P_4). A tram is a rail transit vehicle driven by electricity and running on the track. Because it runs on the street, it is also called road tram, or tram for short.
(5) Mid–Low-Speed Maglev System (P_5). A medium–low-speed maglev is a new technology with independent intellectual property rights in China, and it is also the most advanced technology in urban rail transit. It is applicable to the traffic connection between urban areas, close cities and scenic spots.
(6) Automatic Guided Track System (P_6). Automatic guided track system trains run along special guiding devices. The vehicle operation and stations can be controlled by computer. It can realize full automation and unmanned driving. The automatic guided track system is suitable for urban airport lines and point-to-point transportation lines with relatively concentrated urban passenger flow. When necessary, it can operate with fewer stops in the middle.
(7) Municipal Railway System (P_7). A municipal railway, also known as commuter railway and suburban railway, refers to the passenger rail transit system within the metropolitan area, serving cities and suburbs, central cities and satellite cities, key cities and towns, etc.

4.2. Relevant Criteria

The criteria for urban rail transit system selection are obtained based on literature research and expert consultation summary. The evaluation criteria proposed in this study is from the perspectives of characteristic, technology, economy and environment, with a total of eight criteria, including five benefit criteria and three cost criteria. The detailed description of the criteria is shown in Table 3.

4.3. Method Implementation

In this subsection, based on the above-listed seven urban rail transit system types and eight urban rail transit evaluation criteria, City N is selected as an example to implement the hybrid group decision framework in order to select the most suitable urban rail transit system type for City N. By the end of 2020, the total resident population of city N was 9.404 million, and the population density was 622.52 people per square kilometer. Throughout the year, the whole society completed 75.1333 million passenger trips, including 24.264 million road passenger trips and 40.516 million railway passenger trips. In terms of public transport, at the end of the year, there were 10,035 standard public transport vehicles in the city. Further, 1272 lines were operated, an increase of 8.3%. Rail transit completed 158 million passenger trips in the whole year. At the end of the year, there were

42,000 public bicycles in the city, with a total of 22.597 million car rentals in the whole year. At the end of the year, there were 6281 taxis in the city.

The decision group is still composed of the above four experts, who provide the linguistic evaluation decision matrix and the linguistic attribute direct influence matrix, respectively, as shown in Tables 4 and 5.

Table 4. The linguistic decision-making matrix.

DEs	Urban Rail Transit System	Criteria							
		Q_1	Q_2	Q_3	Q_4	Q_5	Q_6	Q_7	Q_8
D_1	P_1	EH	H	EH	H	VH	VH	VH	H
	P_2	H	H	EH	H	MH	MH	H	VH
	P_3	M	MH	H	VH	L	ML	MH	VH
	P_4	ML	ML	H	VH	ML	L	L	VH
	P_5	M	VH	M	H	H	H	MH	H
	P_6	M	MH	H	H	M	M	M	VH
	P_7	H	EH	VH	H	M	H	H	VH
D_2	P_1	VH	VH	VH	MH	M	M	M	M
	P_2	H	H	VH	MH	M	M	M	M
	P_3	M	M	H	MH	M	ML	M	MH
	P_4	ML	ML	H	MH	M	ML	M	ML
	P_5	M	EH	ML	H	H	H	H	H
	P_6	ML	L	M	MH	M	ML	M	MH
	P_7	VH	VH	VH	MH	M	MH	MH	M
D_3	P_1	VH	VH	VH	MH	M	EH	EH	H
	P_2	MH	MH	VH	MH	M	H	EH	H
	P_3	ML	MH	MH	MH	ML	H	EH	EH
	P_4	ML	ML	MH	ML	H	M	M	M
	P_5	ML	MH	VL	VH	H	H	EH	H
	P_6	VL	ML	VL	MH	ML	M	M	M
	P_7	EH	EH	VH	MH	M	H	VH	H
D_4	P_1	VH	VH	H	M	H	VH	VH	MH
	P_2	MH	MH	H	M	M	MH	ML	H
	P_3	MH	MH	ML	M	ML	M	M	H
	P_4	M	M	ML	M	M	ML	ML	VH
	P_5	H	H	L	MH	MH	H	VH	VH
	P_6	M	M	L	M	MH	VH	VH	MH
	P_7	VH	VH	VH	M	MH	M	M	M

Table 5. The fuzzy direct-influence matrix.

DEs	Criteria	Q_1	Q_2	Q_3	Q_4	Q_5	Q_6	Q_7	Q_8
D_1	Q_1	EL	VL	VL	VL	ML	ML	H	ML
	Q_2	H	EL	VL	VL	MH	MH	H	H
	Q_3	MH	MH	EL	M	VH	VH	VH	H
	Q_4	VL	VL	VL	EL	MH	MH	H	EH
	Q_5	VL	VL	VL	ML	EL	EH	L	L
	Q_6	M	M	VL	VL	MH	EL	L	ML
	Q_7	L	L	VL	VL	VL	VL	EL	VL
	Q_8	VL	VL	VL	VL	VL	VL	VL	EL
D_2	Q_1	EL	MH	ML	MH	ML	ML	ML	MH
	Q_2	M	EL	ML	MH	ML	ML	ML	MH
	Q_3	VH	VH	EL	H	H	M	M	MH
	Q_4	L	L	ML	EL	ML	ML	ML	M
	Q_5	M	H	L	ML	EL	M	ML	M
	Q_6	L	L	H	M	H	EL	ML	ML
	Q_7	ML	M	H	H	MH	M	EL	ML
	Q_8	MH	MH	VH	EH	MH	H	VH	EL

Table 5. *Cont.*

DEs	Criteria	Q_1	Q_2	Q_3	Q_4	Q_5	Q_6	Q_7	Q_8
D_3	Q_1	EL	MH	L	ML	ML	VH	VH	M
	Q_2	VH	EL	MH	ML	VH	MH	VH	M
	Q_3	L	MH	EL	VH	H	M	MH	M
	Q_4	L	ML	VH	EL	H	VH	VH	VH
	Q_5	ML	VH	H	H	EL	VH	VH	VH
	Q_6	VH	MH	M	VH	VH	EL	H	H
	Q_7	VH	VH	MH	VH	H	H	EL	VH
	Q_8	M	M	M	VH	H	H	VH	EL
D_4	Q_1	EL	EL	H	H	VH	VH	VH	M
	Q_2	EL	EL	VH	MH	VH	VH	VH	VH
	Q_3	H	VH	EL	VH	VH	VH	VH	EH
	Q_4	H	MH	VH	EL	M	VH	VH	EH
	Q_5	VH	VH	VH	M	EL	EH	EH	EH
	Q_6	VH	VH	VH	VH	EH	EL	M	M
	Q_7	VH	VH	VH	VH	EH	M	EL	L
	Q_8	M	VH	EH	EH	M	M	L	EL

Then, the linguistic assessment matrix is transformed into a fuzzy evaluation matrix and a fuzzy direct influence matrix represented by intuitionistic fuzzy numbers by the intuitionistic fuzzy scale (adapted from Refs. [33,40]) listed in Tables 1 and 6, as shown in Tables 7 and 8. Then, the expert weight (Table 9) is calculated from Equations (10)–(13), the subjective weight is calculated from Equations (15)–(20), the objective weight is calculated from Equations (21)–(25) and the final comprehensive weight (Table 10) is calculated from Equation (26), and then the ranking of the urban rail transit system most suitable for City N is calculated according to Equations (27)–(32), as shown in Table 11.

Table 6. The transformation relationship of directly affected matrix linguistic variables.

Linguistic Variables	IFNs
Extremely Low (EL)	(0.10, 0.80, 0.10)
Very Low (VL)	(0.20, 0.70, 0.10)
Low (L)	(0.30, 0.60, 0.10)
Medium Low (ML)	(0.40, 0.50, 0.10)
Medium (M)	(0.55, 0.40, 0.05)
Medium High (MH)	(0.65, 0.30, 0.05)
High (H)	(0.75, 0.20, 0.05)
Very High (VH)	(0.90, 0.05, 0.05)
Extremely High (EH)	(1.00, 0.00, 0.00)

Table 7. The fuzzy decision-making matrix.

DEs.	Urban Rail Transit System	Criteria Q_1	Q_2	Q_3	Q_4	Q_5	Q_6	Q_7	Q_8
D_1	P_1	(0.90, 0.10, 0.00)	(0.70, 0.20, 0.10)	(0.90, 0.10, 0.00)	(0.70, 0.20, 0.10)	(0.80, 0.10, 0.10)	(0.80, 0.10, 0.10)	(0.80, 0.10, 0.10)	(0.70, 0.20, 0.10)
	P_2	(0.70, 0.20, 0.10)	(0.70, 0.20, 0.10)	(0.90, 0.10, 0.00)	(0.70, 0.20, 0.10)	(0.60, 0.30, 0.10)	(0.60, 0.30, 0.10)	(0.70, 0.20, 0.10)	(0.80, 0.10, 0.10)
	P_3	(0.50, 0.40, 0.10)	(0.60, 0.30, 0.10)	(0.70, 0.20, 0.10)	(0.80, 0.10, 0.10)	(0.25, 0.60, 0.15)	(0.40, 0.50, 0.10)	(0.60, 0.30, 0.10)	(0.80, 0.10, 0.10)
	P_4	(0.40, 0.50, 0.10)	(0.40, 0.50, 0.10)	(0.70, 0.20, 0.10)	(0.80, 0.10, 0.10)	(0.40, 0.50, 0.10)	(0.25, 0.60, 0.15)	(0.25, 0.60, 0.15)	(0.80, 0.10, 0.10)
	P_5	(0.50, 0.40, 0.10)	(0.80, 0.10, 0.10)	(0.50, 0.40, 0.10)	(0.70, 0.20, 0.10)	(0.70, 0.20, 0.10)	(0.70, 0.20, 0.10)	(0.60, 0.30, 0.10)	(0.70, 0.20, 0.10)
	P_6	(0.50, 0.40, 0.10)	(0.60, 0.30, 0.10)	(0.70, 0.20, 0.10)	(0.70, 0.20, 0.10)	(0.50, 0.40, 0.10)	(0.50, 0.40, 0.10)	(0.50, 0.40, 0.10)	(0.80, 0.10, 0.10)
	P_7	(0.70, 0.20, 0.10)	(0.90, 0.10, 0.00)	(0.80, 0.10, 0.10)	(0.70, 0.20, 0.10)	(0.50, 0.40, 0.10)	(0.70, 0.20, 0.10)	(0.70, 0.20, 0.10)	(0.80, 0.10, 0.10)

Table 7. *Cont.*

DEs.	Urban Rail Transit System	Criteria							
		Q_1	Q_2	Q_3	Q_4	Q_5	Q_6	Q_7	Q_8
D_2	P_1	(0.80, 0.10, 0.10)	(0.80, 0.10, 0.10)	(0.80, 0.10, 0.10)	(0.60, 0.30, 0.10)	(0.50, 0.40, 0.10)	(0.50, 0.40, 0.10)	(0.50, 0.40, 0.10)	(0.50, 0.40, 0.10)
	P_2	(0.70, 0.20, 0.10)	(0.70, 0.20, 0.10)	(0.80, 0.10, 0.10)	(0.60, 0.30, 0.10)	(0.50, 0.40, 0.10)	(0.50, 0.40, 0.10)	(0.50, 0.40, 0.10)	(0.50, 0.40, 0.10)
	P_3	(0.50, 0.40, 0.10)	(0.50, 0.40, 0.10)	(0.70, 0.20, 0.10)	(0.60, 0.30, 0.10)	(0.50, 0.40, 0.10)	(0.40, 0.50, 0.10)	(0.50, 0.40, 0.10)	(0.60, 0.30, 0.10)
	P_4	(0.40, 0.50, 0.10)	(0.40, 0.50, 0.10)	(0.70, 0.20, 0.10)	(0.60, 0.30, 0.10)	(0.50, 0.40, 0.10)	(0.40, 0.50, 0.10)	(0.50, 0.40, 0.10)	(0.40, 0.50, 0.10)
	P_5	(0.50, 0.40, 0.10)	(0.90, 0.10, 0.00)	(0.40, 0.50, 0.10)	(0.70, 0.20, 0.10)	(0.70, 0.20, 0.10)	(0.70, 0.20, 0.10)	(0.70, 0.20, 0.10)	(0.70, 0.20, 0.10)
	P_6	(0.40, 0.50, 0.10)	(0.25, 0.60, 0.15)	(0.50, 0.40, 0.10)	(0.60, 0.30, 0.10)	(0.50, 0.40, 0.10)	(0.40, 0.50, 0.10)	(0.50, 0.40, 0.10)	(0.60, 0.30, 0.10)
	P_7	(0.80, 0.10, 0.10)	(0.80, 0.10, 0.10)	(0.80, 0.10, 0.10)	(0.60, 0.30, 0.10)	(0.50, 0.40, 0.10)	(0.60, 0.30, 0.10)	(0.60, 0.30, 0.10)	(0.50, 0.40, 0.10)
D_3	P_1	(0.80, 0.10, 0.10)	(0.80, 0.10, 0.10)	(0.80, 0.10, 0.10)	(0.60, 0.30, 0.10)	(0.50, 0.40, 0.10)	(0.90, 0.10, 0.00)	(0.90, 0.10, 0.00)	(0.70, 0.20, 0.10)
	P_2	(0.60, 0.30, 0.10)	(0.60, 0.30, 0.10)	(0.80, 0.10, 0.10)	(0.60, 0.30, 0.10)	(0.50, 0.40, 0.10)	(0.70, 0.20, 0.10)	(0.90, 0.10, 0.00)	(0.70, 0.20, 0.10)
	P_3	(0.40, 0.50, 0.10)	(0.60, 0.30, 0.10)	(0.60, 0.30, 0.10)	(0.60, 0.30, 0.10)	(0.40, 0.50, 0.10)	(0.70, 0.20, 0.10)	(0.90, 0.10, 0.00)	(0.90, 0.10, 0.00)
	P_4	(0.40, 0.50, 0.10)	(0.40, 0.50, 0.10)	(0.60, 0.30, 0.10)	(0.40, 0.50, 0.10)	(0.70, 0.20, 0.10)	(0.50, 0.40, 0.10)	(0.50, 0.40, 0.10)	(0.50, 0.40, 0.10)
	P_5	(0.40, 0.50, 0.10)	(0.60, 0.30, 0.10)	(0.10, 0.75, 0.15)	(0.80, 0.10, 0.10)	(0.70, 0.20, 0.10)	(0.70, 0.20, 0.10)	(0.90, 0.10, 0.00)	(0.70, 0.20, 0.10)
	P_6	(0.10, 0.75, 0.15)	(0.40, 0.50, 0.10)	(0.10, 0.75, 0.15)	(0.60, 0.30, 0.10)	(0.40, 0.50, 0.10)	(0.50, 0.40, 0.10)	(0.50, 0.40, 0.10)	(0.50, 0.40, 0.10)
	P_7	(0.90, 0.10, 0.00)	(0.90, 0.10, 0.00)	(0.80, 0.10, 0.10)	(0.60, 0.30, 0.10)	(0.50, 0.40, 0.10)	(0.70, 0.20, 0.10)	(0.80, 0.10, 0.10)	(0.70, 0.20, 0.10)
D_4	P_1	(0.80, 0.10, 0.10)	(0.80, 0.10, 0.10)	(0.70, 0.20, 0.10)	(0.50, 0.40, 0.10)	(0.70, 0.20, 0.10)	(0.80, 0.10, 0.10)	(0.80, 0.10, 0.10)	(0.60, 0.30, 0.10)
	P_2	(0.60, 0.30, 0.10)	(0.60, 0.30, 0.10)	(0.70, 0.20, 0.10)	(0.50, 0.40, 0.10)	(0.50, 0.40, 0.10)	(0.60, 0.30, 0.10)	(0.40, 0.50, 0.10)	(0.70, 0.20, 0.10)
	P_3	(0.60, 0.30, 0.10)	(0.60, 0.30, 0.10)	(0.40, 0.50, 0.10)	(0.50, 0.40, 0.10)	(0.40, 0.50, 0.10)	(0.50, 0.40, 0.10)	(0.50, 0.40, 0.10)	(0.70, 0.20, 0.10)
	P_4	(0.50, 0.40, 0.10)	(0.50, 0.40, 0.10)	(0.40, 0.50, 0.10)	(0.50, 0.40, 0.10)	(0.50, 0.40, 0.10)	(0.40, 0.50, 0.10)	(0.40, 0.50, 0.10)	(0.80, 0.10, 0.10)
	P_5	(0.70, 0.20, 0.10)	(0.70, 0.20, 0.10)	(0.25, 0.60, 0.15)	(0.60, 0.30, 0.10)	(0.60, 0.30, 0.10)	(0.70, 0.20, 0.10)	(0.80, 0.10, 0.10)	(0.80, 0.10, 0.10)
	P_6	(0.50, 0.40, 0.10)	(0.50, 0.40, 0.10)	(0.25, 0.60, 0.15)	(0.50, 0.40, 0.10)	(0.60, 0.30, 0.10)	(0.80, 0.10, 0.10)	(0.80, 0.10, 0.10)	(0.60, 0.30, 0.10)
	P_7	(0.80, 0.10, 0.10)	(0.80, 0.10, 0.10)	(0.80, 0.10, 0.10)	(0.50, 0.40, 0.10)	(0.60, 0.30, 0.10)	(0.50, 0.40, 0.10)	(0.50, 0.40, 0.10)	(0.50, 0.40, 0.10)

Table 8. The intuitionistic fuzzy direct-influence matrix.

DEs	Criteria	Q_1	Q_2	Q_3	Q_4	Q_5	Q_6	Q_7	Q_8
D_1	Q_1	(0.10, 0.80, 0.10)	(0.20, 0.70, 0.10)	(0.20, 0.70, 0.10)	(0.20, 0.70, 0.10)	(0.40, 0.50, 0.10)	(0.40, 0.50, 0.10)	(0.75, 0.20, 0.05)	(0.40, 0.50, 0.10)
	Q_2	(0.75, 0.20, 0.05)	(0.10, 0.80, 0.10)	(0.20, 0.70, 0.10)	(0.20, 0.70, 0.10)	(0.65, 0.30, 0.05)	(0.65, 0.30, 0.05)	(0.75, 0.20, 0.05)	(0.75, 0.20, 0.05)
	Q_3	(0.65, 0.30, 0.05)	(0.65, 0.30, 0.05)	(0.10, 0.80, 0.10)	(0.55, 0.40, 0.05)	(0.90, 0.05, 0.05)	(0.90, 0.05, 0.05)	(0.90, 0.05, 0.05)	(0.75, 0.20, 0.05)
	Q_4	(0.20, 0.70, 0.10)	(0.20, 0.70, 0.10)	(0.20, 0.70, 0.10)	(0.10, 0.80, 0.10)	(0.65, 0.30, 0.05)	(0.65, 0.30, 0.05)	(0.75, 0.20, 0.05)	(1.00, 0.00, 0.00)
	Q_5	(0.20, 0.70, 0.10)	(0.20, 0.70, 0.10)	(0.20, 0.70, 0.10)	(0.40, 0.50, 0.10)	(0.10, 0.80, 0.10)	(1.00, 0.00, 0.00)	(0.30, 0.60, 0.10)	(0.30, 0.60, 0.10)
	Q_6	(0.55, 0.40, 0.05)	(0.55, 0.40, 0.05)	(0.20, 0.70, 0.10)	(0.20, 0.70, 0.10)	(0.65, 0.30, 0.05)	(0.10, 0.80, 0.10)	(0.30, 0.60, 0.10)	(0.40, 0.50, 0.10)
	Q_7	(0.30, 0.60, 0.10)	(0.30, 0.60, 0.10)	(0.20, 0.70, 0.10)	(0.20, 0.70, 0.10)	(0.20, 0.70, 0.10)	(0.20, 0.70, 0.10)	(0.10, 0.80, 0.10)	(0.20, 0.70, 0.10)
	Q_8	(0.20, 0.70, 0.10)	(0.20, 0.70, 0.10)	(0.20, 0.70, 0.10)	(0.20, 0.70, 0.10)	(0.20, 0.70, 0.10)	(0.20, 0.70, 0.10)	(0.20, 0.70, 0.10)	(0.10, 0.80, 0.10)

Table 8. *Cont.*

DEs	Criteria	Q_1	Q_2	Q_3	Q_4	Q_5	Q_6	Q_7	Q_8
D_2	Q_1	(0.10, 0.80, 0.10)	(0.65, 0.30, 0.05)	(0.40, 0.50, 0.10)	(0.65, 0.30, 0.05)	(0.40, 0.50, 0.10)	(0.40, 0.50, 0.10)	(0.40, 0.50, 0.10)	(0.65, 0.30, 0.05)
	Q_2	(0.55, 0.40, 0.05)	(0.10, 0.80, 0.10)	(0.40, 0.50, 0.10)	(0.65, 0.30, 0.05)	(0.40, 0.50, 0.10)	(0.40, 0.50, 0.10)	(0.40, 0.50, 0.10)	(0.65, 0.30, 0.05)
	Q_3	(0.90, 0.05, 0.05)	(0.90, 0.05, 0.05)	(0.10, 0.80, 0.10)	(0.75, 0.20, 0.05)	(0.75, 0.20, 0.05)	(0.55, 0.40, 0.05)	(0.55, 0.40, 0.05)	(0.65, 0.30, 0.05)
	Q_4	(0.30, 0.60, 0.10)	(0.30, 0.60, 0.10)	(0.40, 0.50, 0.10)	(0.10, 0.80, 0.10)	(0.40, 0.50, 0.10)	(0.40, 0.50, 0.10)	(0.40, 0.50, 0.10)	(0.55, 0.40, 0.05)
	Q_5	(0.55, 0.40, 0.05)	(0.75, 0.20, 0.05)	(0.30, 0.60, 0.10)	(0.40, 0.50, 0.10)	(0.10, 0.80, 0.10)	(0.55, 0.40, 0.05)	(0.40, 0.50, 0.10)	(0.55, 0.40, 0.05)
	Q_6	(0.30, 0.60, 0.10)	(0.30, 0.60, 0.10)	(0.75, 0.20, 0.05)	(0.55, 0.40, 0.05)	(0.75, 0.20, 0.05)	(0.10, 0.80, 0.10)	(0.40, 0.50, 0.10)	(0.40, 0.50, 0.10)
	Q_7	(0.40, 0.50, 0.10)	(0.55, 0.40, 0.05)	(0.75, 0.20, 0.05)	(0.75, 0.20, 0.05)	(0.65, 0.30, 0.05)	(0.55, 0.40, 0.05)	(0.10, 0.80, 0.10)	(0.40, 0.50, 0.10)
	Q_8	(0.65, 0.30, 0.05)	(0.65, 0.30, 0.05)	(0.90, 0.05, 0.05)	(1.00, 0.00, 0.00)	(0.65, 0.30, 0.05)	(0.75, 0.20, 0.05)	(0.90, 0.05, 0.05)	(0.10, 0.80, 0.10)
D_3	Q_1	(0.10, 0.80, 0.10)	(0.65, 0.30, 0.05)	(0.30, 0.60, 0.10)	(0.40, 0.50, 0.10)	(0.40, 0.50, 0.10)	(0.90, 0.05, 0.05)	(0.90, 0.05, 0.05)	(0.55, 0.40, 0.05)
	Q_2	(0.90, 0.05, 0.05)	(0.10, 0.80, 0.10)	(0.65, 0.30, 0.05)	(0.40, 0.50, 0.10)	(0.90, 0.05, 0.05)	(0.65, 0.30, 0.05)	(0.90, 0.05, 0.05)	(0.55, 0.40, 0.05)
	Q_3	(0.30, 0.60, 0.10)	(0.65, 0.30, 0.05)	(0.10, 0.80, 0.10)	(0.90, 0.05, 0.05)	(0.75, 0.20, 0.05)	(0.55, 0.40, 0.05)	(0.65, 0.30, 0.05)	(0.55, 0.40, 0.05)
	Q_4	(0.30, 0.60, 0.10)	(0.40, 0.50, 0.10)	(0.90, 0.05, 0.05)	(0.10, 0.80, 0.10)	(0.75, 0.20, 0.05)	(0.90, 0.05, 0.05)	(0.90, 0.05, 0.05)	(0.90, 0.05, 0.05)
	Q_5	(0.40, 0.50, 0.10)	(0.90, 0.05, 0.05)	(0.75, 0.20, 0.05)	(0.75, 0.20, 0.05)	(0.10, 0.80, 0.10)	(0.90, 0.05, 0.05)	(0.90, 0.05, 0.05)	(0.90, 0.05, 0.05)
	Q_6	(0.90, 0.05, 0.05)	(0.65, 0.30, 0.05)	(0.55, 0.40, 0.05)	(0.90, 0.05, 0.05)	(0.90, 0.05, 0.05)	(0.10, 0.80, 0.10)	(0.75, 0.20, 0.05)	(0.75, 0.20, 0.05)
	Q_7	(0.90, 0.05, 0.05)	(0.90, 0.05, 0.05)	(0.65, 0.30, 0.05)	(0.90, 0.05, 0.05)	(0.75, 0.20, 0.05)	(0.75, 0.20, 0.05)	(0.10, 0.80, 0.10)	(0.90, 0.05, 0.05)
	Q_8	(0.55, 0.40, 0.05)	(0.55, 0.40, 0.05)	(0.55, 0.40, 0.05)	(0.90, 0.05, 0.05)	(0.75, 0.20, 0.05)	(0.75, 0.20, 0.05)	(0.90, 0.05, 0.05)	(0.10, 0.80, 0.10)
D_4	Q_1	(0.10, 0.80, 0.10)	(0.10, 0.80, 0.10)	(0.75, 0.20, 0.05)	(0.75, 0.20, 0.05)	(0.90, 0.05, 0.05)	(0.90, 0.05, 0.05)	(0.90, 0.05, 0.05)	(0.55, 0.40, 0.05)
	Q_2	(0.10, 0.80, 0.10)	(0.10, 0.80, 0.10)	(0.90, 0.05, 0.05)	(0.65, 0.30, 0.05)	(0.90, 0.05, 0.05)	(0.90, 0.05, 0.05)	(0.90, 0.05, 0.05)	(0.90, 0.05, 0.05)
	Q_3	(0.75, 0.20, 0.05)	(0.90, 0.05, 0.05)	(0.10, 0.80, 0.10)	(0.90, 0.05, 0.05)	(0.90, 0.05, 0.05)	(0.90, 0.05, 0.05)	(0.90, 0.05, 0.05)	(1.00, 0.00, 0.00)
	Q_4	(0.75, 0.20, 0.05)	(0.65, 0.30, 0.05)	(0.90, 0.05, 0.05)	(0.10, 0.80, 0.10)	(0.55, 0.40, 0.05)	(0.90, 0.05, 0.05)	(0.90, 0.05, 0.05)	(1.00, 0.00, 0.00)
	Q_5	(0.90, 0.05, 0.05)	(0.90, 0.05, 0.05)	(0.90, 0.05, 0.05)	(0.55, 0.40, 0.05)	(0.10, 0.80, 0.10)	(1.00, 0.00, 0.00)	(1.00, 0.00, 0.00)	(1.00, 0.00, 0.00)
	Q_6	(0.90, 0.05, 0.05)	(0.90, 0.05, 0.05)	(0.90, 0.05, 0.05)	(0.90, 0.05, 0.05)	(1.00, 0.00, 0.00)	(0.10, 0.80, 0.10)	(0.55, 0.40, 0.05)	(0.55, 0.40, 0.05)
	Q_7	(0.90, 0.05, 0.05)	(0.90, 0.05, 0.05)	(0.90, 0.05, 0.05)	(0.90, 0.05, 0.05)	(1.00, 0.00, 0.00)	(0.55, 0.40, 0.05)	(0.10, 0.80, 0.10)	(0.30, 0.60, 0.10)
	Q_8	(0.55, 0.40, 0.05)	(0.90, 0.05, 0.05)	(1.00, 0.00, 0.00)	(1.00, 0.00, 0.00)	(0.55, 0.40, 0.05)	(0.55, 0.40, 0.05)	(0.30, 0.60, 0.10)	(0.10, 0.80, 0.10)

Table 9. The weight of DEs.

DEs	λ_k
D_1	0.2548
D_2	0.2521
D_3	0.2431
D_4	0.2499

Table 10. The weight of criteria and ranking.

Criteria	ω_j^s	Ranking	ω_j^o	Ranking	ω_j	Ranking
Q_1	0.0764	8	0.1631	3	0.1198	6
Q_2	0.1066	7	0.1813	2	0.1439	2
Q_3	0.1468	2	0.2039	1	0.1754	1
Q_4	0.1168	6	0.0385	8	0.0776	8
Q_5	0.1522	1	0.1185	5	0.1353	3
Q_6	0.1245	5	0.1168	6	0.1206	5
Q_7	0.1404	3	0.1261	4	0.1332	4
Q_8	0.1363	4	0.0519	7	0.0941	7

Table 11. The ranking of urban rail transit system type.

Urban Rail Transit System	G_i^+	G_i^-	H_i	Ranking
P_1	−0.1409	−1.0936	−0.2057	1
P_2	−0.5312	−0.2748	−0.7893	3
P_3	−1.3059	−0.1643	−1.7376	4
P_4	−1.9230	−0.0440	−3.5346	7
P_5	−1.7623	−0.8585	−1.8449	5
P_6	−2.4013	−0.2118	−2.7362	6
P_7	−0.0762	−0.3434	−0.2828	2

It can be seen from Table 10 that the ranking of the subjective weight and objective weight of criteria are quite different. The weight determination method combining subjective and objective weight can make the evaluation results more objective. The top three final criteria are technology maturity Q_3, transportation speed Q_2 and construction difficulty Q_5. The ranking of criteria may change due to different cities. For City N, the first consideration is the three attributes of technology maturity, transportation speed and construction difficulty.

It can be seen from Table 11 that the ranking of the urban rail transit system in City N can be obtained through the comprehensive evaluation value. Here, the comprehensive evaluation value is negative because the regret theory is considered. P_1 ranks first; that is, the type of urban rail transit most suitable for City N is metro system. City N is the third largest city in Z Province, with a large population and high requirements for transportation capacity. In addition, the metro system has high technical maturity and fast transportation speed. The natural geographical environment of city N also makes the construction of the metro system relatively difficult. Therefore, the metro system is the most suitable urban rail transit for city N. The municipal railway system (P_7) and light rail system (P_2) rank second and third, respectively. These two types are two other options that can be considered for construction in city N in addition to the metro system. They also have the characteristics of high technical maturity and fast transportation speed. Other criteria can be comprehensively considered for selection. The final results of the ranking of the urban rail transit system type can prove the applicability and effectiveness of the evaluation index and evaluation framework proposed in this study.

4.4. Sensitivity Analysis

In this subsection, the stability and robustness of the proposed hybrid intuitionistic fuzzy group decision framework will be explored through sensitivity analysis. The sensitivity analysis of this study is divided into two parts. The first part is the sensitivity analysis of the relative importance coefficient of subjective and objective weights. The second part is the sensitivity analysis of the regret avoidance coefficient of experts.

4.4.1. The Impact Analysis of Parameter φ on Decision Results

The relative importance coefficient φ of subjective and objective weights can express the preference of decision-making experts for weights. In the previous example analysis, the value of φ is 0.5. Next, by changing the value of φ, different criteria weight values are obtained, and then the adjusted criteria ranking results are observed. In this paper, $\varphi \in [0,1]$, first, let $\varphi = 0$, increasing by 0.1; the final ranking results and ranking changes are shown in Table 12 and Figure 2.

Table 12. The ranking of urban rail transit types under different φ values.

	$\varphi = 0$	$\varphi = 0.1$	$\varphi = 0.2$	$\varphi = 0.3$	$\varphi = 0.4$	$\varphi = 0.5$	$\varphi = 0.6$	$\varphi = 0.7$	$\varphi = 0.8$	$\varphi = 0.9$	$\varphi = 1$
P_1	1	1	1	1	1	1	1	1	1	1	1
P_2	3	3	3	3	3	3	3	3	3	3	3
P_3	4	4	4	4	4	4	4	4	4	4	5
P_4	7	7	7	7	7	7	7	7	7	7	7
P_5	5	5	5	5	5	5	5	5	5	5	4
P_6	6	6	6	6	6	6	6	6	6	6	6
P_7	2	2	2	2	2	2	2	2	2	2	2

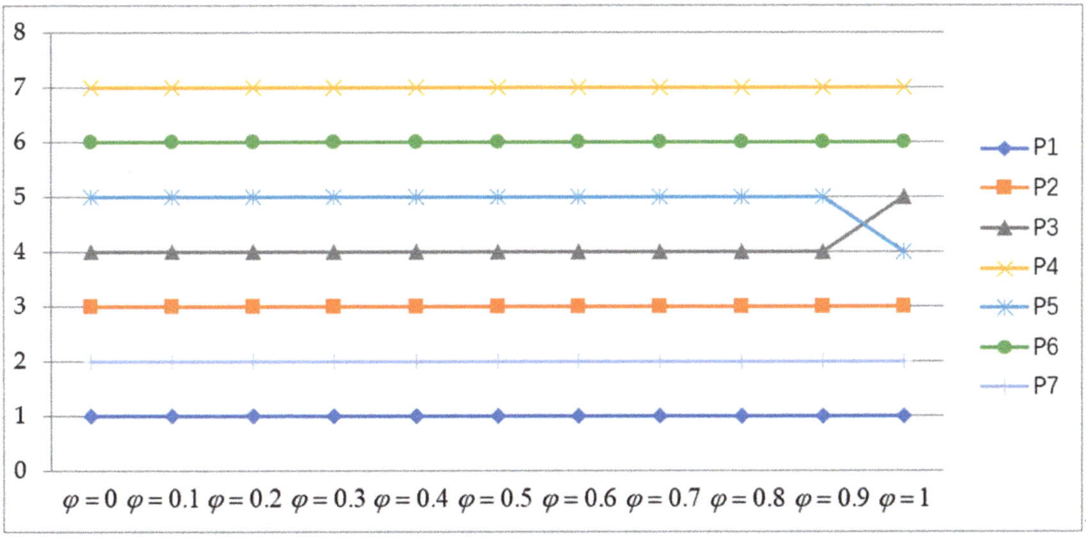

Figure 2. The ranking change of decision results under different parameter φ values.

As can be seen from Figure 2, the final ranking is relatively stable by changing the proportion of subjective and objective weights, and the top three are always $P_1 \succ P_7 \succ P_2$. When $\varphi = 0$ and 1, it means that only objective weight and only subjective weight are considered, respectively. When only the subjective weight is considered, the ranking of the fourth and fifth types will be exchanged and the other rankings will not change. Therefore, comprehensive consideration of the subjective and objective weight can make the decision-making results more stable.

4.4.2. The Impact Analysis of Parameter ϑ on Decision Results

The second part considers the influence of the expert regret avoidance coefficient on the final decision outcome. The larger ϑ is, the higher the degree of the regret of experts. The initial value of ϑ is 5. In the analysis, ϑ takes 1 to 10 and increases by 1. The ranking and changes of the decision results are shown in Table 13 and Figure 3.

Table 13. The ranking of urban rail transit types under different ϑ values.

	$\vartheta = 1$	$\vartheta = 2$	$\vartheta = 3$	$\vartheta = 4$	$\vartheta = 5$	$\vartheta = 6$	$\vartheta = 7$	$\vartheta = 8$	$\vartheta = 9$	$\vartheta = 10$
P_1	1	1	1	1	1	1	1	1	1	1
P_2	3	3	3	3	3	3	3	3	3	3
P_3	5	5	4	4	4	4	4	4	4	4
P_4	7	7	7	7	7	7	7	7	7	7
P_5	4	4	5	5	5	5	5	5	5	5
P_6	6	6	6	6	6	6	6	6	6	6
P_7	2	2	2	2	2	2	2	2	2	2

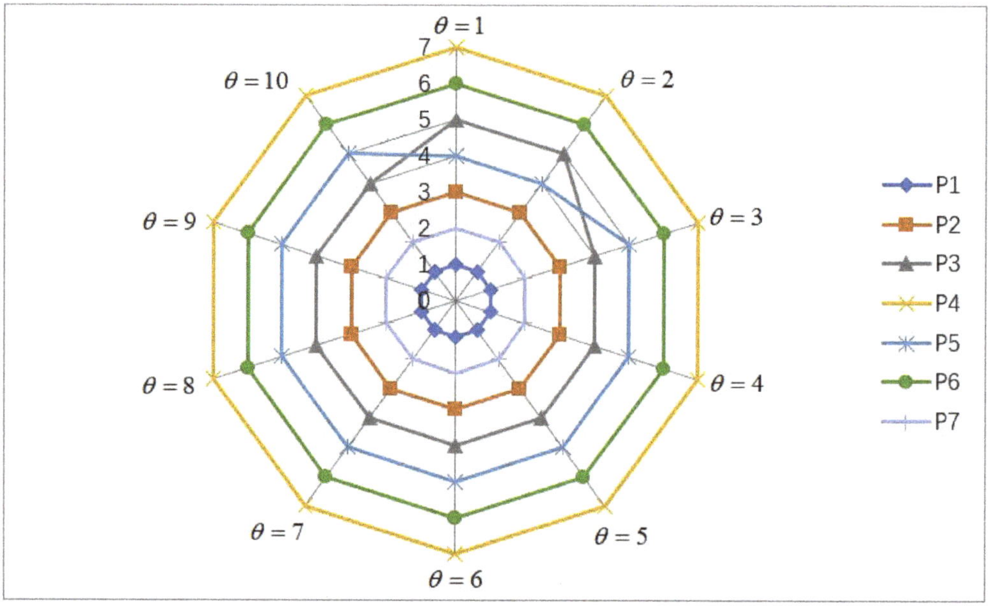

Figure 3. The ranking change in decision results under different parameter ϑ values.

As can be seen from Figure 3, changing the value of the regret avoidance coefficient has little impact on the final ranking result, which is still relatively stable, and the top three are still $P_1 \succ P_7 \succ P_2$; only when the value of $\vartheta = 1$ and 2, the medium–low-speed maglev system (P5) and monorail system (P_3) rank fourth and fifth, respectively. When the value of ϑ is greater than or equal to 3, the rankings of the two types are exchanged. Monorail system (P_3) ranks fourth, while medium–low-speed maglev system (P_5) ranks fifth. From the sensitivity analysis of the above two parts, it can be seen that the model proposed in this paper has strong stability.

4.5. Comparative Analysis

The same as this study uses IFS to deal with the uncertainty and inaccuracy in decisions, the weight determination method remains unchanged based on IFS in the comparative analysis part. Three MCDM methods are selected to compare with the results of this study. The first is the traditional COPRAS method, which does not consider the regret theory. The other two methods are TOPSIS and ARAS. The comparison results are shown in Table 14 and Figure 4.

Table 14. The ranking under different evaluation methods.

Urban Rail Transit System	This Paper	IF-COPRAS	IF-TOPSIS	IF-ARAS
P_1	1	3	1	3
P_2	3	2	4	2
P_3	4	4	6	4
P_4	7	5	7	5
P_5	5	6	3	6
P_6	6	7	5	7
P_7	2	1	2	1

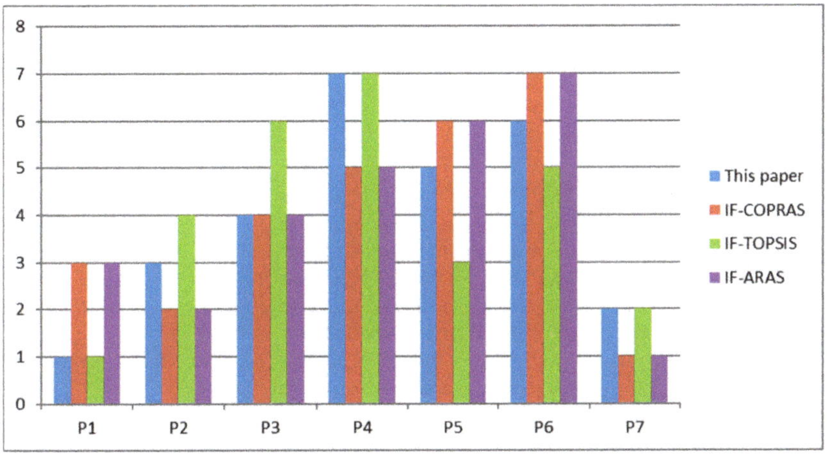

Figure 4. The ranking results based on different evaluation methods.

Figure 4 shows the comparison of the ranking results under the four methods. It can be seen that the ranking results under different evaluation methods are different, but the overall trend is the same. The top three are mainly P_1, P_2 and P_7. The best scheme changes between P_1 and P_7, and the last three are concentrated among P_4, P_5 and P_6. By calculating the Spearman correlation coefficient of the ranking results of the original method and other methods, it can be seen that all the correlation coefficients are greater than 0.78, which shows that the evaluation model proposed in this study is relatively stable. The detailed comparison analyses with other intuitionistic fuzzy decision approaches are illustrated below.

Compared with the results of the traditional IF-COPRAS method, it is found that the results obtained by the two methods are different, and the Spearman correlation coefficient is 0.786. The reason for this difference is that the evaluation model proposed in this study considers the regret theory; that is, the expert risk aversion coefficient and regret aversion coefficient are considered at the same time. The result is the optimization of the traditional IF-COPRAS method.

Compared with the results of the IF-TOPSIS method, the ranking results obtained by the two methods are more consistent, and the ranking of 1, 2 and 7 are the same. This can also be proven by the Spearman correlation coefficient of 0.821. The TOPSIS method is a classical MCDM method, which has wide applicability. Through the consistency of the results of the two methods, it can be seen that the method proposed in this study has stability and robustness.

Compared with the results of the IF-ARAS method, the results of the ARAS method are the same as those of the traditional COPRAS method. Therefore, the Spearman correlation coefficient is also 0.786. This indicates that regret theory will affect the results.

Based on the above discussion and comparative analysis, the proposed hybrid intuitionistic fuzzy group decision framework for the urban rail transit system selection of this paper has the following advantages:

(1) The proposed framework describes the uncertainty and fuzziness in the decision-making process through IFS, which makes the decision-making results closer to the uncertain cognitive thinking of decision-makers.
(2) The proposed framework can effectively solve the decision problem with completely unknown weight information, so it has a wider scope of application.
(3) The proposed framework determines the model through the comprehensive weight, and reasonably considers the subjective and objective factors to make the importance of the criterion more credible.
(4) The proposed framework combines regret theory and the COPRAS method and comprehensively considers the inconsistency of psychological behavior and the attribute transformation process in the process of expert decision-making, so it improves the rationality and reliability of decision outcomes.

5. Conclusions

In view of the shortcomings of the existing research, the main goal of this study is to develop a hybrid MCGDM evaluation model for the selection of an urban rail transit system. In order to overcome the uncertainty and inaccuracy in the process of expert evaluation and make the evaluation information more reliable, this study put forward a hybrid intuitionistic fuzzy group decision framework to select the satisfactory urban rail transit system. The DEMATEL and CRITIC methods were selected to determine the subjective and objective weight of the criteria, and the COPRAS method based on regret theory was used to rank the types of urban rail transit systems and select the optimal urban rail transit system. The sensitivity analysis and comparative analysis prove the stability and robustness of the evaluation model. The results show that, no matter how the coefficient changes, the top three schemes have not changed. Furthermore, the ranking results still have high consistency by a detailed comparison analysis with other prior methodologies. Therefore, the hybrid decision-making framework model proposed in this study has strong practicability. It not only considers the subjective randomness of experts in the decision-making process but also considers the risk preference and regret degree of experts. It is more comprehensive and has more advantages in evaluating the selection of urban rail transit.

The method proposed in this study also has some limitations. For example, when calculating the weights of experts, only the relative distance of expert evaluation information is considered, and the information, such as experts' own experience, is ignored. In the process of expert information fusion, the relationship of decision information under different criteria is not considered. In the future, it can be further studied from the following aspects. Firstly, this research model can be applied to other related MCGDM problems [43,44]. Secondly, this research is based on the intuitionistic fuzzy environment. In the future, different fuzzy linguistic environments and MCDM methods can be applied to this research model. Thirdly, in the face of decision-making experts from different fields, it is difficult to reach a consensus on the preference information provided by different experts, and small-group-decision-making cannot fully ensure the credibility of the final decision-making results; therefore, it is a hot issue to establish a large-scale group consensus decision-making model [45–47] in an intuitionistic fuzzy environment and solve the actual group-decision-making problem combined with big data artificial intelligence technology.

Author Contributions: Conceptualization, B.Y., Y.R. and Y.H.; Formal analysis, B.Y., Y.R. and Y.H.; Funding acquisition, L.Y.; Investigation, Y.R., L.Y. and Y.H.; Methodology, B.Y., Y.R. and Y.H.; Project administration, L.Y.; Visualization, L.Y.; Writing—original draft, B.Y.; Writing—review & editing, B.Y., Y.R. and L.Y. All authors have read and agreed to the published version of the manuscript.

Funding: The research was funded by the General Program of National Natural Science Foundation of China (No: 12071280).

Institutional Review Board Statement: Not applicable.

Informed Consent Statement: Not applicable.

Data Availability Statement: The data presented in this study are available in the article.

Conflicts of Interest: The authors declare that they have no conflict of interest.

Abbreviations

Decision-Making Trial and Evaluation Laboratory	DEMATEL
Criteria Importance Through Inter-criteria Correlation	CRITIC
Complex Proportional Assessment	COPRAS
Intuitionistic Fuzzy Sets	IFS
VlseKriterijuska Optimizacija I Komoromisno Resenje	VIKOR
Technique for Order Preference by Similarity to an Ideal Solution	TOPSIS
Multi-Criteria Group-Decision-Making	MCGDM
Multi-Criteria Decision-Making	MCDM
Multi-Attribute Border Approximation Area Comparison	MABAC
Intuitionistic Fuzzy Number	IFN
Additional Ratio Assessment	ARAS

References

1. Cao, J.; Chen, X.; Qiu, R.; Hou, S. Electric vehicle industry sustainable development with a stakeholder engagement system. *Technol. Soc.* **2021**, *67*, 101771. [CrossRef]
2. Thondoo, M.; Marquet, O.; Marquez, S.; Nieuwenhuijsen, M.J. Small cities, big needs: Urban transport planning in cities of developing countries. *J. Transp. Health* **2020**, *19*, 100944. [CrossRef]
3. Huang, X.; Cao, X.J.; Cao, X.; Yin, J. How does the propensity of living near rail transit moderate the influence of rail transit on transit trip frequency in Xi'an? *J. Transp. Geogr.* **2016**, *54*, 194–204. [CrossRef]
4. Atanassov, K.T. Intuitionistic fuzzy sets. *Fuzzy Sets Syst.* **1986**, *20*, 87–96. [CrossRef]
5. Das, S.; Guha, D.; Mesiar, R. Information measures in the intuitionistic fuzzy framework and their relationships. *IEEE Trans. Fuzzy Syst.* **2017**, *26*, 1626–1637. [CrossRef]
6. Mishra, A.R.; Rani, P. Information measures based TOPSIS method for multicriteria decision making problem in intuitionistic fuzzy environment. *Iran. J. Fuzzy Syst.* **2017**, *14*, 41–63.
7. Ecer, F.; Pamucar, D. MARCOS technique under intuitionistic fuzzy environment for determining the COVID-19 pandemic performance of insurance companies in terms of healthcare services. *Appl. Soft Comput.* **2021**, *104*, 107199. [CrossRef]
8. Schitea, D.; Deveci, M.; Iordache, M.; Bilgili, K.; Akyurt, İ.Z.; Iordache, I. Hydrogen mobility roll-up site selection using intuitionistic fuzzy sets based WASPAS, COPRAS and EDAS. *Int. J. Hydrogen Energy* **2019**, *44*, 8585–8600. [CrossRef]
9. Mishra, A.R.; Mardani, A.; Rani, P.; Kamyab, H.; Alrasheedi, M. A new intuitionistic fuzzy combinative distance-based assessment framework to assess low-carbon sustainable suppliers in the maritime sector. *Energy* **2021**, *237*, 121500. [CrossRef]
10. Zhang, C.; Liao, H.; Luo, L.; Xu, Z. Distance-based consensus reaching process for group decision making with intuitionistic multiplicative preference relations. *Appl. Soft Comput.* **2020**, *88*, 106045. [CrossRef]
11. Meng, F.; Chen, S.M.; Yuan, R. Group decision making with heterogeneous intuitionistic fuzzy preference relations. *Inf. Sci.* **2020**, *523*, 197–219. [CrossRef]
12. Gabus, A.; Fontela, E. *World Problems, an Invitation to Further Thought within the Framework of DEMATEL*; Battelle Geneva Research Center: Geneva, Switzerland, 1972; pp. 1–8.
13. Topgul, M.H.; Kilic, H.S.; Tuzkaya, G. Greenness assessment of supply chains via intuitionistic fuzzy based approaches. *Adv. Eng. Inform.* **2021**, *50*, 101377. [CrossRef]
14. Roostaie, S.; Nawari, N. The DEMATEL approach for integrating resilience indicators into building sustainability assessment frameworks. *Build. Environ.* **2022**, *207*, 108113. [CrossRef]
15. Tseng, M.L.; Ardaniah, V.; Sujanto, R.Y.; Fujii, M.; Lim, M.K. Multicriteria assessment of renewable energy sources under uncertainty: Barriers to adoption. *Technol. Forecast. Soc. Change* **2021**, *171*, 120937. [CrossRef]
16. Liu, Y.; Wood, L.C.; Venkatesh, V.G.; Zhang, A.; Farooque, M. Barriers to sustainable food consumption and production in China: A fuzzy DEMATEL analysis from a circular economy perspective. *Sustain. Prod. Consum.* **2021**, *28*, 1114–1129. [CrossRef]
17. Hosseini, S.M.; Paydar, M.M.; Hajiaghaei-Keshteli, M. Recovery solutions for ecotourism centers during the COVID-19 pandemic: Utilizing Fuzzy DEMATEL and Fuzzy VIKOR methods. *Expert Syst. Appl.* **2021**, *185*, 115594. [CrossRef]
18. Li, H.; Wang, W.; Fan, L.; Li, Q.; Chen, X. A novel hybrid MCDM model for machine tool selection using fuzzy DEMATEL, entropy weighting and later defuzzification VIKOR. *Appl. Soft Comput.* **2020**, *91*, 106207. [CrossRef]

19. Fang, S.; Zhou, P.; Dinçer, H.; Yüksel, S. Assessment of safety management system on energy investment risk using house of quality based on hybrid stochastic interval-valued intuitionistic fuzzy decision-making approach. *Saf. Sci.* **2021**, *141*, 105333. [CrossRef]
20. Liang, W.; Wang, Y.M. A probabilistic interval-valued hesitant fuzzy gained and lost dominance score method based on regret theory. *Comput. Ind. Eng.* **2021**, *159*, 107532. [CrossRef]
21. Loomes, G.; Sugden, R. Regret theory: An alternative theory of rational choice under uncertainty. *Econ. J.* **1982**, *92*, 805–824. [CrossRef]
22. Bell, D.E. Regret in decision making under uncertainty. *Oper. Res.* **1982**, *30*, 961–981. [CrossRef]
23. Tian, X.; Xu, Z.; Gu, J.; Herrera, F. A consensus process based on regret theory with probabilistic linguistic term sets and its application in venture capital. *Inf. Sci.* **2021**, *562*, 347–369. [CrossRef]
24. Huang, L.; Mao, L.X.; Chen, Y.; Liu, H.C. New method for emergency decision making with an integrated regret theory-EDAS method in 2-tuple spherical linguistic environment. *Appl. Intell.* **2022**. [CrossRef] [PubMed]
25. Zhang, K.; Wang, Y.M.; Zheng, J. Regret Theory-Based Case-Retrieval Method with Multiple Heterogeneous Attributes and Incomplete Weight Information. *Int. J. Comput. Intell. Syst.* **2021**, *14*, 1022–1033. [CrossRef]
26. Liu, P.; Cheng, S. An improved MABAC group decision-making method using regret theory and likelihood in probability multi-valued neutrosophic sets. *Int. J. Inf. Technol. Decis. Mak.* **2020**, *19*, 1353–1387.
27. Huang, X.; Zhan, J. TWD-R: A three-way decision approach based on regret theory in multi-scale decision information systems. *Inf. Sci.* **2021**, *581*, 711–739. [CrossRef]
28. Liu, Z.; Wang, D.; Wang, W.; Liu, P. An integrated group decision—Making framework for selecting cloud service providers based on regret theory and EVAMIX with hybrid information. *Int. J. Intell. Syst.* **2021**, *37*, 3480–3513. [CrossRef]
29. Zavadskas, E.K.; Kaklauskas, A.; Sarka, V. The new method of multicriteria complex proportional assessment of projects. *Technol. Econ. Dev. Econ.* **1994**, *1*, 131–139.
30. Organ, A.; Yalçın, E. Performance evaluation of research assistants by COPRAS method. *Eur. Sci. J.* **2016**, *12*, 102–109.
31. Buyukozkan, G.; Gocer, F. A novel approach integrating AHP and COPRAS under pythagorean fuzzy sets for digital supply chain partner selection. *IEEE Trans. Eng. Manag.* **2019**, *68*, 1486–1503. [CrossRef]
32. Balali, A.; Valipour, A.; Edwards, R.; Moehler, R. Ranking effective risks on human resources threats in natural gas supply projects using ANP-COPRAS method: Case study of Shiraz. *Reliab. Eng. Syst. Saf.* **2021**, *208*, 107442. [CrossRef]
33. Mishra, A.R.; Rani, P.; Pandey, K.; Mardani, A.; Streimikis, J.; Streimikiene, D.; Alrasheedi, M. Novel multi-criteria intuitionistic fuzzy SWARA–COPRAS approach for sustainability evaluation of the bioenergy production process. *Sustainability* **2020**, *12*, 4155.
34. Alipour, M.; Hafezi, R.; Rani, P.; Hafezi, M.; Mardani, A. A new Pythagorean fuzzy-based decision-making method through entropy measure for fuel cell and hydrogen components supplier selection. *Energy* **2021**, *234*, 121208. [CrossRef]
35. Yuan, Y.; Xu, Z.; Zhang, Y. The DEMATEL–COPRAS hybrid method under probabilistic linguistic environment and its application in Third Party Logistics provider selection. *Fuzzy Optim. Decis. Mak.* **2022**, *21*, 137–156. [CrossRef]
36. Narayanamoorthy, S.; Ramya, L.; Kalaiselvan, S.; Kureethara, J.V.; Kang, D. Use of DEMATEL and COPRAS method to select best alternative fuel for control of impact of greenhouse gas emissions. *Socio-Econ. Plan. Sci.* **2021**, *76*, 100996. [CrossRef]
37. Zeng, S.; Chen, S.M.; Kuo, L.W. Multiattribute decision making based on novel score function of intuitionistic fuzzy values and modified VIKOR method. *Inf. Sci.* **2019**, *488*, 76–92. [CrossRef]
38. Xu, Z. Intuitionistic fuzzy aggregation operators. *IEEE Trans. Fuzzy Syst.* **2007**, *15*, 1179–1187.
39. Zhang, S.; Zhu, J.; Liu, X.; Chen, Y. Regret theory-based group decision-making with multidimensional preference and incomplete weight information. *Inf. Fusion* **2016**, *31*, 1–13. [CrossRef]
40. Yurdakul, M.; İç, Y.T.; Atalay, K.D. Development of an intuitionistic fuzzy ranking model for nontraditional machining processes. *Soft Comput.* **2020**, *24*, 10095–10110. [CrossRef]
41. Quiggin, J. Regret theory with general choice sets. *J. Risk Uncertain.* **1994**, *8*, 153–165. [CrossRef]
42. Chen, W.J.; Goh, M.; Zou, Y. Logistics provider selection for omni-channel environment with fuzzy axiomatic design and extended regret theory. *Appl. Soft Comput.* **2018**, *71*, 353–363. [CrossRef]
43. Rong, Y.; Liu, Y.; Pei, Z. Complex q-rung orthopair fuzzy 2-tuple linguistic Maclaurin symmetric mean operators and its application to emergency program selection. *Int. J. Intell. Syst.* **2020**, *35*, 1749–1790. [CrossRef]
44. Rong, Y.; Liu, Y.; Pei, Z. A novel multiple attribute decision-making approach for evaluation of emergency man-agement schemes under picture fuzzy environment. *Int. J. Mach. Learn. Cybern.* **2021**, *13*, 633–661. [CrossRef]
45. Lu, Y.; Xu, Y.; Huang, J.; Wei, J.; Herrera-Viedma, E. Social network clustering and consensus-based distrust be-haviors management for large-scale group decision-making with incomplete hesitant fuzzy preference relations. *Appl. Soft Comput.* **2022**, *117*, 108373. [CrossRef]
46. Rodríguez, R.M.; Labella, Á.; Nunez-Cacho, P.; Molina-Moreno, V.; Martínez, L. A comprehensive minimum cost consensus model for large scale group decision making for circular economy measurement. *Technol. Forecast. Soc. Change* **2022**, *175*, 121391. [CrossRef]
47. Wan, S.P.; Yan, J.; Dong, J.Y. Personalized individual semantics based consensus reaching process for large-scale group decision making with probabilistic linguistic preference relations and application to COVID-19 surveillance. *Expert Syst. Appl.* **2022**, *191*, 116328. [CrossRef]

mathematics

Article

A Ship Fire Escape Speed Correction Method Considering the Influence of Crowd Interaction

Jingyuan Li [1], Weile Liu [1], Fangwei Zhang [1,2,*], Taiyang Li [1] and Rui Wang [1]

[1] School of Navigation and Shipping, Shandong Jiaotong University, Weihai 264209, China; jy971221@163.com (J.L.); lwlglxxtl@163.com (W.L.); litaiyang0115@163.com (T.L.); wr19991127@163.com (R.W.)
[2] College of Transport and Communications, Shanghai Maritime University, Shanghai 201306, China
* Correspondence: fangweirzhang@163.com

Abstract: The aim of this study is to explore the effect of different personnel attributes and the relationship between different people on evacuation efficiency in the case of a passenger ship fire. As such, this study proposes a speed correction method that considers human attributes and interactions between different populations. Firstly, a hesitant fuzzy set and hesitant fuzzy average operator are adopted to quantify four kinds of personnel attributes. Secondly, considering the influence of different people, this study extracts the formula for acceleration under the interactive influence of different groups of people. At the same time, based on the first-order linear relationship between velocity and acceleration, an interactive velocity correction method is presented in the evacuation of ship personnel. Finally, this study uses the personnel evacuation simulation software Pathfinder to conduct experiments, and introduces the corrected speed and the uncorrected speed into the evacuation simulation process, respectively. The results show that the simulation results of the revised speed plan are more consistent with reality.

Keywords: ship fire; emergency evacuation; hesitation fuzzy sets

MSC: 41-02

Citation: Li, J.; Liu, W.; Zhang, F.; Li, T.; Wang, R. A Ship Fire Escape Speed Correction Method Considering the Influence of Crowd Interaction. *Mathematics* **2022**, *10*, 2749. https://doi.org/10.3390/math10152749

Academic Editor: Yanhui Guo

Received: 27 June 2022
Accepted: 20 July 2022
Published: 3 August 2022

1. Introduction

In recent years, the density of maritime navigation has been increasing. The number of ships, as the main means of water transportation, is also increasing year by year. During navigation, ship fire is one of the main threats to navigation safety. In shipwreck incidents over the years, accidents caused by fire account for about 11% [1]. Fire on ships is particularly dangerous because rescue is more difficult and fires spread fast, and the internal structure of ships is complex. Therefore, it is difficult to evacuate and put out fires. Once a fire occurs, it will cause a lot of economic losses and seriously threaten the safety of people's lives [2]. The occurrence of such accidents is closely related to the evacuation behavior and evacuation time [3]. Therefore, the behavior mechanism and rules of different groups in the evacuation process are studied extensively. It is beneficial to formulate scientific and efficient emergency evacuation measures. It is also of great significance to ensure the personal safety of evacuees in case of fire on ships.

Relevant studies have found that in the event of a fire, passengers will not only be affected by personal attributes such as their emergency ability, cognitive ability, psychological endurance, and value orientation, but also by different groups of people [4]. Therefore, in the ship scenario, the interactive hesitancy fuzzy integration operator is used to integrate the information of different groups. At the same time, the velocity and acceleration formulas under the influence of different crowd interaction are extracted. Then the cognitive ability and emergency response ability of the population at the fire scene are quantitatively analyzed. This study revises and supplements the effects of capacity, value orientation, psychological bearing capacity and group effect on evacuation efficiency. The models of escape acceleration and escape speed are also established, which provide theoretical basis and decision support for the evacuation of ship fire personnel.

1.1. Literature Review

Ensuring the safety of people in a fire is the fundamental goal of evacuating people. To achieve this goal, it is necessary to study the behavioral laws of evacuated people. The study of evacuation behavior is one of the nine key research directions of fire science. At present, the research on pedestrian evacuation mainly focuses on three aspects: evacuation model construction, evacuation decision-making and personnel evacuation efficiency.

Concerning the construction of the evacuation model, Treuille et al. [5] proposed a real-time crowd model with congestion in public places in multiple cities as the research object. Helbing et al. [6,7] established a social model to describe the walking behavior of the crowd in evacuation according to the calculation formula of Newton's second theorem. Wang et al. [8] integrated human factors into emergency evacuation and analyzed the influence of various factors on evacuation behavior in different stages by building an evacuation model. Wang et al. [9] constructed an evacuation model that considers Openness, Conscientiousness, Extroversion, Agreeableness, and Neuroticism (OCEAN) to analyze the impact of passengers' personality traits on evacuation behavior. Hu et al. [10] considered the interaction between the fire environment and evacuees from the perspective of the system and established a manual evacuation procedure.

Concerning the evacuation decision part, Feng et al. [11] proposed an evacuation decision-making model consisting of three parts: pedestrian distribution prediction model, pedestrian flow calculation model and path situation and feedback correction model. Lovreglio et al. [12] introduced an evacuation decision model predicting pre-evacuation behavior, and the model simulates the probability of evacuees' behavioral state. Peng et al. [13] established a two-level decision-making model for emergency evacuation paths of high-rise buildings based on BIM, which realized the optimal planning of emergency evacuation paths. Sun et al. [14] used game-based theory in a small-world network context and built an evolutionary game model of evacuation decision diffusion between evacuees in the context of a complex network. Tian et al. [15] have designed a mobile-based system to collect medical and temporal data produced during an emergency response to mass casualty incidents.

Concerning the evacuation efficiency, Koo et al. [16] studied the psychological panic effect coefficient of evacuation speed by combining theoretical derivation with three-dimensional simulation technology. Chen et al. [17] adopted an improved social force model to study the influences of the total number of pedestrians, required speed, and specific location of obstacles on the evacuation efficiency of multi-exit configuration. Jeon [18] studied the impact of escape routes and emergency exits on evacuation speed under different environmental conditions. Yu et al. [19] conducted experimental and numerical simulation study on evacuation time and average evacuation speed of personnel in railway tunnels under train fire conditions.

The above research provides a theoretical basis for this research. At present, however, there are few studies on fire evacuation in the scenario of ships. The research on the behavior of personnel evacuation is not refined enough. Therefore, this study takes ships as the research scene and adopts the method of questionnaire survey to collect data. Then, the factors affecting evacuation efficiency of different groups are quantified. The relationship between the behavior and psychological characteristics of people in a fire and the evacuation speed are difficult to directly quantify into an accurate mathematical relationship, therefore, this paper uses fuzzy logic to quantify their influence, and selects the classical hesitant fuzzy weighted average operator for information integration.

1.2. Objective Contribution

The purpose of this study is to explore the influence of the interaction between different attribute groups on the evacuation efficiency in the ship fire scenario. The main research contributions are as follows.

Firstly, this study is based on fuzzy mathematics theory, using the classical hesitant fuzzy average operator. Then, the four attributes that affect crew escape on board are integrated. Through the quantification of four objective influencing factors, evacuation research is more realistic.

Secondly, a speed correction model considering the interaction of different populations is developed. The reduction in evacuation speed caused by different crowd interaction is quantified. It provides a reference for realizing evacuation research under the influence of multiple factors.

Finally, this study collects data through questionnaires and calculates the model. The simulation software is used to compare the modified speed plan with the unmodified speed plan, and more realistic simulation results are obtained.

The contents of this study are arranged as follows. In Section 2, the research ideas are described and the factors affecting the escape of personnel on board are analyzed. In Section 3, a fuzzy set containing four kinds of people is established by using fuzzy mathematics theory. Then, a speed correction method considering personnel attributes and the interaction between different groups is developed. Section 4 verifies the validity of the revised velocity model through simulation.

2. Research Foundation

2.1. Research Idea

The ship fire evacuation efficiency is affected by many factors. This study mainly considers the influence of the interaction between different groups of people on the evacuation efficiency. Then, a more realistic fire evacuation velocity model is extracted. The specific research ideas are as follows.

Based on the above research goals, this study has carried out the following work. Firstly, this study establishes the hesitant fuzzy sets of four kinds of people. The influence of the attributes of emergency response ability, cognitive ability, psychological bearing ability and value orientation is quantitatively analyzed. Secondly, this study introduces the interaction between the target population and other groups of people, combined with the classical universal gravitation formula. The acceleration formula of interaction between the influence of different attributes and the influence of different people is extracted. Finally, this study collects data through questionnaires, and uses simulation software to compare the revised speed plan with the uncorrected speed plan to verify the validity of the model (please, see Figure 1 for the research process).

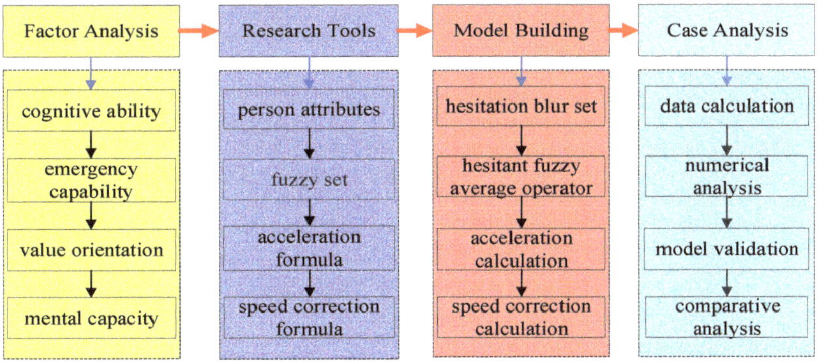

Figure 1. Research process.

2.2. *Analysis of Key Factors for the Escape of Personnel on Board*

In the process of fire evacuation, emergency ability, cognitive ability, psychological endurance and value orientation are the key factors that affect the survival of people [20]. The sudden stimuli of fire make the crowd react instantaneously, and the instantaneous response is closely related to the above four abilities of different people. The specific explanations of emergency ability, cognitive ability [21], psychological bearing ability and value orientation are as follows.

(i) Emergency ability: When people encounter an emergency, the brain immediately deals with it based on past experience and the ability to think for itself. Self-thinking is a subconscious response. Because children and the elderly have far less physical function than adults, once a fire breaks out, children and the elderly will become vulnerable groups. Their evacuation speed is also significantly lower than that of adults.

(ii) Cognitive ability: People with higher education levels have weaker fear, faster reactions and stronger ability to escape. On the other hand, people with lower education levels have slower reactions and weaker escape ability in the face of fire.

(iii) Psychological endurance: When a fire occurs on a ship, people will have a fear of fire due to a lack of understanding of fire. In this state, people are prone to irrational behavior. Adults have a strong psychological bearing capacity, while the elderly and children have a weak psychological bearing capacity. There is a certain gap in the psychological response of different groups of people in terms of psychological bearing capacity.

(iv) Value orientation: When a ship fire occurs, the value orientation of the elderly is conservative, which greatly affects the escape ability of the elderly. Therefore, value orientation is also one of the key factors affecting the escape speed of the crew on board.

2.3. *Research Tools*

When a fire occurs on a ship, panic and chaotic behavior are bound to occur in the crowd. In a ship with concentrated personnel, the personnel's emergency ability, cognitive ability, psychological bearing ability and value orientation vary greatly. In addition, under fire conditions, the four abilities of different groups are complex and abstract, and the relationship with evacuation speed cannot be directly quantified. Therefore, this study proposes a fire escape velocity correction model based on hesitant fuzzy sets.

Due to the complexity and uncertainty of objective information and the ambiguity of human thinking, Zadeh introduced the concept of fuzzy sets [22]. Hesitant fuzzy sets are fuzzy set extensions to handle hesitant situations that were not well handled by previous tools [23]. Operator theory is an important part of fuzzy theory. Based on the arithmetic ensemble method, this study uses the classical hesitancy fuzzy weighted average operator (HFWA) and the classical hesitant fuzzy average (HFA). Then, the cognitive ability, emergency response ability, value orientation, psychological bearing ability and group effect of people in the fire scene are integrated. Furthermore, the objective information of different groups is quantified [24]. The relevant definitions of the hesitant fuzzy set weighted average operator, acceleration and velocity formulas used in this study are as follows.

Definition 1. *Let X be a given finite set, then* $E = \{\langle x, h_E(x)\rangle | x \in X\}$ *is called hesitant fuzzy set. Among them,* $h_E(x)$ *represents the possible membership degree of x belonging to X, which is a subset of the interval* $[0,1]$*, and let* h_1, h_2, h_3*, be three hesitant fuzzy elements, then their basic operations are as follows*

$$
\begin{aligned}
h_1 \cap h_2 &= \cup_{\gamma_1 \subset h_1, \gamma_2 \subset h_2}\{\min(\gamma_1, \gamma_2)\};\\
h_1 \cup h_2 &= \cup_{\gamma_1 \subset h_1, \gamma_2 \subset h_2}\{\max(\gamma_1, \gamma_2)\};\\
h_1 \oplus h_2 &= \cup_{\gamma_1 \subset h_1, \gamma_2 \subset h_2}\{\gamma_1 + \gamma_2 - \gamma_1\gamma_2\};\\
h_1 \otimes h_2 &= \cup_{\gamma_1 \subset h_1, \gamma_2 \subset h_2}\{\gamma_1\gamma_2\}.
\end{aligned}
$$

In addition, $h_j(j = 1, 2\cdots, n)$ is a set of hesitant fuzzy elements, and the operation of hesitant fuzzy weighted average operator $H^n \to H$ is the mapping of $H^n \to H$, and the specific operation is as follows [25]

$$HFWA(h_1, h_2, \cdots h_n) = \underset{j=1}{\overset{n}{\otimes}} (w_j h_j) = \cup_{\gamma_1 \subset h_1, \gamma_2 \subset h_2} \left\{ 1 - \prod_{j=1}^{n} (1 - \gamma_j)^{w_j} \right\}; \tag{1}$$

where $w = (w_1, w_2, \cdots w_n)^T$ is the weight vector of h_j, $w_j > 0$, $\sum_{j=1}^{n} w_j = 1$.

Definition 2. *To calculate the escape speed of people on board, this study introduces relevant acceleration formulas. For research convenience, the probability that the target group perceives the other group is denoted as θ. The influence of the other group on the target group is denoted as μ, and the influence direction of the other group on the target group is denoted as $sgn(v_j - v_i)$, When $v_j < v_i$, its value is -1, $v_j > v_i$, its value is +1. t_{act} is the instantaneous reaction time of personnel escape. The acceleration equation is obtained as*

$$a = \frac{\Delta v}{\Delta t} = \frac{\theta_i \cdot \mu \cdot \text{sgn}(v_j - v_i)}{t_{act}}. \tag{2}$$

Definition 3. *Considering the influence of hesitancy fuzzy average operator and based on the first-order linear relationship between velocity and acceleration, this study gives the velocity correction formula of different people under different influences when fire occurs. The influence of all attributes on a single population was denoted as Q_i, v_i is the expected speed of group i. The corrected speed v' can be obtained as*

$$v' = (1 - Q_i) \cdot v_i + a \cdot t_{act}. \tag{3}$$

3. Model Formulation

Based on the above analysis of human behavior characteristics in a passenger ship fire, this study constructs its model as follows. Firstly, fuzzy mathematical theory is applied to acquire fuzzy sets including different groups of people. Secondly, an ensemble operator with different attributes is obtained by using the classical hesitant fuzzy average operator. Thirdly, the escape speed of each crowd is obtained by combining the universal gravitation formula and the relationship between velocity and acceleration. Finally, considering various special cases, the properties and inferences are acquired. The model building process is shown in Figure 2.

3.1. Consider Different Attributes and the Speed Correction Model of the Crowd

The steps of model construction are as follows. Step 1 is to construct the hesitant fuzzy sets including an adult male, adult female, children and the elderly, and comprehensively consider the factors affecting the escape speed of people on the ship. The classical hesitancy fuzzy integration operator is used to consider and quantify various factors in Step 2. Step 3 is inspired by the classical universal gravitation formula, and extracts the formula of escape acceleration of people on board under the interaction of two factors, i.e., different attributes and different people [26]. Step 4, combined with the relationship between acceleration and velocity, further extracts the escape speed of people on board under this interactive influence [27].

Step 1. This study establishes hesitant fuzzy sets of four groups of people. For the convenience of research, in this study, $\{H_{i1}, H_{i2}, H_{i3}, H_{i4}\}$ represents the hesitant fuzzy sets under the four attributes of the four groups of people. N_{ijk} is the hesitant fuzzy elements under the four attributes of the four groups of people. ρ_{ijk} is the percentage of i crowd, j attribute, and k ability judgment options. Among them, i represents the four groups of people, respectively; $i \in \{1,2,3,4\}$. j represents the four attributes,$j \in \{1,2,3,4\}$. k represents the evaluation options of the four attributes; $k \in \{1,2,3,4\}$. The hesitant fuzzy sets under the four attributes of the four groups of people are expressed as follows:

$$H_{i1} = \left\{ \overbrace{\rho_{i11}, \rho_{i11} \cdots \rho_{i11}}^{N_{i11}}, \overbrace{\rho_{i12}, \rho_{i12} \cdots \rho_{i12}}^{N_{i12}}, \overbrace{\rho_{i13}, \rho_{i13} \cdots \rho_{i13}}^{N_{i13}}, \overbrace{\rho_{i14}, \rho_{i14} \cdots \rho_{i14}}^{N_{i14}} \right\},$$

$$H_{i2} = \left\{ \overbrace{\rho_{i21}, \rho_{i21} \cdots \rho_{i21}}^{N_{i21}}, \overbrace{\rho_{i22}, \rho_{i22} \cdots \rho_{i22}}^{N_{i22}}, \overbrace{\rho_{i23}, \rho_{i23} \cdots \rho_{i23}}^{N_{i23}}, \overbrace{\rho_{i24}, \rho_{i24} \cdots \rho_{i24}}^{N_{i24}} \right\},$$

$$H_{i3} = \left\{ \overbrace{\rho_{i31}, \rho_{i31} \cdots \rho_{i31}}^{N_{i31}}, \overbrace{\rho_{i32}, \rho_{i32} \cdots \rho_{i32}}^{N_{i32}}, \overbrace{\rho_{i33}, \rho_{i33} \cdots \rho_{i33}}^{N_{i33}}, \overbrace{\rho_{i34}, \rho_{i34} \cdots \rho_{i34}}^{N_{i34}} \right\},$$

$$H_{i4} = \left\{ \overbrace{\rho_{i41}, \rho_{i41} \cdots \rho_{i41}}^{N_{i41}}, \overbrace{\rho_{i42}, \rho_{i42} \cdots \rho_{i42}}^{N_{i42}}, \overbrace{\rho_{i43}, \rho_{i43} \cdots \rho_{i43}}^{N_{i43}}, \overbrace{\rho_{i44}, \rho_{i44} \cdots \rho_{i44}}^{N_{i44}} \right\}.$$

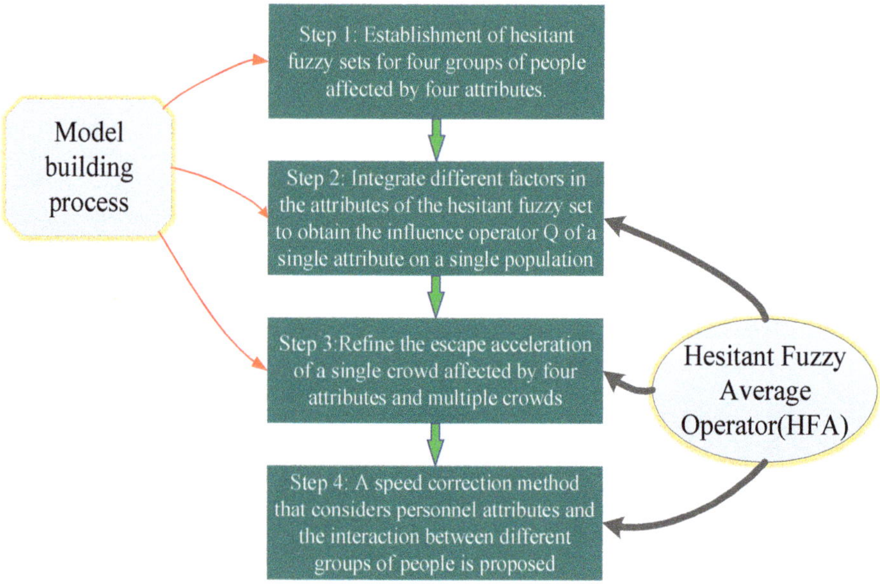

Figure 2. Model building process.

Step 2. Based on the fuzzy mathematics theory, this study transforms the qualitative problem into the quantitative problem. According to the above hesitant fuzzy sets, the objective factors affecting crowd evacuation efficiency are integrated. Then, the influence of a single attribute on a single population can be denoted as Q_{ij}. The influence of all attributes on a single population was denoted as Q_i by integrating Q_{ij}. Based on experience, the literature, and current research, this study assumes that the weight of the four groups is equal. According to the classical hesitant fuzzy average operator [24], Q_i is obtained as

$$Q_{ij} = \left(1 - \prod_{i=1}^{4}\left(1 - \rho_{ijk}\right)^{\frac{1}{4}}\right), \tag{4}$$

$$Q_i = \frac{\sum\limits_{j=1}^{4} Q_{ij}}{4}, \tag{5}$$

respectively, where $j = 1,2,3,4; i = 1,2,3,4$.

Step 3. Consider that there is only a single target population, its escape speed is only affected by four attributes and its expected speed. However, in a case of multiple groups, the escape speed of the target group will be affected not only by cognitive ability, emergency response ability, value orientation and psychological endurance but also by other groups. To simplify the research work, this study divides the influences from other groups into three categories, as follows. The probability that the target group perceives the other group is denoted as θ, the magnitude of the influence of the other group on the target group is denoted as μ, and the direction of the influence of the other group on the target group is denoted as $\mathrm{sgn}(v_j - v_i)$. Since this study only considers the influence between the four groups of people, other influencing factors are not considered. Therefore, in this study, other unconsidered factors are denoted as λ^* and defaulted to 0.375 [28]. M_i is used to represent the number of single people, with v_i representing the expectations of a single population escape velocity, with v_i' representing a single population after a reaction time of the final velocity. λ_{ij} denotes mutual influence between the two groups, t_{act} instantaneous response time for escape, ω_1 is the weight of the probability that the target group feels other groups of people, and ω_2 is the weight of the influence of other groups on the target population. Based on experience, the literature, and current research [29], this study considers that the instantaneous reaction time of adult men and women is $t_{1act} = t_{2act} = 2$ s, and the elderly and children is $t_{3act} = t_{4act} = 3$ s. Thus, the formula of escape acceleration (a_i) of a single crowd is obtained as

$$\begin{aligned} a_i &= \sum_{i=1}^{4} (\lambda_{ij}) \\ &= \sum_{j=1}^{4} \frac{\frac{4}{\pi} \cdot \arctan\left(\left(\frac{M_j}{M_i+M_j}\right)^{w_1} \cdot \left(\frac{|v_i-v_j|}{\max\{v_i,v_j\}}\right)^{w_2} \right) \mathrm{sgn}(v_j-v_i) \cdot |v_j-v_i|}{t_{iact}}. \end{aligned} \tag{6}$$

Among them, $\mathrm{sgn}(v_j - v_i)$ represents the influence direction of other groups on the target group. In $i = \{1,2,3,4\}$, $j \in \{1,2,3,4\}$, 1, 2, 3, and 4 represent adult males, adult females, the elderly, and children, respectively. When $v_j > v_i$, the value of $\mathrm{sgn}(v_j - v_i)$ is $+1$, indicating that other groups have a positive influence on the target group. When $v_j < v_i$, the value of $\mathrm{sgn}(v_j - v_i)$ is -1, indicating that other groups have a negative influence on the target group. When $v_j = v_i$, the value of $\mathrm{sgn}(v_j - v_i)$ is 0, and other groups are in the same direction as the target group.

Step 4. This study integrates the influence of four attributes and other populations on the target population, substitute it into Equation (3), v_i' is obtained as

$$v'_i = (1 - Q_i) \cdot v_i + \sum_{j=1}^{4} (\lambda_{ij}) \cdot \lambda^* \cdot t_{iact}, \tag{7}$$

where, $i = 1,2,3,4$.

Then, substitute Equation (7) into Equation (6), and it is obtained as

$$\begin{aligned} v'_i &= (1 - Q_i) \cdot v_i + \\ &\sum_{j=1}^{4} \frac{\frac{4}{\pi} \cdot \arctan\left(\left(\frac{M_j}{M_i+M_j}\right)^{w_1} \cdot \left(\frac{|v_i-v_j|}{\max\{v_i,v_j\}}\right)^{w_2} \right) \mathrm{sgn}(v_j-v_i) \cdot |v_j-v_i|}{t_{iact}} \cdot \lambda^* \cdot t_{iact} \end{aligned} \tag{8}$$

where, $i = 1,2,3,4$.

3.2. Supplement and Description

The corollary and properties of Equation (8) are as follows.

Corollary 1. *When M_i is much larger than M_j, M_i is regarded as the maximum value and M_j as the minimum value, thus, which is substituted into Equation (8) to obtain $v'_i = (1 - Q_i) \cdot v_i$.*

Theorem 1. *When the number of the target population is considered larger than that of other groups, the speed of the group is not affected by other groups but is only related to its expected speed and its cognitive ability, emergency response-ability, value orientation and psychological bearing capacity.*

Corollary 2. *When M_j is much larger than M_i, M_j is regarded as the maximum value and M_i as the minimum value, thus $\frac{M_j}{M_i+M_j}= 1$, which is substituted into Equation (8) to obtain the simplified Equation (9).*

$$v'_i = (1 - Q_i) \cdot v_i + \sum_{j=1}^{4} \frac{\frac{4}{\pi}\cdot\arctan\left(\frac{|v_i-v_j|}{\max\{v_i,v_j\}}\right)^{w_2} \operatorname{sgn}(v_j-v_i)\cdot|v_j-v_i|}{t_{iact}} \cdot \lambda^* \cdot t_{iact} \quad (9)$$

where, $i = 1, 2, 3, 4$.

Theorem 2. *When the number of other groups is much larger than the number of target groups, the impact of the number of groups can be ignored.*

Corollary 3. *When v_i and v_j differs greatly, there will be the following two situations.*

(i) v_j is regarded as the maximum value and v_i as the minimum value, so $\frac{|v_i-v_j|}{\max\{v_i,v_j\}} = 1$ are substituted into Equation (8) to obtain simplified Equation (10).

$$v'_i = (1 - Q_i) \cdot v_i + \sum_{j=1}^{4} \frac{\frac{4}{\pi} \cdot \arctan\left(\frac{M_j}{M_i+M_j}\right)^{w_1} \operatorname{sgn}(v_j - v_i) \cdot v_i}{t_{iact}} \cdot \lambda^* \cdot t_{iact}, \quad (10)$$

where, $i = 1, 2, 3, 4$.

(ii) Similarly, v_i is regarded as the maximum value and v_j as the minimum value, so $\frac{|v_i-v_j|}{\max\{v_i,v_j\}} = 1$ can be substituted into Equation (8) to obtain simplified Equation (10).

Theorem 3. *When the speed of the target crowd differs considerably from that of another crowd, the evacuation speed of the target crowd is not affected by the speed of others.*

Corollary 4. *When $v_i = v_j$, then $\frac{|v_i-v_j|}{max\{v_i,v_j\}} = |v_i - v_j| = 0$, substitute the sub-data into Equation (8) to obtain $v'_i = (1 - Q_i) \cdot v_i$.*

Theorem 4. *When the speed of the target group is the same as that of other groups, the speed of the group is not affected by other groups, but is only related to its own speed and the impact of cognitive ability, emergency response-ability, psychological bearing capacity and value orientation.*

4. Simulation Example

In order to verify the effectiveness of the interactive speed correction method, a comparative simulation experiment is carried out in this study. Firstly, a questionnaire survey was conducted. Based on the results of the questionnaire survey, the expected speed of four different groups of people is revised by using the interactive speed modification method. Secondly, the single deck of a ro-ro passenger ship is selected as a simulation example. The deck is then modeled by Pathfinder evacuation software. Finally, this study sets up two evacuation plans, namely, ordinary evacuation and evacuation under the speed correction of interactive influence. Through the comparison of simulation results, it is concluded that the speed correction method of interactive influence proposed in this study is in line with reality.

4.1. Personnel Evacuation Speed Correction

The age, gender, cultural background, and other factors of the people on board will not only affect their judgment of the degree of fire risk, but also affect the evacuation speed. Therefore, this study combines the previous research results to design a questionnaire on ship fire evacuation behavior [30]. The questionnaire structure of personnel evacuation behavior in a ship fire situation is shown in Table 1. The questionnaire topic is mainly set to investigate the cognitive ability, emergency response ability, value orientation, and psychological endurance of different groups of people. Questionnaires were randomly distributed on an online questionnaire survey platform. The respondents were then divided into different age groups. A total of 129 questionnaires were distributed, and 105 questionnaires were finally effectively recovered. Finally, the reliability of the questionnaire was tested. The Cronbach reliability coefficient α of the questionnaire was 0.67. This indicates that the data reliability of the questionnaire is good and meets the requirements of usability. Among them, the proportion of adult men, adult women, the elderly, and children surveyed is 8:8:3:2. The information summary is shown in Table 2.

Table 1. Questionnaire structure of personnel evacuation behavior in a ship fire situation.

Variable	Problem Setting	Options
basic information of personnel	gender	men; women
	age	under 20 years old; 20–60 years old: over 60 years old
cognitive ability	education level	high school diploma and below; college degree; bachelor degree; Master degree and above
emergency capability	escape response when hearing a fire alarm	look around and judge for yourself;
		ask others to determine the direction of escape;
		escape immediately;
		observe the behavior of others
value orientation	will you escape with valuables	no; possibly; not sure, must
psychological endurance	the level of panic at hearing a fire alarm	no panic; low panic; moderate panic; extreme panic

Table 2. Summary of questionnaire information.

Influencing Factors		Adult Male (20–60 Years Old)	Adult Female (20–60 Years Old)	Elderly (Over 60 Years Old)	Children (Under 20 Years Old)
Cognitive ability	high school diploma or below	3%	18%	14%	0%
	college degree	54%	53%	43%	54%
	bachelor degree	33%	24%	29%	46%
	master degree and above	10%	5%	14%	0%

Table 2. *Cont.*

Influencing Factors		Adult Male (20–60 Years Old)	Adult Female (20–60 Years Old)	Elderly (Over 60 Years Old)	Children (Under 20 Years Old)
The emergency ability	look around and judge for yourself	29%	22%	43%	38%
	ask others to determine the direction of escape	37%	36%	14%	31%
	escape immediately	20%	31%	43%	23%
	observe other people's behavior	14%	11%	0%	8%
The value orientation	no	58%	58%	68%	70%
	probably	15%	10%	8%	10%
	not sure	19%	22%	16%	10%
	must	8%	10%	8%	10%
Mental endurance	no panic	17%	22%	29%	8%
	low panic	32%	29%	43%	31%
	moderate panic	29%	29%	14%	38%
	extreme panic	12%	20%	14%	23%

Based on the questionnaire data, this study brings the above data into the speed correction model and obtains the expected correction speed of four groups. The specific model calculation steps are as follows.

Step 1. Bring ρ_{ijk} into a fuzzy set based on the questionnaire data in Table 2. It can be obtained that the impact of a single attribute on a single population is recorded as Q_{ij}. Then, Q_{ij} is integrated to calculate the impact of all attributes on a single population, which is recorded as Q_i. Among them, M_1= 40, M_2= 40, M_3= 15, M_4= 10. The effects of all attributes on adult men, adult women, the elderly, and children are

$$Q_1 = 0.246,\ Q_2 = 0.266,\ Q_3 = 0.275,\ Q_4 = 0.277$$

Step 2. In this study, the expected speeds of adult men, adult women, children and the elderly are set to be 1.5 m/s, 1.3 m/s, 1.1 m/s, and 0.9 m/s, respectively. Combined with the questionnaire data in Table 2 and substituted into Formula (6), the escape acceleration a_i of the four groups is, respectively,

$$a_1 = 0.205\ \mathrm{m/s^2},\ a_2 = 0.056\ \mathrm{m/s^2},\ a_3 = -0.076\ \mathrm{m/s^2},\ a_4 = -0.236\ \mathrm{m/s^2}.$$

Step 3. Substituting v_i, Q_i, a_i and t_{iact} into Formula (8), the correction speed of adult males, adult females, the elderly, and children can be obtained as

$$v_1 = 1.28\ \mathrm{m/s},\ v_2 = 1\ \mathrm{m/s},\ v_3 = 0.71\ \mathrm{m/s},\ v_4 = 0.38\mathrm{m/s}.$$

The above results are used as examples to simulate the modified expected speed of adult men, adult women, children, and the elderly.

4.2. Simulation Model Construction

This study takes a ro-ro passenger ship as the simulation object. The ship has a length of 196.27 m, a width of 28.60 m, a seating capacity of 1588 people, a passenger quota of 1500 people, and 10 decks. The seven and eight decks of the ship belong to the passenger activity area, and both decks have independent evacuation assembly areas, vertical single channel marine evacuation system, and fully enclosed lifeboats. In this study, the eight decks of the ship are selected as the simulation model for the evacuation of ship fire personnel.

In this study, Pathfinder software is used to model the above simulation examples. The deck model is 196 m long, 28.6 m wide and 3 m high. A total of 620 people need to be evacuated on the deck, and the proportion of adult men, adult women, children, and the elderly is 8:8:3:2, who are randomly distributed on the deck. According to the internal structure of the eighth deck, the evacuation routes and exits shall be set. Considering that the vertical single channel marine evacuation system and the fully enclosed lifeboat are difficult to set up in Pathfinder software, this paper sets the personnel evacuation deck as the form of successful personnel evacuation. The 3D model of personnel evacuation on the ship's deck is shown in Figure 3. The green column represents adult men, the blue column represents adult women, the yellow column represents children, and the black column represents the elderly. The green line indicates the evacuation exit, that is, the evacuation from the green line indicates the successful evacuation.

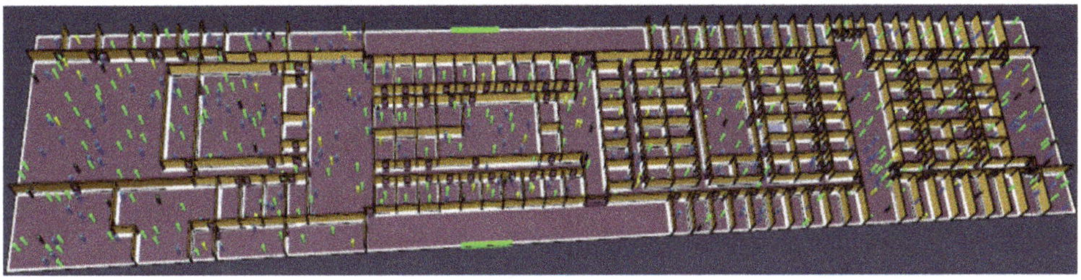

Figure 3. Ship Deck Evacuation 3D model.

4.3. *Comparative Analysis of Evacuation Results*

In order to verify the effectiveness of the correction method, this study sets up two evacuation plans, namely ordinary evacuation and speed correction method of interactive influence. At the same time, the velocity correction methods of ordinary evacuation and interactive influence are compared and analyzed experimentally. In addition, the specific details of the plan are as follows.

Plan 1: ordinary evacuation. The range of passenger evacuation speed is set to be 0.51~1.50 m/s, and remains unchanged. The evacuation path is uniform evacuation at the exits on both sides of the deck.

Plan 2: speed correction method of interactive influence. Based on the calculation results in Section 4.3, the expected evacuation speeds of adult men, adult women, the elderly, and children are set to 1.28 m/s, 1 m/s, 0.71 m/s, and 0.38 m/s, respectively. The evacuation speed obeys the normal distribution. The expected speed of each group is compared with the corrected speed (please see Figure 4 for details). The experimental results of the two evacuation plans are compared, as shown in Figure 5.

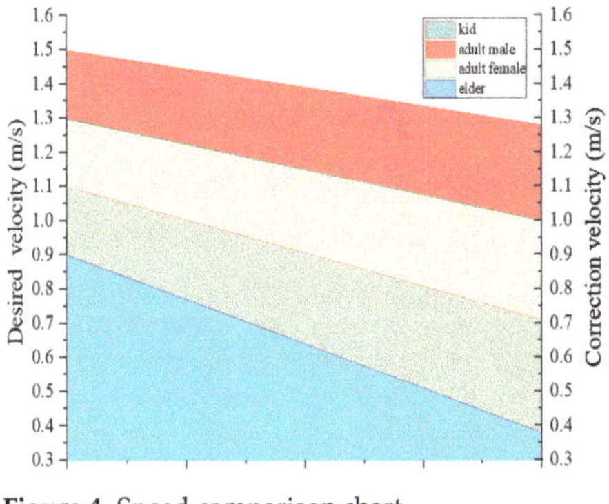

Figure 4. Speed comparison chart.

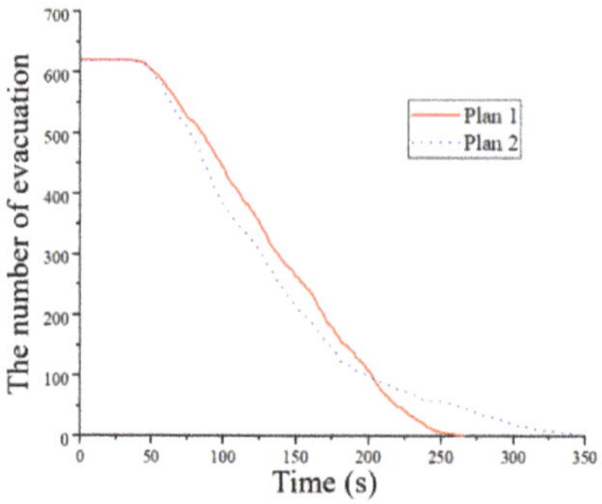

Figure 5. Experimental results of two plans.

According to Figure 5 and Table 3, the total evacuation time of ordinary evacuation is 265.28 s, and the overall evacuation efficiency is high. The evacuation time of the interactive speed correction method is 349.03 s, and the overall evacuation efficiency is low. With the development of time, the evacuation efficiency decreases significantly. After 325 s, the evacuation efficiency is close to 0. The evacuation efficiency (curve slope) of the speed correction method of interactive influence is significantly lower than that of ordinary evacuation, and the evacuation efficiency decreases significantly with the development of time. This is due to the consideration of the negative psychology of personnel in the fire and the influence of fire smoke. As time goes on, smoke concentration and temperature will gradually increase, causing harm to the human body and resulting in reduced evacuation efficiency. After 325 s, the evacuation efficiency of the interactive speed correction method is close to 0. This is because the concentration and temperature of fire smoke are enough to threaten the lives of people at 325 s. However, considering the mutual assistance behavior of the group, people around will actively help those who are slow and unconscious, so that they can keep the same speed and continue to move.

Table 3. Data comparison between two plans.

Plan	Evacuation Time (s)	Evacuation Efficiency (Rate of the Curve)	Remaining Staff
1	265.28	high	0
2	349.03	low	0

To sum up, ordinary evacuation oversimplifies the evacuation behavior of people. It is assumed that people only evacuate evenly according to the evacuation path during the evacuation process, and the impact of fire smoke and group behavior on people is not considered, which is not in line with reality and there is a large error. The aim of the speed correction method of interactive influence is to study evacuation from the human point of view. Through the analysis of the complex psychology and group effect on evacuation behavior, the evacuation path and evacuation speed are modified and supplemented. This is more realistic.

5. Conclusions

This study adopts the HFA to research evacuation from the perspective of the people on board, analyzes the influence of cognitive ability, emergency response ability, value orientation, psychological tolerance and group effect on evacuation behavior of the people in the fire scene. Then the escape acceleration and escape speed are corrected and supplanted to make them more realistic. The innovation of this study is mainly reflected in the following three points.

First, this study considers the interaction among fire escape personnel. The main influencing factors among escaped groups are summarized as emergency ability, cognitive ability, psychological bearing ability and value orientation. The interactive effects of the four attributes of the four groups of people are introduced into the evacuation model, which makes the evacuation research more realistic.

Secondly, this study uses the hesitant fuzzy integration operator to integrate the four attributes of the four groups of people, and realizes the quantification of the interaction between the groups. Then, the acceleration formula and velocity correction modulus formula considering the interaction effect of different groups of people are extracted, and the influence from other types of people is introduced into the evacuation research.

Finally, this study collects data through questionnaires and calculates the revised speed for different populations. Then, simulation software is used to compare the revised speed plan with the uncorrected speed plan, it is concluded that the revised speed plan is more realistic and provides a reference for subsequent evacuation research.

The shortcoming of this study is that it fails to take into account the emotional contagion of panic among pedestrians, the influence of fire smoke toxicity and fire temperature as well as other factors. Further study will be carried out by combining the above influencing factors with simulation examples.

Author Contributions: Conceptualization, F.Z.; Data curation, W.L.; Formal analysis, T.L.; Investigation, R.W.; Writing—original draft, J.L. All authors have read and agreed to the published version of the manuscript.

Funding: The Fangwei Zhang's work is partially supported by Shanghai Pujiang Program (No. 2019PJC062), the Natural Science Foundation of Shandong Province (No. ZR2021MG003), the Research Project on Undergraduate Teaching Reform of Higher Education in Shandong Province (No. Z2021046).

Institutional Review Board Statement: Not applicable.

Informed Consent Statement: Informed consent was obtained from all subjects involved in the study.

Data Availability Statement: Not applicable.

Conflicts of Interest: The authors declare no conflict of interest.

References

1. Zhang, T. Investigation and Research Report on Ship Fire. *WSRJ* **2020**, *6*, 16–23.
2. Kim, B.; Hwang, K.I. Text mining techniques to identify causes and hazards of ship fire accidents. *J. Mar. Eng. Technol.* **2020**, *44*, 189–195. [CrossRef]
3. Tong, D.; Canter, D. The decision to evacuate: A study of the motivations which contribute to evacuation in the event of fire. *Fire Saf. J.* **1985**, *9*, 257–265. [CrossRef]
4. Kinateder, M.T.; Kuligowski, E.D.; Reneke, P.A.; Peacock, R.D. Risk perception in fire evacuation behavior revisited: Definitions, related concepts, and empirical evidence. *Fire Sci. Rev.* **2015**, *4*, 1. [CrossRef] [PubMed]
5. Treuille, A.; Cooper, S.Z.; Popović, Z. Continuum crowds. *ACM Trans. Graph.* **2006**, *25*, 1160–1168. [CrossRef]
6. Helbing, D.; Farkas, I.; Vicsek, T. Simulating dynamical features of escape panic. *Nature* **2000**, *407*, 487–490. [CrossRef]
7. Helbing, D.; Farkas, I.J.; Vicsek, T. Freezing by heating in a driven mesoscopic system. *Phys. Rev. Lett.* **2000**, *84*, 1240. [CrossRef]
8. Wang, Y.; Kyriakidis, M.; Dang, V.N. Incorporating human factors in emergency evacuation–An overview of behavioral factors and models. *Int. J. Disaster Risk Reduct.* **2021**, *60*, 102254. [CrossRef]
9. Wang, H.; Xu, T.; Li, F. A novel emergency evacuation model of subway station passengers considering personality traits. *Sustainablity* **2021**, *13*, 10463. [CrossRef]
10. Hu, Y.L.; Wang, X.; Wang, F.Y. A quantitative study of factors influence on evacuation in building fire emergencies. *IEEE Trans. Comput. Soc. Syst.* **2018**, *5*, 544–552. [CrossRef]
11. Feng, J.N.; Wang, Q.F. Emergency safety evacuation decision based on dynamic Gaussian Bayesian network. *Mater. Sci. Eng.* **2019**, *688*, 055076. [CrossRef]
12. Lovreglio, R.; Ronchi, E.; Nilsson, D. An Evacuation Decision Model based on perceived risk, social influence and behavioural uncertainty. *Simul. Model. Pract. Theory* **2016**, *66*, 226–242. [CrossRef]
13. Peng, M.L.; Liu, X.X.; Chen, Q. Two-layer decision-making model for HRB emergency evacuation route based on BIM. *Comput. Simul.* **2021**, *38*, 471–475.
14. Sun, D.; Zhang, L.P.; Su, Z.F. Evacuate or Stay? A Typhoon Evacuation Decision Model in China Based on the Evolutionary Game Theory in Complex Networks. *Int. J. Environ. Res. Public Health* **2020**, *17*, 682. [CrossRef]

15. Tian, Y.; Zhou, T.S.; Yao, Q.; Zhang, M.; Li, J.S. Use of an agent-based simulation model to evaluate a mobile-based system for supporting emergency evacuation decision making. *J. Med. Syst.* **2014**, *38*, 149. [CrossRef] [PubMed]
16. Koo, J.; Kim, B.I.; Kim, Y.S. Estimating the effects of mental disorientation and physical fatigue in a semi-panic evacuation. *Expert. Syst. Appl.* **2014**, *41*, 2379–2390. [CrossRef]
17. Chen, L.; Zheng, Q.; Li, K.; Li, Q.R. Emergency evacuation from multi-exits rooms in the presence of obstacles. *Phys. Scripta* **2021**, *96*, 115208. [CrossRef]
18. Jeon, G.Y.; Na, W.J.; Hong, W.H.; Lee, J.K. Influence of design and installation of emergency exit signs on evacuation speed. *J. Asian Archit. Build. Eng.* **2019**, *18*, 104–111. [CrossRef]
19. Yu, L.; Deng, T.; Wang, M.; Li, Q.; Xu, X.X. Passengers' evacuation from a fire train in railway tunnel. *Int. J. Rail. Transp.* **2019**, *7*, 159–172. [CrossRef]
20. Jia, L.; Yun, C. Simulation on staffs evacuation behavior in plant fire emergencies. *Syst. Res. Behav. Sci.* **2014**, *31*, 527–536. [CrossRef]
21. Kinsey, M.J.; Gwynne, S.M.V.; Kuligowski, E.D.; Kinateder, M. Cognitive biases within decision making during fire evacuations. *Fire Technol.* **2019**, *55*, 465–485. [CrossRef]
22. Zadeh, L.A. Fuzzy sets. *Inf. Control* **1965**, *8*, 338–353. [CrossRef]
23. Torra, V. Hesitant fuzzy sets. *Int. J. Intell. Syst.* **2010**, *25*, 529–539. [CrossRef]
24. Rodríguez, R.M.; Martínez, L.; Torra, V.; Xu, Z.S.; Herrera, F. Hesitant fuzzy sets: State of the art and future directions. *Int. J. Intell. Syst.* **2014**, *29*, 495–524. [CrossRef]
25. Xia, M.; Xu, Z. Hesitant fuzzy information aggregation in decision making. *Int. J. Approx. Reason.* **2011**, *52*, 395–407. [CrossRef]
26. No, H.; Cho, A.; Kee, C. Attitude estimation method for small UAV under accelerative environment. *GPS Solut.* **2015**, *19*, 343–355. [CrossRef]
27. Xu, M.; Wang, G. Research on the applicability of ship motion and acceleration formulas to fatigue strength assessment of container ships. *China Ship Build.* **2022**, *63*, 166–175.
28. Yuan, C.Y.; Wang, K.; Chen, H.Y.; Liu, X.; Lang, Y.J. Research on fire evacuation speed correction method based on personnel psychological-environmental factors. *China Saf. Prod. Sci. Technol.* **2020**, *16*, 112–118.
29. Xu, Z.; Yang, X.K.; Zhao, X.H.; Li, L.J. Driver perception reaction time under the emergency evacuation situation. *J. Chongqing Univ.* **2011**, *34*, 54–60.
30. Koshiba, Y.; Suzuki, Y. Factors affecting post-evacuation behaviors following an earthquake: A questionnaire-based survey. *Int. J. Disaster Risk Reduct.* **2018**, *31*, 548–554. [CrossRef]

mathematics MDPI

Article

Medical Diagnosis and Pattern Recognition Based on Generalized Dice Similarity Measures for Managing Intuitionistic Hesitant Fuzzy Information

Majed Albaity [1,*] and Tahir Mahmood [2,*]

1 Department of Mathematics, Faculty of Science, King Abdulaziz University, P.O. Box 80348, Jeddah 22254, Saudi Arabia
2 Department of Mathematics and Statistics, International Islamic University, Islamabad 44000, Pakistan
* Correspondence: malbaity@kau.edu.sa (M.A.); tahirbakhat@yahoo.com or tahirbakhat@iiu.edu.pk (T.M.)

Abstract: Pattern recognition is the computerized identification of shapes, designs, and reliabilities in information. It has applications in information compression, machine learning, statistical information analysis, signal processing, image analysis, information retrieval, bioinformatics, and computer graphics. Similarly, a medical diagnosis is a procedure to illustrate or identify diseases or disorders, which would account for a person's symptoms and signs. Moreover, to illustrate the relationship between any two pieces of intuitionistic hesitant fuzzy (IHF) information, the theory of generalized dice similarity (GDS) measures played an important and valuable role in the field of genuine life dilemmas. The main influence of GDS measures is that we can easily obtain a lot of measures by using different values of parameters, which is the main part of every measure, called DGS measures. The major influence of this theory is to utilize the well-known and valuable theory of dice similarity measures (DSMs) (four different types of DSMs) under the assumption of the IHF set (IHFS), because the IHFS covers the membership grade (MG) and non-membership grade (NMG) in the form of a finite subset of [0, 1], with the rule that the sum of the supremum of the duplet is limited to [0, 1]. Furthermore, we pioneered the main theory of generalized DSMs (GDSMs) computed based on IHFS, called the IHF dice similarity measure, IHF weighted dice similarity measure, IHF GDS measure, and IHF weighted GDS measure, and computed their special cases with the help of parameters. Additionally, to evaluate the proficiency and capability of pioneered measures, we analyzed two different types of applications based on constructed measures, called medical diagnosis and pattern recognition problems, to determine the supremacy and consistency of the presented approaches. Finally, based on practical application, we enhanced the worth of the evaluated measures with the help of a comparative analysis of proposed and existing measures.

Keywords: intuitionistic hesitant fuzzy sets; generalized dice similarity measures; medical diagnosis; pattern recognition; artificial intelligence

MSC: 03B52; 68T27; 68T37; 94D05; 03E72

Citation: Albaity, M.; Mahmood, T. Medical Diagnosis and Pattern Recognition Based on Generalized Dice Similarity Measures for Managing Intuitionistic Hesitant Fuzzy Information. *Mathematics* **2022**, *10*, 2815. https://doi.org/10.3390/math10152815

Academic Editor: Pasi Luukka

Received: 21 June 2022
Accepted: 1 August 2022
Published: 8 August 2022

1. Introduction

The decision-making procedure covers four main stages: intelligence, design, choice, and implementation. The principle of the decision-making technique begins with the intelligence stage. In this stage, the intellectual determines reality and identifies and explains the troubles. However, before 1965, no one had utilized or studied the decision-making troubles in the environment of the fuzzy set (FS) theory. For this, the well-known idea of FS was initiated by Zadeh [1] by modifying the technique of crisp set into FS, which covers the MG belonging to [0, 1]. FS has received considerable attention from the distinct intellectual, and certain applications have been carried out by different scholars. For example, Aydin [2] proposed the fuzzy multicriteria decision-making technique by using the Fermatean fuzzy sets, John [3] discussed the certain application of the type-2 FSs, Mandel and John [4] explored the type-2 fuzzy sets made simple, and Mahmood [5] initiated the idea of a bipolar soft set, discussed operational laws, and applied it in decision-making problems.

FS has received attention from the distinct intellectual, and certain applications have been carried out by different scholars. However, if an intellectual faces information in the shape of {0.8,0.9,0.7}, then the principle of FS has been neglected. For this, the well-known idea of hesitant FS (HFS) was initiated by Torra [6] by modifying the technique of FS into HFS, which covers the MG, whose supremum value is belonging to [0, 1]. HFS is a modified version of FS and has received attention from the distinct intellectual; certain applications have been performed by different scholars. For example, Meng and Chen [7] developed the correlation measures for HFSs, Li et al. [8] investigated the distance and similarity measures for HFSs, Su et al. [9] proposed certain measures based on dual HFSs, and Wei et al. [10] investigated the entropy and certain types of measures based on HFSs.

If a piece of intellectual faced information in the shape of "yes" or "no", then the principle of FS has been neglected. For this, the well-known idea of intuitionistic FS (IFS) was initiated by Atanassov [11] by modifying the technique of FS into IFS, which covers the MG and NMG, whose sum is belonging to [0, 1]. IFS is a modified version of FS and has received attention from the distinct intellectual; certain applications have been carried out by different scholars. For example, Ye [12] initiated the certain cosine measure by using IFSs, Rani and Garg [13] developed the distance measures by using complex IFSs, Liang and Shi [14] also explored certain measures based on IFSs, Xu and Chen [15] examined the distance and similarity measures for IFSs, Xu [16] proposed the intuitionistic fuzzy similarity measures, Garg and Rani [17] presented the correlation among any number of complex IFSs, Zeshui [18] utilized certain measures for interval-valued IFSs, Wei et al. [19] investigated the entropy and similarity measures for interval-valued IFSS, and Wang and Xin [20] proposed the distance measures for IFSs.

It was demonstrated that the prevailing information computed based on FSs, HFSs, and IFSs has a variety of applications in many different fields, for instance, computer science, economics and finance, engineering sciences, and road signals. However, it is also clear that they have many limitations and restrictions. For instance, we know that IFS has managed only with two-dimensional information in a singleton set, and each dimension of information can express only one value, but what if someone provided two-dimensional information in the shape of singleton sets, and each dimension of information could represent more than one value? In such a situation, experts noticed that the theory of IFS was not able to proceed with the above information accurately. For this, the well-known idea of an intuitionistic hesitant fuzzy set (IHFS) was initiated by Beg and Rashid [21] by modifying the technique of IFS into IHFS, which covers the MG and NMG in the form of a finite subset of [0, 1], whose sum of the supremum of the duplet is belonging to [0, 1]. IHFS is a modified version of IFS and HFS to cope with complicated and unreliable information in genuine life troubles, and it has gotten massive attraction from the distinct intellectual. Certain applications have been carried out by different scholars. For example, Peng et al. [22] initiated the cross-entropy measures by using the IHFSs, and Zhai et al. [23] examined probabilistic interval-valued IHFSs.

In statistics and related theories, a similarity function, i.e., similarity metric or similarity measure, is a real-valued function that computes the similarity among two terms. Even though no single idea of similarity exists, generally such measures are, in a particular sense, the inverse of distance metrics. Cosine similarity, Tangent similarity, hamming similarity, Euclidean similarity, dice similarity, and generalized dice similarity measures are the commonly employed types of similarity measures for real-valued vectors, used in data retravel to score the similarity of documents in the vector space model. In machine learning, common kernel mappings such as the Radial based function kernel can be observed as similarity measures. In all these measures, we noticed that the GDS measures are massively valuable and effective, as they are more generalized than the prevailing studied measures.

Furthermore, GDS measures are a very significant part of the decision-making technique to determine the closeness between any number of attributes. A certain application has been performed by different scholars. By using different values of the parameter, we can easily obtain the prevailing measures of cosine similarity, tangent similarity, hamming similarity, Euclidean similarity, dice similarity. However, the principle of dice and GDS measures are not implemented in the environment of IHFSs. The main goal of this study is to utilize the principle of GDS measures in the environment of IHFS to improve the quality of the research. We propose this theory, due to the following reasons:

1. How do we find the relation between two objects?
2. How do we propose new types of measures based on IHF information?
3. How do we find our required result?

To handle the above questions, we aim to illustrate the following investigations, which are briefly explained in the form of certain points below:

1. To diagnose certain dice similarity measures based on IHF information.
2. To evaluate different types of GDS measures based on IHF information.
3. To investigate many cases of the investigated measures in order to improve the worth of the evaluated measures.
4. To utilize two different types of applications, called medical diagnosis and pattern recognition, based on pioneered measures.
5. To describe the sensitive analysis, advantages, and geomatical expressions of the evaluated theories to determine the partibility of the investigated measures.

The main contribution of this study is constructed as follows: In Section 2, we briefly recall the idea of IFSs, HFSs, and IHFSs. The main idea of dice similarity measure (DSM) is also revised. In Section 3, we propose certain types of DSM measures based on IHFSs. In Section 4, we explore the IHF GDS measure and IHF-weighted GDS measure. Based on the investigated measures, certain special cases are also evaluated. In Section 5, we utilize two different types of applications, called medical diagnosis and pattern recognition, based on pioneered measures and discuss their comparative analysis. The conclusion of this study is discussed in Section 6.

2. Preliminaries

The theory of IFSs, HFSs, IHFSs, and DSMs are the parts of this section. Further, the mathematical term X, represented as a universal set with MG "$\mathfrak{M}_I$" and NMG "$\mathfrak{N}_I$".

Definition 1 ([11]). *An IFS I is investigated by:*

$$I = \{(\mathfrak{X}, \mathfrak{M}_I(\mathfrak{X}), \mathfrak{N}_I(\mathfrak{X})) : \mathfrak{X} \in X\}$$

with a rule: $0 \leq \mathfrak{M}_I(\mathfrak{X}) + \mathfrak{N}_I(\mathfrak{X}) \leq 1$. *Moreover, the hesitancy degree is shown by:* $d_I(\mathfrak{X}) = 1 - (\mathfrak{M}_I(\mathfrak{X}) + \mathfrak{N}_I(\mathfrak{X}))$. *During this study, the IFN is elaborated by* $I = (\mathfrak{M}, \mathfrak{N})$.

Definition 2 ([6]). *A HFS I is investigated by:*

$$I = \{(\mathfrak{X}, \mathfrak{M}_I(\mathfrak{X})) : \mathfrak{X} \in X\}$$

where $\mathfrak{M}_I = \{\mathfrak{M}_1, \mathfrak{M}_2, \ldots, \mathfrak{M}_n\}$ *with a rule:* $0 \leq \sup(\mathfrak{M}_I) \leq 1$.

Definition 3 ([21]). *An IHFS* Ξ *is investigated by:*

$$\Xi = \{(\mathfrak{X}, \mathfrak{M}_\Xi(\mathfrak{X}), \mathfrak{N}_\Xi(\mathfrak{X})) : \mathfrak{X} \in X\}$$

where $\mathfrak{M}_\Xi(\mathfrak{X})$ *and* $\mathfrak{N}_\Xi(\mathfrak{X})$ *are expressed the hesitant fuzzy numbers (HFNs), with a rule:* $0 \leq \mathfrak{MAX}(\mathfrak{M}_\Xi(\mathfrak{X})) + max(\mathfrak{N}_\Xi(\mathfrak{X})) \leq 1$. *Moreover, the refusal grade is initiated by:* $\pi_\Xi(\mathfrak{X}) =$

$1-(\mathfrak{MAX}(\mathfrak{M}_{\Xi}(\mathfrak{X}))+max(\mathfrak{N}_{\Xi}(\mathfrak{X})))$. *The intuitionistic hesitant fuzzy number is expressed by:* $\Xi=\left(\mathfrak{M}^j_{\Xi},\mathfrak{N}^j_{\Xi}\right)$.

Definition 4 ([24]). *For any two-positive vector X and Y, the DSM is initiated by:*

$$D(X,Y)=\frac{2X.Y}{\|X\|_2^2+\|Y\|_2^2}=\frac{2\sum_{j=1}^{l}\mathfrak{X}_j y_j}{\sum_{j=1}^{l}\mathfrak{X}_j^2+\sum_{j=1}^{l}y_j^2}$$

where $X.Y=\sum_{j=1}^{l}\mathfrak{X}_j y_j$ *is expressed as the inner product and* $\|X\|_2=\sqrt{\sum_{j=1}^{l}\mathfrak{X}_j^2}$ *and* $\|Y\|_2=\sqrt{\sum_{j=1}^{l}y_j^2}$ *is expressed in the Euclidean or* L_2 *norms of X and Y.*

3. DSM for IHFSs

To illustrate the relationship between any two pieces of IHF information, the theory of DSMs played an important and valuable role in the field of genuine life dilemmas. The main influence of GDS measures is that we can easily obtain many measures by using different values of parameters, which is the main part of every measure, called DGS measures. In this study, we chose one of the most flexible and genuine principles, called the IHFS, which covers the MG and NMG in the form of a finite subset of [0, 1], with the rule that the sum of the supremum of the duplet is limited to [0, 1] and GDS measures are to develop the four sorts of IHF dice similarity measure and IHF weighted dice similarity measure. Based on the investigated measures, certain special cases were also evaluated.

Definition 5. *By using any two IHFNS* Ξ *and* Ξ', *a DSM* $D^1{}_{P\Xi F}(\Xi,\Xi')$ *is investigated by:*

$$D^1{}_{P\Xi F}(\Xi,\Xi')=\frac{1}{M}\sum_{i=1}^{M}\frac{2\left(\frac{1}{\mathfrak{M}}\sum_{j=1}^{l}\mathfrak{M}^j_{\Xi}(\mathfrak{X}_i)\mathfrak{M}^j_{\Xi'}(\mathfrak{X}_i)+\frac{1}{\mathfrak{N}}\sum_{j=1}^{l}\mathfrak{N}^j_{\Xi}(\mathfrak{X}_i)\mathfrak{N}^j_{\Xi'}(\mathfrak{X}_i)\right)}{\left(\frac{1}{L_{\mathfrak{M}_{\Xi}(\mathfrak{X})}}\sum_{j=1}^{l}\left(\mathfrak{M}^j_{\Xi}(\mathfrak{X}_i)\right)^2+\frac{1}{L_{\mathfrak{N}_{\Xi}(\mathfrak{X})}}\sum_{j=1}^{l}\left(\mathfrak{N}^j_{\Xi}(\mathfrak{X}_i)\right)^2+\frac{1}{L_{\mathfrak{M}_{\Xi'}(\mathfrak{X})}}\sum_{j=1}^{l}\left(\mathfrak{M}^j_{\Xi'}(\mathfrak{X}_i)\right)^2+\frac{1}{L_{\mathfrak{N}_{\Xi'}(\mathfrak{X})}}\sum_{j=1}^{l}\left(\mathfrak{N}^j_{\Xi'}(\mathfrak{X}_i)\right)^2\right)}$$

which holds the necessary rules:

1. $0\le D^1{}_{P\Xi F}(\Xi,\Xi')\le 1$
2. $D^1{}_{P\Xi F}(\Xi,\Xi')=D^1{}_{P\Xi F}(\Xi',\Xi)$
3. $D^1{}_{P\Xi F}(\Xi,\Xi')=1\Leftrightarrow \Xi=\Xi'$

Using some conditions, we can easily obtain further particular cases from the above theory; for instance, to put $\mathfrak{N}^j_{\Xi}(\mathfrak{X}_i)=\mathfrak{N}^j_{\Xi'}(\mathfrak{X}_i)=0$ in $D^1{}_{P\Xi F}(\Xi,\Xi')$, then $D^1{}_{P\Xi F}(\Xi,\Xi')$ will change for HFSs. Furthermore, to put $\mathfrak{M}^j_{\Xi}(\mathfrak{X}_i),\mathfrak{M}^j_{\Xi'}(\mathfrak{X}_i)$ and $\mathfrak{N}^j_{\Xi}(\mathfrak{X}_i),\mathfrak{N}^j_{\Xi'}(\mathfrak{X}_i)$ as a singleton set, then $D^1{}_{P\Xi F}(\Xi,\Xi')$ will change for IFSs, meaning the theory diagnosed in this study is massively powerful and dominant compared to others.

Definition 6. *By using any two IHFNS* Ξ *and* Ξ', *a WDSM* $WD^1{}_{P\Xi F}(\Xi,\Xi')$ *is investigated by:*

$$WD^1{}_{P\Xi F}(\Xi,\Xi')=\sum_{i=1}^{M}w_i\frac{2\left(\frac{1}{\mathfrak{M}}\sum_{j=1}^{l}\mathfrak{M}^j_{\Xi}(\mathfrak{X}_i)\mathfrak{M}^j_{\Xi'}(\mathfrak{X}_i)+\frac{1}{\mathfrak{N}}\sum_{j=1}^{l}\mathfrak{N}^j_{\Xi}(\mathfrak{X}_i)\mathfrak{N}^j_{\Xi'}(\mathfrak{X}_i)\right)}{\left(\frac{1}{L_{\mathfrak{M}_{\Xi}(\mathfrak{X})}}\sum_{j=1}^{l}\left(\mathfrak{M}^j_{\Xi}(\mathfrak{X}_i)\right)^2+\frac{1}{L_{\mathfrak{N}_{\Xi}(\mathfrak{X})}}\sum_{j=1}^{l}\left(\mathfrak{N}^j_{\Xi}(\mathfrak{X}_i)\right)^2+\frac{1}{L_{\mathfrak{M}_{\Xi'}(\mathfrak{X})}}\sum_{j=1}^{l}\left(\mathfrak{M}^j_{\Xi'}(\mathfrak{X}_i)\right)^2+\frac{1}{L_{\mathfrak{N}_{\Xi'}(\mathfrak{X})}}\sum_{j=1}^{l}\left(\mathfrak{N}^j_{\Xi'}(\mathfrak{X}_i)\right)^2\right)}$$

which holds the necessary rules of Definition 5.

Using some conditions, we can easily obtain further particular cases from the above theory, for instance, to put $\mathfrak{N}^j_{\Xi}(\mathfrak{X}_i)=\mathfrak{N}^j_{\Xi'}(\mathfrak{X}_i)=0$ in $WD^1{}_{P\Xi F}(\Xi,\Xi')$, $WD^1{}_{P\Xi F}(\Xi,\Xi')$ will change for HFSs. Furthermore, to put $\mathfrak{M}^j_{\Xi}(\mathfrak{X}_i),\mathfrak{M}^j_{\Xi'}(\mathfrak{X}_i)$ and $\mathfrak{N}^j_{\Xi}(\mathfrak{X}_i),\mathfrak{N}^j_{\Xi'}(\mathfrak{X}_i)$ as a singleton set, $WD^1{}_{P\Xi F}(\Xi,\Xi')$ will change for IFSs, meaning the theory diagnosed in this manuscript is massively powerful and dominant as compared to others. For $w = \left(\frac{1}{M}, \frac{1}{M}, \ldots, \frac{1}{M}\right)^T$, the WDSM is converted for DSM based on IHFS such that $WD^1{}_{P\Xi F}(\Xi,\Xi') = D^1{}_{P\Xi F}(\Xi,\Xi')$.

Definition 7. *By using any two IHFNS Ξ and Ξ', a DSM $D^2{}_{P\Xi F}(\Xi,\Xi')$ is investigated by:*

$$D^2{}_{P\Xi F}(\Xi,\Xi') = \frac{1}{M}\sum_{i=1}^{M} \frac{2\left(\frac{1}{\mathfrak{M}}\sum_{j=1}^{l}\mathfrak{M}^j_{\Xi}(\mathfrak{X}_i)\mathfrak{M}^j_{\Xi'}(\mathfrak{X}_i) + \frac{1}{\mathfrak{N}}\sum_{j=1}^{l}\mathfrak{N}^j_{\Xi}(\mathfrak{X}_i)\mathfrak{N}^j_{\Xi'}(\mathfrak{X}_i) + \frac{1}{\pi}\sum_{j=1}^{l}\pi^j_{\Xi}(\mathfrak{X}_i)\pi^j_{\Xi'}(\mathfrak{X}_i)\right)}{\left(\begin{array}{c}\frac{1}{L_{\mathfrak{M}_{\Xi}(\mathfrak{X})}}\sum_{j=1}^{l}\left(\mathfrak{M}^j_{\Xi}(\mathfrak{X}_i)\right)^2 + \frac{1}{L_{\mathfrak{N}_{\Xi}(\mathfrak{X})}}\sum_{j=1}^{l}\left(\mathfrak{N}^j_{\Xi}(\mathfrak{X}_i)\right)^2 + \frac{1}{L_{\pi_{\Xi}(\mathfrak{X})}}\sum_{j=1}^{l}\left(\pi^j_{\Xi}(\mathfrak{X}_i)\right)^2 + \\ \frac{1}{L_{\mathfrak{M}_{\Xi'}(\mathfrak{X})}}\sum_{j=1}^{l}\left(\mathfrak{M}^j_{\Xi'}(\mathfrak{X}_i)\right)^2 + \frac{1}{L_{\mathfrak{N}_{\Xi'}(\mathfrak{X})}}\sum_{j=1}^{l}\left(\mathfrak{N}^j_{\Xi'}(\mathfrak{X}_i)\right)^2 + \frac{1}{L_{\pi_{\Xi'}(\mathfrak{X})}}\sum_{j=1}^{l}\left(\pi^j_{\Xi'}(\mathfrak{X}_i)\right)^2\end{array}\right)}$$

which holds the necessary rules of Definition 5.

Using some conditions, we can easily obtain a lot of further particular cases from the above theory; for instance, to put $\mathfrak{N}^j_{\Xi}(\mathfrak{X}_i) = \mathfrak{N}^j_{\Xi'}(\mathfrak{X}_i) = 0$ in $D^2{}_{P\Xi F}(\Xi,\Xi')$, then $D^2{}_{P\Xi F}(\Xi,\Xi')$ will change for HFSs. Furthermore, to put $\mathfrak{M}^j_{\Xi}(\mathfrak{X}_i), \mathfrak{M}^j_{\Xi'}(\mathfrak{X}_i)$ and $\mathfrak{N}^j_{\Xi}(\mathfrak{X}_i)$, $\mathfrak{N}^j_{\Xi'}(\mathfrak{X}_i)$ as a singleton set, $D^2{}_{P\Xi F}(\Xi,\Xi')$ will change for IFSs, meaning the theory diagnosed in this study is massively powerful and dominant compared to others.

Definition 8. *By using any two IHFNS Ξ and Ξ', a WDSM $WD^2{}_{P\Xi F}(\Xi,\Xi')$ is investigated by:*

$$WD^2{}_{P\Xi F}(\Xi,\Xi') = \sum_{i=1}^{M} w_i \frac{2\left(\frac{1}{\mathfrak{M}}\sum_{j=1}^{l}\mathfrak{M}^j_{\Xi}(\mathfrak{X}_i)\mathfrak{M}^j_{\Xi'}(\mathfrak{X}_i) + \frac{1}{\mathfrak{N}}\sum_{j=1}^{l}\mathfrak{N}^j_{\Xi}(\mathfrak{X}_i)\mathfrak{N}^j_{\Xi'}(\mathfrak{X}_i) + \frac{1}{\pi}\sum_{j=1}^{l}\pi^j_{\Xi}(\mathfrak{X}_i)\pi^j_{\Xi'}(\mathfrak{X}_i)\right)}{\left(\begin{array}{c}\frac{1}{L_{\mathfrak{M}_{\Xi}(\mathfrak{X})}}\sum_{j=1}^{l}\left(\mathfrak{M}^j_{\Xi}(\mathfrak{X}_i)\right)^2 + \frac{1}{L_{\mathfrak{N}_{\Xi}(\mathfrak{X})}}\sum_{j=1}^{l}\left(\mathfrak{N}^j_{\Xi}(\mathfrak{X}_i)\right)^2 + \frac{1}{L_{\pi_{\Xi}(\mathfrak{X})}}\sum_{j=1}^{l}\left(\pi^j_{\Xi}(\mathfrak{X}_i)\right)^2 + \\ \frac{1}{L_{\mathfrak{M}_{\Xi'}(\mathfrak{X})}}\sum_{j=1}^{l}\left(\mathfrak{M}^j_{\Xi'}(\mathfrak{X}_i)\right)^2 + \frac{1}{L_{\mathfrak{N}_{\Xi'}(\mathfrak{X})}}\sum_{j=1}^{l}\left(\mathfrak{N}^j_{\Xi'}(\mathfrak{X}_i)\right)^2 + \frac{1}{L_{\pi_{\Xi'}(\mathfrak{X})}}\sum_{j=1}^{l}\left(\pi^j_{\Xi'}(\mathfrak{X}_i)\right)^2\end{array}\right)}$$

which holds the necessary rules of Definition 5.

Using some conditions, we can easily obtain a lot of further particular cases from the above theory; for instance, to put $\mathfrak{N}^j_{\Xi}(\mathfrak{X}_i) = \mathfrak{N}^j_{\Xi'}(\mathfrak{X}_i) = 0$ in $WD^2{}_{P\Xi F}(\Xi,\Xi')$, $WD^2{}_{P\Xi F}(\Xi,\Xi')$ will change for HFSs. Furthermore, to put $\mathfrak{M}^j_{\Xi}(\mathfrak{X}_i), \mathfrak{M}^j_{\Xi'}(\mathfrak{X}_i)$ and $\mathfrak{N}^j_{\Xi}(\mathfrak{X}_i), \mathfrak{N}^j_{\Xi'}(\mathfrak{X}_i)$ as a singleton set, $WD^2{}_{P\Xi F}(\Xi,\Xi')$ will change for IFSs, meaning that the theory diagnosed in this study is massively powerful and dominant compared to others. For $w = \left(\frac{1}{M}, \frac{1}{M}, \ldots, \frac{1}{M}\right)^T$ then $WD^2{}_{P\Xi F}(\Xi,\Xi') = D^2{}_{P\Xi F}(\Xi,\Xi')$.

Definition 9. *By using any two IHFNS Ξ and Ξ', a DSM $D^3{}_{P\Xi F}(\Xi,\Xi')$ is investigated by:*

$$D^3{}_{P\Xi F}(\Xi,\Xi') = \frac{\sum_{i=1}^{M} 2\left(\frac{1}{\mathfrak{M}}\sum_{j=1}^{l}\mathfrak{M}^j_{\Xi}(\mathfrak{X}_i)\mathfrak{M}^j_{\Xi'}(\mathfrak{X}_i) + \frac{1}{\mathfrak{N}}\sum_{j=1}^{l}\mathfrak{N}^j_{\Xi}(\mathfrak{X}_i)\mathfrak{N}^j_{\Xi'}(\mathfrak{X}_i)\right)}{\begin{array}{c}\sum_{i=1}^{M}\left(\frac{1}{L_{\mathfrak{M}_{\Xi}(\mathfrak{X})}}\sum_{j=1}^{l}\left(\mathfrak{M}^j_{\Xi}(\mathfrak{X}_i)\right)^2 + \frac{1}{L_{\mathfrak{N}_{\Xi}(\mathfrak{X})}}\sum_{j=1}^{l}\left(\mathfrak{N}^j_{\Xi}(\mathfrak{X}_i)\right)^2\right) + \\ \sum_{i=1}^{M}\left(\frac{1}{L_{\mathfrak{M}_{\Xi'}(\mathfrak{X})}}\sum_{j=1}^{l}\left(\mathfrak{M}^j_{\Xi'}(\mathfrak{X}_i)\right)^2 + \frac{1}{L_{\mathfrak{N}_{\Xi'}(\mathfrak{X})}}\sum_{j=1}^{l}\left(\mathfrak{N}^j_{\Xi'}(\mathfrak{X}_i)\right)^2\right)\end{array}}$$

which holds the necessary rules of Definition 5.

Using some conditions, we can easily obtain further particular cases from the above theory; for instance, to put $\mathfrak{N}^j_{\Xi}(\mathfrak{X}_i) = \mathfrak{N}^j_{\Xi'}(\mathfrak{X}_i) = 0$ in $D^3{}_{P\Xi F}(\Xi,\Xi')$, $D^3{}_{P\Xi F}(\Xi,\Xi')$ will change for HFSs. Furthermore, to put $\mathfrak{M}^j_{\Xi}(\mathfrak{X}_i), \mathfrak{M}^j_{\Xi'}(\mathfrak{X}_i)$ and $\mathfrak{N}^j_{\Xi}(\mathfrak{X}_i), \mathfrak{N}^j_{\Xi'}(\mathfrak{X}_i)$ as a singleton set, $D^3{}_{P\Xi F}(\Xi,\Xi')$ will change for IFSs, meaning that the theory diagnosed in this study is massively powerful and dominant compared to others.

Definition 10. *By using any two IHFNS Ξ and Ξ', a WDSM $WD^3{}_{P\Xi F}(\Xi,\Xi')$ is investigated by:*

$$WD^3{}_{P\Xi F}(\Xi,\Xi')=\frac{\sum_{i=1}^{M}2w_i^2\left(\frac{1}{\mathfrak{M}}\sum_{j=1}^{l}\mathfrak{M}_{\Xi}^{j}(\mathfrak{X}_i)\mathfrak{M}_{\Xi'}^{j}(\mathfrak{X}_i)+\frac{1}{\mathfrak{N}}\sum_{j=1}^{l}\mathfrak{N}_{\Xi}^{j}(\mathfrak{X}_i)\mathfrak{N}_{\Xi'}^{j}(\mathfrak{X}_i)\right)}{\begin{array}{c}\sum_{i=1}^{M}w_i^2\left(\frac{1}{L_{\mathfrak{M}_{\Xi}(\mathfrak{X})}}\sum_{j=1}^{l}\left(\mathfrak{M}_{\Xi}^{j}(\mathfrak{X}_i)\right)^2+\frac{1}{L_{\mathfrak{N}_{\Xi}(\mathfrak{X})}}\sum_{j=1}^{l}\left(\mathfrak{N}_{\Xi}^{j}(\mathfrak{X}_i)\right)^2\right)+\\ \sum_{i=1}^{M}w_i^2\left(\frac{1}{L_{\mathfrak{M}_{\Xi'}(\mathfrak{X})}}\sum_{j=1}^{l}\left(\mathfrak{M}_{\Xi'}^{j}(\mathfrak{X}_i)\right)^2+\frac{1}{L_{\mathfrak{N}_{\Xi'}(\mathfrak{X})}}\sum_{j=1}^{l}\left(\mathfrak{N}_{\Xi'}^{j}(\mathfrak{X}_i)\right)^2\right)\end{array}}$$

which holds the necessary rules of Definition 5.

Using some conditions, we can easily obtain further particular cases from the above theory; for instance, to put $\mathfrak{N}_{\Xi}^{j}(\mathfrak{X}_i)=\mathfrak{N}_{\Xi'}^{j}(\mathfrak{X}_i)=0$ in $WD^3{}_{P\Xi F}(\Xi,\Xi')$, $WD^3{}_{P\Xi F}(\Xi,\Xi')$ will change for HFSs. Furthermore, to put $\mathfrak{M}_{\Xi}^{j}(\mathfrak{X}_i),\mathfrak{M}_{\Xi'}^{j}(\mathfrak{X}_i)$ and $\mathfrak{N}_{\Xi}^{j}(\mathfrak{X}_i),\mathfrak{N}_{\Xi'}^{j}(\mathfrak{X}_i)$ as a singleton set, $WD^3{}_{P\Xi F}(\Xi,\Xi')$ will change for IFSs, meaning that the theory diagnosed in this study is massively powerful and dominant compared to others. For $w=\left(\frac{1}{M},\frac{1}{M},\ldots,\frac{1}{M}\right)^T$ then $WD^3{}_{P\Xi F}(\Xi,\Xi')=D^3{}_{P\Xi F}(\Xi,\Xi')$.

Definition 11. *By using any two IHFNS Ξ and Ξ', a DSM $D^4{}_{P\Xi F}(\Xi,\Xi')$ is investigated by:*

$$D^4{}_{P\Xi F}(\Xi,\Xi')=\frac{2\sum_{i=1}^{M}\left(\frac{1}{\mathfrak{M}}\sum_{j=1}^{l}\mathfrak{M}_{\Xi}^{j}(\mathfrak{X}_i)\mathfrak{M}_{\Xi'}^{j}(\mathfrak{X}_i)+\frac{1}{\mathfrak{N}}\sum_{j=1}^{l}\mathfrak{N}_{\Xi}^{j}(\mathfrak{X}_i)\mathfrak{N}_{\Xi'}^{j}(\mathfrak{X}_i)+\frac{1}{\pi}\sum_{j=1}^{l}\pi_{\Xi}^{j}(\mathfrak{X}_i)\pi_{\Xi'}^{j}(\mathfrak{X}_i)\right)}{\begin{array}{c}\sum_{i=1}^{M}\left(\frac{1}{L_{\mathfrak{M}_{\Xi}(\mathfrak{X})}}\sum_{j=1}^{l}\left(\mathfrak{M}_{\Xi}^{j}(\mathfrak{X}_i)\right)^2+\frac{1}{L_{\mathfrak{N}_{\Xi}(\mathfrak{X})}}\sum_{j=1}^{l}\left(\mathfrak{N}_{\Xi}^{j}(\mathfrak{X}_i)\right)^2+\frac{1}{L_{\pi_{\Xi}(\mathfrak{X})}}\sum_{j=1}^{l}\left(\pi_{\Xi}^{j}(\mathfrak{X}_i)\right)^2\right)+\\ \sum_{i=1}^{M}\left(\frac{1}{L_{\mathfrak{M}_{\Xi'}(\mathfrak{X})}}\sum_{j=1}^{l}\left(\mathfrak{M}_{\Xi'}^{j}(\mathfrak{X}_i)\right)^2+\frac{1}{L_{\mathfrak{N}_{\Xi'}(\mathfrak{X})}}\sum_{j=1}^{l}\left(\mathfrak{N}_{\Xi'}^{j}(\mathfrak{X}_i)\right)^2+\frac{1}{L_{\pi_{\Xi'}(\mathfrak{X})}}\sum_{j=1}^{l}\left(\pi_{\Xi'}^{j}(\mathfrak{X}_i)\right)^2\right)\end{array}}$$

which holds the necessary rules of Definition 5.

Using some conditions, we can easily obtain further particular cases from the above theory; for instance, to put $\mathfrak{N}_{\Xi}^{j}(\mathfrak{X}_i)=\mathfrak{N}_{\Xi'}^{j}(\mathfrak{X}_i)=0$ in $D^4{}_{P\Xi F}(\Xi,\Xi')$, $D^4{}_{P\Xi F}(\Xi,\Xi')$ will change for HFSs. Furthermore, to put $\mathfrak{M}_{\Xi}^{j}(\mathfrak{X}_i),\mathfrak{M}_{\Xi'}^{j}(\mathfrak{X}_i)$ and $\mathfrak{N}_{\Xi}^{j}(\mathfrak{X}_i),\mathfrak{N}_{\Xi'}^{j}(\mathfrak{X}_i)$ as a singleton set, $D^4{}_{P\Xi F}(\Xi,\Xi')$ will change for IFSs, meaning the theory diagnosed in this study is massively powerful and dominant compared to others.

Definition 12. *By using any two IHFNS Ξ and Ξ', a WDSM $WD^4{}_{P\Xi F}(\Xi,\Xi')$ is investigated by:*

$$WD^4{}_{P\Xi F}(\Xi,\Xi')=\frac{2\sum_{i=1}^{M}w_i^2\left(\frac{1}{\mathfrak{M}}\sum_{j=1}^{l}\mathfrak{M}_{\Xi}^{j}(\mathfrak{X}_i)\mathfrak{M}_{\Xi'}^{j}(\mathfrak{X}_i)+\frac{1}{\mathfrak{N}}\sum_{j=1}^{l}\mathfrak{N}_{\Xi}^{j}(\mathfrak{X}_i)\mathfrak{N}_{\Xi'}^{j}(\mathfrak{X}_i)+\frac{1}{\pi}\sum_{j=1}^{l}\pi_{\Xi}^{j}(\mathfrak{X}_i)\pi_{\Xi'}^{j}(\mathfrak{X}_i)\right)}{\begin{array}{c}\sum_{i=1}^{M}w_i^2\left(\frac{1}{L_{\mathfrak{M}_{\Xi}(\mathfrak{X})}}\sum_{j=1}^{l}\left(\mathfrak{M}_{\Xi}^{j}(\mathfrak{X}_i)\right)^2+\frac{1}{L_{\mathfrak{N}_{\Xi}(\mathfrak{X})}}\sum_{j=1}^{l}\left(\mathfrak{N}_{\Xi}^{j}(\mathfrak{X}_i)\right)^2+\frac{1}{L_{\pi_{\Xi}(\mathfrak{X})}}\sum_{j=1}^{l}\left(\pi_{\Xi}^{j}(\mathfrak{X}_i)\right)^2\right)+\\ \sum_{i=1}^{M}w_i^2\left(\frac{1}{L_{\mathfrak{M}_{\Xi'}(\mathfrak{X})}}\sum_{j=1}^{l}\left(\mathfrak{M}_{\Xi'}^{j}(\mathfrak{X}_i)\right)^2+\frac{1}{L_{\mathfrak{N}_{\Xi'}(\mathfrak{X})}}\sum_{j=1}^{l}\left(\mathfrak{N}_{\Xi'}^{j}(\mathfrak{X}_i)\right)^2+\frac{1}{L_{\pi_{\Xi'}(\mathfrak{X})}}\sum_{j=1}^{l}\left(\pi_{\Xi'}^{j}(\mathfrak{X}_i)\right)^2\right)\end{array}}$$

which holds the necessary rules of Definition 5.

Using some conditions, we can easily obtain further particular cases from the above theory; for instance, to put $\mathfrak{N}_{\Xi}^{j}(\mathfrak{X}_i)=\mathfrak{N}_{\Xi'}^{j}(\mathfrak{X}_i)=0$ in $WD^4{}_{P\Xi F}(\Xi,\Xi')$, $WD^4{}_{P\Xi F}(\Xi,\Xi')$ will change for HFSs. Furthermore, to put $\mathfrak{M}_{\Xi}^{j}(\mathfrak{X}_i),\mathfrak{M}_{\Xi'}^{j}(\mathfrak{X}_i)$ and $\mathfrak{N}_{\Xi}^{j}(\mathfrak{X}_i),\mathfrak{N}_{\Xi'}^{j}(\mathfrak{X}_i)$ as a singleton set, $WD^4{}_{P\Xi F}(\Xi,\Xi')$ will change for IFSs, meaning the theory diagnosed in this manuscript is massively powerful and dominant compared to others. For $w=\left(\frac{1}{M},\frac{1}{M},\ldots,\frac{1}{M}\right)^T$, $WD^4{}_{P\Xi F}(\Xi,\Xi')=D^4{}_{P\Xi F}(\Xi,\Xi')$.

4. GDSM for IHFSs

To illustrate the relationship between any two pieces of IHF information, the theory of GDS measures played an important and valuable role in the field of genuine life dilemmas. The main influence of GDS measures is that we can easily obtain a large number of measures by using different values of parameters, which is the main part of every measure, called DGS measures. In this study, we chose one of the most flexible and genuine principles, called the IHFS, which covers the MG and NMG in the form of a finite subset of [0, 1], with the rule that the sum of the supremum of the duplet is limited to [0, 1], GDS measures are to develop the four sorts of IHF GDS measure, and IHF weighted GDS measure. Based on the investigated measures, certain special cases are also evaluated, with $0 \leq \rho \leq 1$.

Definition 13. *By using any two IHFNS Ξ and Ξ', a GDSM $GD^1{}_{P\Xi F}(\Xi, \Xi')$ is investigated by:*

$$GD^1{}_{P\Xi F}(\Xi,\Xi')= \frac{1}{M}\sum_{i=1}^{M}\frac{\left(\frac{1}{\mathfrak{M}}\sum_{j=1}^{l}\mathfrak{M}^j_{\Xi}(\mathfrak{X}_i)\mathfrak{M}^j_{\Xi'}(\mathfrak{X}_i)+\frac{1}{\mathfrak{N}}\sum_{j=1}^{l}\mathfrak{N}^j_{\Xi}(\mathfrak{X}_i)\mathfrak{N}^j_{\Xi'}(\mathfrak{X}_i)\right)}{\left(\begin{array}{c}\gamma\left(\frac{1}{L_{\mathfrak{M}_{\Xi}(\mathfrak{X})}}\sum_{j=1}^{l}\left(\mathfrak{M}^j_{\Xi}(\mathfrak{X}_i)\right)^2+\frac{1}{L_{\mathfrak{N}_{\Xi}(\mathfrak{X})}}\sum_{j=1}^{l}\left(\mathfrak{N}^j_{\Xi}(\mathfrak{X}_i)\right)^2\right)+\\(1-\gamma)\left(\frac{1}{L_{\mathfrak{M}_{\Xi'}(\mathfrak{X})}}\sum_{j=1}^{l}\left(\mathfrak{M}^j_{\Xi'}(\mathfrak{X}_i)\right)^2+\frac{1}{L_{\mathfrak{N}_{\Xi'}(\mathfrak{X})}}\sum_{j=1}^{l}\left(\mathfrak{N}^j_{\Xi'}(\mathfrak{X}_i)\right)^2\right)\end{array}\right)}$$

which holds the necessary rules of Definition 5.

Using some conditions, we can easily obtain further particular cases from the above theory; for instance, to put $\mathfrak{N}^j_{\Xi}(\mathfrak{X}_i)=\mathfrak{N}^j_{\Xi'}(\mathfrak{X}_i)=0$ in $GD^1{}_{P\Xi F}(\Xi,\Xi')$, $GD^1{}_{P\Xi F}(\Xi,\Xi')$ will change for HFSs. Furthermore, to put $\mathfrak{M}^j_{\Xi}(\mathfrak{X}_i), \mathfrak{M}^j_{\Xi'}(\mathfrak{X}_i)$ and $\mathfrak{N}^j_{\Xi}(\mathfrak{X}_i), \mathfrak{N}^j_{\Xi'}(\mathfrak{X}_i)$ as a singleton set, $GD^1{}_{P\Xi F}(\Xi,\Xi')$ will change for IFSs, meaning the theory diagnosed in this study is massively powerful and dominant compared to others.

Definition 14. *By using any two IHFNS Ξ and Ξ', a WGDSM $WGD^1{}_{P\Xi F}(\Xi, \Xi')$ is investigated by:*

$$WGD^1{}_{P\Xi F}(\Xi,\Xi')= \sum_{i=1}^{M}w_i\frac{\left(\frac{1}{\mathfrak{M}}\sum_{j=1}^{l}\mathfrak{M}^j_{\Xi}(\mathfrak{X}_i)\mathfrak{M}^j_{\Xi'}(\mathfrak{X}_i)+\frac{1}{\mathfrak{N}}\sum_{j=1}^{l}\mathfrak{N}^j_{\Xi}(\mathfrak{X}_i)\mathfrak{N}^j_{\Xi'}(\mathfrak{X}_i)\right)}{\left(\begin{array}{c}\gamma\left(\frac{1}{L_{\mathfrak{M}_{\Xi}(\mathfrak{X})}}\sum_{j=1}^{l}\left(\mathfrak{M}^j_{\Xi}(\mathfrak{X}_i)\right)^2+\frac{1}{L_{\mathfrak{N}_{\Xi}(\mathfrak{X})}}\sum_{j=1}^{l}\left(\mathfrak{N}^j_{\Xi}(\mathfrak{X}_i)\right)^2\right)+\\(1-\gamma)\left(\frac{1}{L_{\mathfrak{M}_{\Xi'}(\mathfrak{X})}}\sum_{j=1}^{l}\left(\mathfrak{M}^j_{\Xi'}(\mathfrak{X}_i)\right)^2+\frac{1}{L_{\mathfrak{N}_{\Xi'}(\mathfrak{X})}}\sum_{j=1}^{l}\left(\mathfrak{N}^j_{\Xi'}(\mathfrak{X}_i)\right)^2\right)\end{array}\right)}$$

which holds the necessary rules of Definition 5.

Using some conditions, we can easily obtain further particular cases from the above theory; for instance, to put $\mathfrak{N}^j_{\Xi}(\mathfrak{X}_i)=\mathfrak{N}^j_{\Xi'}(\mathfrak{X}_i)=0$ in $WGD^1{}_{P\Xi F}(\Xi,\Xi')$, $WGD^1{}_{P\Xi F}(\Xi,\Xi')$ will change for HFSs. Furthermore, to put $\mathfrak{M}^j_{\Xi}(\mathfrak{X}_i), \mathfrak{M}^j_{\Xi'}(\mathfrak{X}_i)$ and $\mathfrak{N}^j_{\Xi}(\mathfrak{X}_i), \mathfrak{N}^j_{\Xi'}(\mathfrak{X}_i)$ as a singleton set, $WGD^1{}_{P\Xi F}(\Xi,\Xi')$ will change for IFSs, meaning the theory diagnosed in this study is massively powerful and dominant compared to others. For $w=\left(\frac{1}{M},\frac{1}{M},\ldots,\frac{1}{M}\right)^T$, $WGD^1{}_{P\Xi F}(\Xi,\Xi')=GD^1{}_{P\Xi F}(\Xi,\Xi')$.

Definition 15. *By using any two IHFNS Ξ and Ξ', a GDSM $GD^2{}_{P\Xi F}(\Xi, \Xi')$ is investigated by:*

$$GD^2{}_{P\Xi F}(\Xi,\Xi')= \frac{1}{M}\sum_{i=1}^{M}\frac{\left(\frac{1}{\mathfrak{M}}\sum_{j=1}^{l}\mathfrak{M}^j_{\Xi}(\mathfrak{X}_i)\mathfrak{M}^j_{\Xi'}(\mathfrak{X}_i)+\frac{1}{\mathfrak{N}}\sum_{j=1}^{l}\mathfrak{N}^j_{\Xi}(\mathfrak{X}_i)\mathfrak{N}^j_{\Xi'}(\mathfrak{X}_i)+\frac{1}{\pi}\sum_{j=1}^{l}\pi^j_{\Xi}(\mathfrak{X}_i)\pi^j_{\Xi'}(\mathfrak{X}_i)\right)}{\left(\begin{array}{c}\gamma\left(\frac{1}{L_{\mathfrak{M}_{\Xi}(\mathfrak{X})}}\sum_{j=1}^{l}\left(\mathfrak{M}^j_{\Xi}(\mathfrak{X}_i)\right)^2+\frac{1}{L_{\mathfrak{N}_{\Xi}(\mathfrak{X})}}\sum_{j=1}^{l}\left(\mathfrak{N}^j_{\Xi}(\mathfrak{X}_i)\right)^2+\frac{1}{L_{\pi_{\Xi}(\mathfrak{X})}}\sum_{j=1}^{l}\left(\pi^j_{\Xi}(\mathfrak{X}_i)\right)^2\right)+\\(1-\gamma)\left(\frac{1}{L_{\mathfrak{M}_{\Xi'}(\mathfrak{X})}}\sum_{j=1}^{l}\left(\mathfrak{M}^j_{\Xi'}(\mathfrak{X}_i)\right)^2+\frac{1}{L_{\mathfrak{N}_{\Xi'}(\mathfrak{X})}}\sum_{j=1}^{l}\left(\mathfrak{N}^j_{\Xi'}(\mathfrak{X}_i)\right)^2+\frac{1}{L_{\pi_{\Xi'}(\mathfrak{X})}}\sum_{j=1}^{l}\left(\pi^j_{\Xi'}(\mathfrak{X}_i)\right)^2\right)\end{array}\right)}$$

which holds the necessary rules of Definition 5.

Using some conditions, we can easily obtain further particular cases from the above theory; for instance, to put $\mathfrak{N}^j_{\Xi}(\mathfrak{X}_i) = \mathfrak{N}^j_{\Xi'}(\mathfrak{X}_i) = 0$ in $GD^2{}_{P\Xi F}(\Xi, \Xi')$, $GD^2{}_{P\Xi F}(\Xi, \Xi')$ will change for HFSs. Furthermore, to put $\mathfrak{M}^j_{\Xi}(\mathfrak{X}_i), \mathfrak{M}^j_{\Xi'}(\mathfrak{X}_i)$ and $\mathfrak{N}^j_{\Xi}(\mathfrak{X}_i), \mathfrak{N}^j_{\Xi'}(\mathfrak{X}_i)$ as a singleton set, $GD^2{}_{P\Xi F}(\Xi, \Xi')$ will change for IFSs, meaning the theory diagnosed in this study is massively powerful and dominant compared to others.

Definition 16. *By using any two IHFNS Ξ and Ξ', a WGDSM $WGD^2{}_{P\Xi F}(\Xi, \Xi')$ is investigated by:*

$$WGD^2{}_{P\Xi F}(\Xi, \Xi') = \sum_{i=1}^{M} w_i \frac{2\left(\frac{1}{\mathfrak{M}}\sum_{j=1}^{l} \mathfrak{M}^j_{\Xi}(\mathfrak{X}_i)\mathfrak{M}^j_{\Xi'}(\mathfrak{X}_i) + \frac{1}{\mathfrak{N}}\sum_{j=1}^{l} \mathfrak{N}^j_{\Xi}(\mathfrak{X}_i)\mathfrak{N}^j_{\Xi'}(\mathfrak{X}_i) + \frac{1}{\pi}\sum_{j=1}^{l} \pi^j_{\Xi}(\mathfrak{X}_i)\pi^j_{\Xi'}(\mathfrak{X}_i)\right)}{\left(\begin{array}{c}\gamma\left(\frac{1}{L_{\mathfrak{M}_{\Xi}(\mathfrak{X})}}\sum_{j=1}^{l}\left(\mathfrak{M}^j_{\Xi}(\mathfrak{X}_i)\right)^2 + \frac{1}{L_{\mathfrak{N}_{\Xi}(\mathfrak{X})}}\sum_{j=1}^{l}\left(\mathfrak{N}^j_{\Xi}(\mathfrak{X}_i)\right)^2 + \frac{1}{L_{\pi_{\Xi}(\mathfrak{X})}}\sum_{j=1}^{l}\left(\pi^j_{\Xi}(\mathfrak{X}_i)\right)^2\right) + \\ (1-\gamma)\left(\frac{1}{L_{\mathfrak{M}_{\Xi'}(\mathfrak{X})}}\sum_{j=1}^{l}\left(\mathfrak{M}^j_{\Xi'}(\mathfrak{X}_i)\right)^2 + \frac{1}{L_{\mathfrak{N}_{\Xi'}(\mathfrak{X})}}\sum_{j=1}^{l}\left(\mathfrak{N}^j_{\Xi'}(\mathfrak{X}_i)\right)^2 + \frac{1}{L_{\pi_{\Xi'}(\mathfrak{X})}}\sum_{j=1}^{l}\left(\pi^j_{\Xi'}(\mathfrak{X}_i)\right)^2\right)\end{array}\right)}$$

which holds the necessary rules of Definition 5.

Using some conditions, we can easily obtain further particular cases from the above theory; for instance, to put $\mathfrak{N}^j_{\Xi}(\mathfrak{X}_i) = \mathfrak{N}^j_{\Xi'}(\mathfrak{X}_i) = 0$ in $WGD^2{}_{P\Xi F}(\Xi, \Xi')$, $WGD^2{}_{P\Xi F}(\Xi, \Xi')$ will change for HFSs. Furthermore, to put $\mathfrak{M}^j_{\Xi}(\mathfrak{X}_i), \mathfrak{M}^j_{\Xi'}(\mathfrak{X}_i)$ and $\mathfrak{N}^j_{\Xi}(\mathfrak{X}_i), \mathfrak{N}^j_{\Xi'}(\mathfrak{X}_i)$ as a singleton set, $WGD^2{}_{P\Xi F}(\Xi, \Xi')$ will change for IFSs, meaning the theory diagnosed in this study is massively powerful and dominant compared to others. For $w = \left(\frac{1}{M}, \frac{1}{M}, \ldots, \frac{1}{M}\right)^T$, $WD^2{}_{P\Xi F}(\Xi, \Xi') = D^2{}_{P\Xi F}(\Xi, \Xi')$.

Definition 17. *By using any two IHFNS Ξ and Ξ', a GDSM $GD^3{}_{P\Xi F}(\Xi, \Xi')$ is investigated by:*

$$GD^3{}_{P\Xi F}(\Xi, \Xi') = \frac{\sum_{i=1}^{M}\left(\frac{1}{\mathfrak{M}}\sum_{j=1}^{l} \mathfrak{M}^j_{\Xi}(\mathfrak{X}_i)\mathfrak{M}^j_{\Xi'}(\mathfrak{X}_i) + \frac{1}{\mathfrak{N}}\sum_{j=1}^{l} \mathfrak{N}^j_{\Xi}(\mathfrak{X}_i)\mathfrak{N}^j_{\Xi'}(\mathfrak{X}_i)\right)}{\begin{array}{c}\gamma\sum_{i=1}^{M}\left(\frac{1}{L_{\mathfrak{M}_{\Xi}(\mathfrak{X})}}\sum_{j=1}^{l}\left(\mathfrak{M}^j_{\Xi}(\mathfrak{X}_i)\right)^2 + \frac{1}{L_{\mathfrak{N}_{\Xi}(\mathfrak{X})}}\sum_{j=1}^{l}\left(\mathfrak{N}^j_{\Xi}(\mathfrak{X}_i)\right)^2\right) + \\ (1-\gamma)\sum_{i=1}^{M}\left(\frac{1}{L_{\mathfrak{M}_{\Xi'}(\mathfrak{X})}}\sum_{j=1}^{l}\left(\mathfrak{M}^j_{\Xi'}(\mathfrak{X}_i)\right)^2 + \frac{1}{L_{\mathfrak{N}_{\Xi'}(\mathfrak{X})}}\sum_{j=1}^{l}\left(\mathfrak{N}^j_{\Xi'}(\mathfrak{X}_i)\right)^2\right)\end{array}}$$

which holds the necessary rules of Definition 5.

Using some conditions, we can easily obtain further particular cases from the above theory; for instance, to put $\mathfrak{N}^j_{\Xi}(\mathfrak{X}_i) = \mathfrak{N}^j_{\Xi'}(\mathfrak{X}_i) = 0$ in $GD^3{}_{P\Xi F}(\Xi, \Xi')$, $GD^3{}_{P\Xi F}(\Xi, \Xi')$ will change for HFSs. Furthermore, to put $\mathfrak{M}^j_{\Xi}(\mathfrak{X}_i), \mathfrak{M}^j_{\Xi'}(\mathfrak{X}_i)$ and $\mathfrak{N}^j_{\Xi}(\mathfrak{X}_i), \mathfrak{N}^j_{\Xi'}(\mathfrak{X}_i)$ as a singleton set, $GD^3{}_{P\Xi F}(\Xi, \Xi')$ will change for IFSs, meaning the theory diagnosed in this study is massively powerful and dominant compared to others.

Definition 18. *By using any two IHFNS Ξ and Ξ', a WGDSM $WGD^3{}_{P\Xi F}(\Xi, \Xi')$ is investigated by:*

$$WGD^3{}_{P\Xi F}(\Xi, \Xi') = \frac{\sum_{i=1}^{M} w_i^2\left(\frac{1}{\mathfrak{M}}\sum_{j=1}^{l} \mathfrak{M}^j_{\Xi}(\mathfrak{X}_i)\mathfrak{M}^j_{\Xi'}(\mathfrak{X}_i) + \frac{1}{\mathfrak{N}}\sum_{j=1}^{l} \mathfrak{N}^j_{\Xi}(\mathfrak{X}_i)\mathfrak{N}^j_{\Xi'}(\mathfrak{X}_i)\right)}{\begin{array}{c}\gamma\sum_{i=1}^{M} w_i^2\left(\frac{1}{L_{\mathfrak{M}_{\Xi}(\mathfrak{X})}}\sum_{j=1}^{l}\left(\mathfrak{M}^j_{\Xi}(\mathfrak{X}_i)\right)^2 + \frac{1}{L_{\mathfrak{N}_{\Xi}(\mathfrak{X})}}\sum_{j=1}^{l}\left(\mathfrak{N}^j_{\Xi}(\mathfrak{X}_i)\right)^2\right) + \\ (1-\gamma)\sum_{i=1}^{M} w_i^2\left(\frac{1}{L_{\mathfrak{M}_{\Xi'}(\mathfrak{X})}}\sum_{j=1}^{l}\left(\mathfrak{M}^j_{\Xi'}(\mathfrak{X}_i)\right)^2 + \frac{1}{L_{\mathfrak{N}_{\Xi'}(\mathfrak{X})}}\sum_{j=1}^{l}\left(\mathfrak{N}^j_{\Xi'}(\mathfrak{X}_i)\right)^2\right)\end{array}}$$

which holds the necessary rules of Definition 5.

Using some conditions, we can easily obtain further particular cases from the above theory; for instance, to put $\mathfrak{N}^j_{\Xi}(\mathfrak{X}_i) = \mathfrak{N}^j_{\Xi'}(\mathfrak{X}_i) = 0$ in $WGD^3{}_{P\Xi F}(\Xi, \Xi')$, $WGD^3{}_{P\Xi F}(\Xi, \Xi')$ will change for HFSs. Furthermore, to put $\mathfrak{M}^j_{\Xi}(\mathfrak{X}_i), \mathfrak{M}^j_{\Xi'}(\mathfrak{X}_i)$ and $\mathfrak{N}^j_{\Xi}(\mathfrak{X}_i), \mathfrak{N}^j_{\Xi'}(\mathfrak{X}_i)$ as a singleton set, $WGD^3{}_{P\Xi F}(\Xi, \Xi')$ will change for IFSs, meaning the theory diagnosed in this study is massively powerful and dominant compared to others. For $w = \left(\frac{1}{M}, \frac{1}{M}, \ldots, \frac{1}{M}\right)^T$, $WGD^3{}_{P\Xi F}(\Xi, \Xi') = GD^3{}_{P\Xi F}(\Xi, \Xi')$.

Definition 19. *By using any two IHFNS Ξ and Ξ', a GDSM $GD^4{}_{P\Xi F}(\Xi, \Xi')$ is investigated by:*

$$GD^4{}_{P\Xi F}(\Xi, \Xi') = \frac{\sum_{i=1}^{M}\left(\frac{1}{\mathfrak{M}}\sum_{j=1}^{l}\mathfrak{M}^j_{\Xi}(\mathfrak{X}_i)\mathfrak{M}^j_{\Xi'}(\mathfrak{X}_i) + \frac{1}{\mathfrak{N}}\sum_{j=1}^{l}\mathfrak{N}^j_{\Xi}(\mathfrak{X}_i)\mathfrak{N}^j_{\Xi'}(\mathfrak{X}_i) + \frac{1}{\pi}\sum_{j=1}^{l}\pi^j_{\Xi}(\mathfrak{X}_i)\pi^j_{\Xi'}(\mathfrak{X}_i)\right)}{\begin{array}{c}\gamma\sum_{i=1}^{M}\left(\frac{1}{L_{\mathfrak{M}_{\Xi}(\mathfrak{X})}}\sum_{j=1}^{l}\left(\mathfrak{M}^j_{\Xi}(\mathfrak{X}_i)\right)^2 + \frac{1}{L_{\mathfrak{N}_{\Xi}(\mathfrak{X})}}\sum_{j=1}^{l}\left(\mathfrak{N}^j_{\Xi}(\mathfrak{X}_i)\right)^2 + \frac{1}{L_{\pi_{\Xi}(\mathfrak{X})}}\sum_{j=1}^{l}\left(\pi^j_{\Xi}(\mathfrak{X}_i)\right)^2\right) + \\ (1-\gamma)\sum_{i=1}^{M}\left(\frac{1}{L_{\mathfrak{M}_{\Xi'}(\mathfrak{X})}}\sum_{j=1}^{l}\left(\mathfrak{M}^j_{\Xi'}(\mathfrak{X}_i)\right)^2 + \frac{1}{L_{\mathfrak{N}_{\Xi'}(\mathfrak{X})}}\sum_{j=1}^{l}\left(\mathfrak{N}^j_{\Xi'}(\mathfrak{X}_i)\right)^2 + \frac{1}{L_{\pi_{\Xi'}(\mathfrak{X})}}\sum_{j=1}^{l}\left(\pi^j_{\Xi'}(\mathfrak{X}_i)\right)^2\right)\end{array}}$$

which holds the necessary rules of Definition 5.

Using some conditions, we can easily obtain further particular cases from the above theory; for instance, to put $\mathfrak{N}^j_{\Xi}(\mathfrak{X}_i) = \mathfrak{N}^j_{\Xi'}(\mathfrak{X}_i) = 0$ in $GD^4{}_{P\Xi F}(\Xi, \Xi')$, $GD^4{}_{P\Xi F}(\Xi, \Xi')$ will change for HFSs. Furthermore, to put $\mathfrak{M}^j_{\Xi}(\mathfrak{X}_i), \mathfrak{M}^j_{\Xi'}(\mathfrak{X}_i)$ and $\mathfrak{N}^j_{\Xi}(\mathfrak{X}_i), \mathfrak{N}^j_{\Xi'}(\mathfrak{X}_i)$ as a singleton set, $GD^4{}_{P\Xi F}(\Xi, \Xi')$ will change for IFSs, meaning the theory diagnosed in this study is massively powerful and dominant compared to others.

Definition 20. *By using any two IHFNS Ξ and Ξ', a WGDSM $WGD^4{}_{P\Xi F}(\Xi, \Xi')$ is investigated by:*

$$WGD^4{}_{P\Xi F}(\Xi, \Xi') = \frac{\sum_{i=1}^{M} w_i^2\left(\frac{1}{\mathfrak{M}}\sum_{j=1}^{l}\mathfrak{M}^j_{\Xi}(\mathfrak{X}_i)\mathfrak{M}^j_{\Xi'}(\mathfrak{X}_i) + \frac{1}{\mathfrak{N}}\sum_{j=1}^{l}\mathfrak{N}^j_{\Xi}(\mathfrak{X}_i)\mathfrak{N}^j_{\Xi'}(\mathfrak{X}_i) + \frac{1}{\pi}\sum_{j=1}^{l}\pi^j_{\Xi}(\mathfrak{X}_i)\pi^j_{\Xi'}(\mathfrak{X}_i)\right)}{\begin{array}{c}\gamma\sum_{i=1}^{M} w_i^2\left(\frac{1}{L_{\mathfrak{M}_{\Xi}(\mathfrak{X})}}\sum_{j=1}^{l}\left(\mathfrak{M}^j_{\Xi}(\mathfrak{X}_i)\right)^2 + \frac{1}{L_{\mathfrak{N}_{\Xi}(\mathfrak{X})}}\sum_{j=1}^{l}\left(\mathfrak{N}^j_{\Xi}(\mathfrak{X}_i)\right)^2 + \frac{1}{L_{\pi_{\Xi}(\mathfrak{X})}}\sum_{j=1}^{l}\left(\pi^j_{\Xi}(\mathfrak{X}_i)\right)^2\right) + \\ (1-\gamma)\sum_{i=1}^{M} w_i^2\left(\frac{1}{L_{\mathfrak{M}_{\Xi'}(\mathfrak{X})}}\sum_{j=1}^{l}\left(\mathfrak{M}^j_{\Xi'}(\mathfrak{X}_i)\right)^2 + \frac{1}{L_{\mathfrak{N}_{\Xi'}(\mathfrak{X})}}\sum_{j=1}^{l}\left(\mathfrak{N}^j_{\Xi'}(\mathfrak{X}_i)\right)^2 + \frac{1}{L_{\pi_{\Xi'}(\mathfrak{X})}}\sum_{j=1}^{l}\left(\pi^j_{\Xi'}(\mathfrak{X}_i)\right)^2\right)\end{array}}$$

which holds the necessary rules of Definition 5.

Using some conditions, we can easily obtain further particular cases from the above theory; for instance, to put $\mathfrak{N}^j_{\Xi}(\mathfrak{X}_i) = \mathfrak{N}^j_{\Xi'}(\mathfrak{X}_i) = 0$ in $WGD^4{}_{P\Xi F}(\Xi, \Xi')$, $WGD^4{}_{P\Xi F}(\Xi, \Xi')$ will change for HFSs. Furthermore, to put $\mathfrak{M}^j_{\Xi}(\mathfrak{X}_i), \mathfrak{M}^j_{\Xi'}(\mathfrak{X}_i)$ and $\mathfrak{N}^j_{\Xi}(\mathfrak{X}_i), \mathfrak{N}^j_{\Xi'}(\mathfrak{X}_i)$ as a singleton set, $WGD^4{}_{P\Xi F}(\Xi, \Xi')$ will change for IFSs, meaning the theory diagnosed in this study is massively powerful and dominant compared to others. For $w = \left(\frac{1}{M}, \frac{1}{M}, \ldots, \frac{1}{M}\right)^T$, $WGD^4{}_{P\Xi F}(\Xi, \Xi') = GD^4{}_{P\Xi F}(\Xi, \Xi')$.

By using the investigated measures, we discussed certain special cases of the DSM, WDSM, GDSM, and WGDSM.

For $\gamma = 0$, in $GD^1{}_{P\Xi F}(\Xi, \Xi')$, we obtained

$$GD^1{}_{P\Xi F}(\Xi, \Xi') = \frac{1}{M}\sum_{i=1}^{M}\frac{\left(\frac{1}{\mathfrak{M}}\sum_{j=1}^{l}\mathfrak{M}^j_{\Xi}(\mathfrak{X}_i)\mathfrak{M}^j_{\Xi'}(\mathfrak{X}_i) + \frac{1}{\mathfrak{N}}\sum_{j=1}^{l}\mathfrak{N}^j_{\Xi}(\mathfrak{X}_i)\mathfrak{N}^j_{\Xi'}(\mathfrak{X}_i)\right)}{\left(\frac{1}{L_{\mathfrak{M}_{\Xi'}(\mathfrak{X})}}\sum_{j=1}^{l}\left(\mathfrak{M}^j_{\Xi'}(\mathfrak{X}_i)\right)^2 + \frac{1}{L_{\mathfrak{N}_{\Xi'}(\mathfrak{X})}}\sum_{j=1}^{l}\left(\mathfrak{N}^j_{\Xi'}(\mathfrak{X}_i)\right)^2\right)}$$

Similarly, for $\gamma = 0.5$,

$$GD^1{}_{P\Xi F}(\Xi,\Xi')=\frac{1}{M}\sum_{i=1}^{M}\frac{2\left(\frac{1}{\mathfrak{M}}\sum_{j=1}^{l}\mathfrak{M}^j_{\Xi}(\mathfrak{x}_i)\mathfrak{M}^j_{\Xi'}(\mathfrak{x}_i)+\frac{1}{\mathfrak{A}}\sum_{j=1}^{l}\mathfrak{A}^j_{\Xi}(\mathfrak{x}_i)\mathfrak{A}^j_{\Xi'}(\mathfrak{x}_i)+\frac{1}{\mathfrak{N}}\sum_{j=1}^{l}\mathfrak{N}^j_{\Xi}(\mathfrak{x}_i)\mathfrak{N}^j_{\Xi'}(\mathfrak{x}_i)\right)}{\left(\begin{array}{c}\left(\frac{1}{L_{\mathfrak{M}_{\Xi}(\mathfrak{x})}}\sum_{j=1}^{l}\left(\mathfrak{M}^j_{\Xi}(\mathfrak{x}_i)\right)^2+\frac{1}{L_{\mathfrak{A}_{\Xi}(\mathfrak{x})}}\sum_{j=1}^{l}\left(\mathfrak{A}^j_{\Xi}(\mathfrak{x}_i)\right)^2+\frac{1}{L_{\mathfrak{N}_{\Xi}(\mathfrak{x})}}\sum_{j=1}^{l}\left(\mathfrak{N}^j_{\Xi}(\mathfrak{x}_i)\right)^2\right)+\\\left(\frac{1}{L_{\mathfrak{M}_{\Xi'}(\mathfrak{x})}}\sum_{j=1}^{l}\left(\mathfrak{M}^j_{\Xi'}(\mathfrak{x}_i)\right)^2+\frac{1}{L_{\mathfrak{A}_{\Xi'}(\mathfrak{x})}}\sum_{j=1}^{l}\left(\mathfrak{A}^j_{\Xi'}(\mathfrak{x}_i)\right)^2+\frac{1}{L_{\mathfrak{N}_{\Xi'}(\mathfrak{x})}}\sum_{j=1}^{l}\left(\mathfrak{N}^j_{\Xi'}(\mathfrak{x}_i)\right)^2\right)\end{array}\right)}=D^1{}_{P\Xi F}(\Xi,\Xi')$$

For $\gamma=0.5$, in $GD^1{}_{P\Xi F}(\Xi,\Xi')$, we obtained

$$GD^1{}_{P\Xi F}(\Xi,\Xi')=\frac{1}{M}\sum_{i=1}^{M}\frac{\left(\frac{1}{\mathfrak{M}}\sum_{j=1}^{l}\mathfrak{M}^j_{\Xi}(\mathfrak{x}_i)\mathfrak{M}^j_{\Xi'}(\mathfrak{x}_i)+\frac{1}{\mathfrak{N}}\sum_{j=1}^{l}\mathfrak{N}^j_{\Xi}(\mathfrak{x}_i)\mathfrak{N}^j_{\Xi'}(\mathfrak{x}_i)\right)}{\left(\frac{1}{L_{\mathfrak{M}_{\Xi}(\mathfrak{x})}}\sum_{j=1}^{l}\left(\mathfrak{M}^j_{\Xi}(\mathfrak{x}_i)\right)^2+\frac{1}{L_{\mathfrak{N}_{\Xi}(\mathfrak{x})}}\sum_{j=1}^{l}\left(\mathfrak{N}^j_{\Xi}(\mathfrak{x}_i)\right)^2\right)}$$

For $\gamma=0$ and 0.5, $GD^2{}_{P\Xi F}(\Xi,\Xi')$, $GD^3{}_{P\Xi F}(\Xi,\Xi')$, and $GD^4{}_{P\Xi F}(\Xi,\Xi')$ are similar.

5. Decision-Making Processes

Pattern recognition is the computerized identification of shapes, designs, and reliabilities in information. It has applications in information compression, machine learning, statistical information analysis, signal processing, image analysis, information retrieval, bioinformatics, and computer graphics. Similarly, a medical diagnosis is a procedure to illustrate or identify diseases or disorders, which would account for a person's symptoms and signs. The decision-making procedure covers four main stages: intelligence, design, choice, and implementation. The principle of decision-making technique begins with the intelligence stage. In this stage, the intellectual determines reality and identifies and explains the troubles. The main influence of this theory is to explore the main idea of medical diagnosis and pattern recognition under the consideration of IHF information. The main importance and briefing explanation about every application is available below. These applications are taken from Ref. [17]. By using the proposed measures, the applications of medical diagnosis and pattern recognition are discussed below.

5.1. Medical Diagnosis

Certain sorts of diseases have distinct symptoms and different affection; the medical diagnosis procedure is determined by the distinct symptoms of the required diseases of the intellectual which is safer from them. The diseases are expressed using the symbols $\Xi_1,\Xi_2,\ldots,\Xi_n$, and their symptoms are expressed by the values of universal sets. Using the proposed measures, the numerical example is discussed below.

Example 1. *For any set of diseases whose expressions are in the form of* $\Xi=\left\{\begin{array}{c}\Xi_1(Typhoid),\Xi_2(Flu),\Xi_3(Heart\ Probelms),\\\Xi_4(Pneumonia),\Xi_5(coronavirus)\end{array}\right\}$ *and their symptoms whose expressions are in the form of* $X=\left\{\begin{array}{c}Fever,\ Cough,\ Heart\ pain,\\Loss\ of\ appetite, Short\ of\ breath\end{array}\right\}$. *The symptoms of the distinct diseases are discussed below in the form of unknown diseases:*

$$\Xi_1=\left\{\begin{array}{c}(\{0.1,0.2\},\{0.2,0.3,0.4\}),\\(\{0.11,0.21\},\{0.21,0.31,0.41\}),\\(\{0.12,0.22\},\{0.22,0.32,0.42\}),\\(\{0.13,0.23\},\{0.23,0.33,0.43\}),\\(\{0.14,0.24\},\{0.24,0.34,0.44\})\end{array}\right\},\ \Xi_2=\left\{\begin{array}{c}(\{0.2,0.3\},\{0.1,0.3,0.2\}),\\(\{0.21,0.31\},\{0.11,0.31,0.21\}),\\(\{0.22,0.32\},\{0.12,0.32,0.22\}),\\(\{0.23,0.33\},\{0.13,0.33,0.23\}),\\(\{0.24,0.34\},\{0.14,0.34,0.24\})\end{array}\right\},$$

$$\Xi_3=\left\{\begin{array}{c}(\{0.3,0.1\},\{0.5,0.2,0.1\}),\\(\{0.31,0.11\},\{0.51,0.21,0.11\}),\\(\{0.32,0.12\},\{0.52,0.22,0.12\}),\\(\{0.33,0.13\},\{0.53,0.23,0.13\}),\\(\{0.34,0.14\},\{0.54,0.24,0.14\})\end{array}\right\},\ \Xi_4=\left\{\begin{array}{c}(\{0.1,0.1\},\{0.2,0.2,0.4\}),\\(\{0.11,0.11\},\{0.21,0.21,0.41\}),\\(\{0.12,0.12\},\{0.22,0.22,0.42\}),\\(\{0.13,0.13\},\{0.23,0.23,0.43\}),\\(\{0.14,0.14\},\{0.24,0.24,0.44\})\end{array}\right\},\ \Xi_5=$$

$\left\{\begin{array}{c}(\{0.3,0.5\},\{0.1,0.2,0.3\}),\\(\{0.31,0.51\},\{0.11,0.21,0.31\}),\\(\{0.32,0.52\},\{0.12,0.22,0.32\}),\\(\{0.33,0.53\},\{0.13,0.23,0.33\}),\\(\{0.34,0.54\},\{0.14,0.24,0.34\})\end{array}\right\}$. *For this, we choose the known diseases* $\Xi' = \left\{\begin{array}{c}(\{1,1\},\{0.0,0.0.0.0\}),\\(\{1,1\},\{0.0,0.0.0.0\}),(\{1,1\},\{0.0,0.0.0.0\}),\\(\{1,1\},\{0.0,0.0.0.0\}),(\{1,1\},\{0.0,0.0.0.0\})\end{array}\right\}$. *Then, by using the* $GD^1{}_{P\Xi F}(\Xi,\Xi')$, $WGD^1{}_{P\Xi F}(\Xi,\Xi')$, $GD^2{}_{P\Xi F}(\Xi,\Xi')$, *and* $WGD^2{}_{P\Xi F}(\Xi,\Xi')$, *the examined measures are discussed in the form of Table 1 by using the weight vector 0.2, 0.3, 0.2, 0.2, and 0.1. For this, we chose the value of* $\gamma = 1$, *then*

Table 1. Expressions of the measured values by using different measures.

Methods	Values
$GD^1{}_{P\Xi F}(\Xi,\Xi')$	0.5056, 0.8248, 0.542, 0.4772, 0.7232
$WGD^1{}_{P\Xi F}(\Xi,\Xi')$	0.0595, 0.1036, 0.0640, 0.0553, 0.0866
$GD^2{}_{P\Xi F}(\Xi,\Xi')$	0.4649, 0.7866, 0.4981, 0.4396, 0.6635
$WGD^2{}_{P\Xi F}(\Xi,\Xi')$	0.0473, 0.0821, 0.0509, 0.0441, 0.0687

Further, information computed in Table 2 is constructed based on the information available in Table 1.

Table 2. Contained ranking analysis of the information in Table 1.

Methods	Values
$GD^1{}_{P\Xi F}(\Xi,\Xi')$	$\Xi_{\cdot 2} \geq \Xi_{\cdot 5} \geq \Xi_{\cdot 3} \geq \Xi_{\cdot 1} \geq \Xi_{\cdot 4}$
$WGD^1{}_{P\Xi F}(\Xi,\Xi')$	$\Xi_{\cdot 2} \geq \Xi_{\cdot 5} \geq \Xi_{\cdot 3} \geq \Xi_{\cdot 1} \geq \Xi_{\cdot 4}$
$GD^2{}_{P\Xi F}(\Xi,\Xi')$	$\Xi_{\cdot 2} \geq \Xi_{\cdot 5} \geq \Xi_{\cdot 3} \geq \Xi_{\cdot 1} \geq \Xi_{\cdot 4}$
$WGD^2{}_{P\Xi F}(\Xi,\Xi')$	$\Xi_{\cdot 2} \geq \Xi_{\cdot 5} \geq \Xi_{\cdot 3} \geq \Xi_{\cdot 1} \geq \Xi_{\cdot 4}$

From Table 2, all sorts of measures are provided with the same ranking results. the best optimal is Ξ_2. Additionally, by using distinct types of measures based on IFSs and IHFSs, the comparative analysis of the elaborated measures with certain prevailing measures are discussed in the form of Table 3. The information related to prevailing measures is as follows: Ye [12] initiated certain cosine measures based on IFSs, Beg and Rashid [21] proposed certain measures based on IHFSs, and Peng et al. [22] proposed the cross-entropy measures based on IHFSs. By using the information in Section 5.1, the comparative analysis is discussed in the form of Table 3.

Table 3. Contained comparative information.

Methods	Values	Ranking Results
Ye [12]	*Cannot be Calculated*	*Cannot be Calculated*
Beg and Rashid [21]	0.0484, 0.1025, 0.0530, 0.0442, 0.0755	$\Xi_{\cdot 2} \geq \Xi_{\cdot 5} \geq \Xi_{\cdot 3} \geq \Xi_{\cdot 1} \geq \Xi_{\cdot 4}$
Peng et al. [22]	0.3538, 0.6755, 0.3870, 0.3285, 0.5524	$\Xi_{\cdot 2} \geq \Xi_{\cdot 5} \geq \Xi_{\cdot 3} \geq \Xi_{\cdot 1} \geq \Xi_{\cdot 4}$
$GD^1{}_{P\Xi F}(\Xi,\Xi')$	0.5056, 0.8248, 0.542, 0.4772, 0.7232	$\Xi_{\cdot 2} \geq \Xi_{\cdot 5} \geq \Xi_{\cdot 3} \geq \Xi_{\cdot 1} \geq \Xi_{\cdot 4}$
$WGD^1{}_{P\Xi F}(\Xi,\Xi')$	0.0595, 0.1036, 0.0640, 0.0553, 0.0866	$\Xi_{\cdot 2} \geq \Xi_{\cdot 5} \geq \Xi_{\cdot 3} \geq \Xi_{\cdot 1} \geq \Xi_{\cdot 4}$
$GD^2{}_{P\Xi F}(\Xi,\Xi')$	0.4649, 0.7866, 0.4981, 0.4396, 0.6635	$\Xi_{\cdot 2} \geq \Xi_{\cdot 5} \geq \Xi_{\cdot 3} \geq \Xi_{\cdot 1} \geq \Xi_{\cdot 4}$
$WGD^2{}_{P\Xi F}(\Xi,\Xi')$	0.0473, 0.0821, 0.0509, 0.0441, 0.0687	$\Xi_{\cdot 2} \geq \Xi_{\cdot 5} \geq \Xi_{\cdot 3} \geq \Xi_{\cdot 1} \geq \Xi_{\cdot 4}$

From Table 2, all sorts of measures are provided with the same ranking results. The best optimal is Ξ_2.

5.2. Pattern Recognition

By using the elaborated measures, we aimed to use a practical application called pattern recognition and try to evaluate it by using pioneered information.

Example 2. *Without any complication or difficulty, the construction of any building is very complicated. For this, a decision-maker collects the information for different places and resolves it using the elaborated measures; then a very safe decision can be made. For this, we chose the different types of building material, the information associated with which is discussed below.*

$$\Xi_1 = \left\{\begin{array}{c}(\{0.1,0.2\},\{0.1,0.2,0.3\}),\\(\{0.11,0.21\},\{0.11,0.21,0.31\}),\\(\{0.12,0.22\},\{0.12,0.22,0.32\}),\\(\{0.13,0.23\},\{0.13,0.23,0.33\}),\\(\{0.14,0.24\},\{0.14,0.24,0.34\})\end{array}\right\}, \Xi_2 = \left\{\begin{array}{c}(\{0.2,0.3\},\{0.2,0.3,0.4\}),\\(\{0.21,0.31\},\{0.21,0.31,0.41\}),\\(\{0.22,0.32\},\{0.22,0.32,0.42\}),\\(\{0.23,0.33\},\{0.23,0.33,0.43\}),\\(\{0.24,0.34\},\{0.24,0.34,0.44\})\end{array}\right\},$$

$$\Xi_3 = \left\{\begin{array}{c}(\{0.1,0.3\},\{0.2,0.1,0.1\}),\\(\{0.11,0.31\},\{0.21,0.11,0.11\}),\\(\{0.12,0.32\},\{0.22,0.12,0.12\}),\\(\{0.13,0.33\},\{0.23,0.13,0.13\}),\\(\{0.14,0.34\},\{0.24,0.14,0.14\})\end{array}\right\}, \Xi_4 = \left\{\begin{array}{c}(\{0.1,0.2\},\{0.3,0.2,0.4\}),\\(\{0.11,0.21\},\{0.31,0.21,0.41\}),\\(\{0.12,0.22\},\{0.32,0.22,0.42\}),\\(\{0.13,0.23\},\{0.33,0.23,0.43\}),\\(\{0.14,0.24\},\{0.34,0.24,0.44\})\end{array}\right\},$$

$$\Xi_5 = \left\{\begin{array}{c}(\{0.4,0.5\},\{0.1,0.1,0.1\}),\\(\{0.41,0.51\},\{0.11,0.11,0.11\}),\\(\{0.42,0.52\},\{0.12,0.12,0.12\}),\\(\{0.43,0.53\},\{0.13,0.13,0.13\}),\\(\{0.44,0.54\},\{0.14,0.14,0.14\})\end{array}\right\}$$

For this, we choose the known diseases, which are expressed below:

$$\Xi' = \left\{\begin{array}{c}(\{1,1\},\{0.0,0.0.0.0\}),\\(\{1,1\},\{0.0,0.0.0.0\}),(\{1,1\},\{0.0,0.0.0.0\}),\\(\{1,1\},\{0.0,0.0.0.0\}),(\{1,1\},\{0.0,0.0.0.0\})\end{array}\right\}$$

Then, by using the $GD^1{}_{P\Xi F}(\Xi,\Xi')$, $WGD^1{}_{P\Xi F}(\Xi,\Xi')$, $GD^2{}_{P\Xi F}(\Xi,\Xi')$, and $WGD^2{}_{P\Xi F}(\Xi$ the examined measures are discussed in the form of Table 4 by using the weight vector 0.2, 0.3, 0.2, 0.2, and 0.1. For this, we chose the value of $\gamma = 1$.

Table 4. Expressions of the measured values using different measures.

Methods	Values
$GD^1{}_{P\Xi F}(\Xi,\Xi')$	0.8166, 0.5859, 0.8271, 0.5057, 0.8219
$WGD^1{}_{P\Xi F}(\Xi,\Xi')$	0.098, 0.0728, 0.1313, 0.0596, 0.0984
$GD^2{}_{P\Xi F}(\Xi,\Xi')$	0.749, 0.5589, 0.9916, 0.4649, 0.7514
$WGD^2{}_{P\Xi F}(\Xi,\Xi')$	0.0777, 0.0578, 0.104, 0.0474, 0.078

Further, the information computed in Table 5 was constructed based on the information available in Table 4.

Table 5. Contained ranking analysis.

Methods	Values
$GD^1{}_{P\Xi F}(\Xi, \Xi')$	$\Xi_{\cdot 5} \geq \Xi_{\cdot 3} \geq \Xi_{\cdot 1} \geq \Xi_{\cdot 2} \geq \Xi_{\cdot 4}$
$WGD^1{}_{P\Xi F}(\Xi, \Xi')$	$\Xi_{\cdot 3} \geq \Xi_{\cdot 5} \geq \Xi_{\cdot 1} \geq \Xi_{\cdot 2} \geq \Xi_{\cdot 4}$
$GD^2{}_{P\Xi F}(\Xi, \Xi')$	$\Xi_{\cdot 5} \geq \Xi_{\cdot 3} \geq \Xi_{\cdot 1} \geq \Xi_{\cdot 2} \geq \Xi_{\cdot 4}$
$WGD^2{}_{P\Xi F}(\Xi, \Xi')$	$\Xi_{\cdot 5} \geq \Xi_{\cdot 3} \geq \Xi_{\cdot 1} \geq \Xi_{\cdot 2} \geq \Xi_{\cdot 4}$

From Table 5, all sorts of measures are provided with the different ranking results. the best optimal is Ξ_5 and Ξ_3. Additionally, by using distinct types of measures based on IFSs and IHFSs, the comparative analysis of the elaborated measures with certain prevailing measures are discussed in the form of Table 6. The information related to prevailing measures is as follows: Ye [12] initiated certain cosine measures based on IFSs, Beg and Rashid [21] proposed certain measures based on IHFSs, and Peng et al. [22] proposed the cross-entropy measures based on IHFSs. By using the information from Example 1, the comparative analysis is discussed in the form of Table 6.

Table 6. Contained comparative analysis.

Methods	Values	Ranking Results
Ye [12]	*Cannot be Calculated*	*Cannot be Calculated*
Beg and Rashid [21]	0.638, 0.4478, 0.8805, 0.3538, 0.6403	$\Xi_5 \geq \Xi_3 \geq \Xi_1 \geq \Xi_2 \geq \Xi_4$
Peng et al. [22]	0.0666, 0.0467, 0.103, 0.0363, 0.067	$\Xi_5 \geq \Xi_3 \geq \Xi_1 \geq \Xi_2 \geq \Xi_4$
$GD^1{}_{P\Xi F}(\Xi, \Xi')$	0.8166, 0.5859, 0.8271, 0.5057, 0.8219	$\Xi_5 \geq \Xi_3 \geq \Xi_1 \geq \Xi_2 \geq \Xi_4$
$WGD^1{}_{P\Xi F}(\Xi, \Xi')$	0.098, 0.0728, 0.1313, 0.0596, 0.0984	$\Xi_3 \geq \Xi_5 \geq \Xi_1 \geq \Xi_2 \geq \Xi_4$
$GD^2{}_{P\Xi F}(\Xi, \Xi')$	0.749, 0.5589, 0.9916, 0.4649, 0.7514	$\Xi_5 \geq \Xi_3 \geq \Xi_1 \geq \Xi_2 \geq \Xi_4$
$WGD^2{}_{P\Xi F}(\Xi, \Xi')$	0.0777, 0.0578, 0.104, 0.0474, 0.078	$\Xi_5 \geq \Xi_3 \geq \Xi_1 \geq \Xi_2 \geq \Xi_4$

From Table 2, all sorts of measures are provided with the different ranking results. the best optimal is Ξ_5 and Ξ_3. In the future, we will utilize different types of operators, methods, and measures in the environment of picture hesitant fuzzy sets and neutrosophic hesitant fuzzy sets [24–31] to improve the quality of the proposed works. Therefore, the elaborated measures based on IHFS are more powerful and more fixable than the prevailing ideas [23–31].

6. Conclusions

The main and major features of this analysis are described below:

1. We pioneered the main theory of DSM based on IHFS and evaluated their particular cases.
2. We established the GDS measures in the environment of IHFSs and discussed IHFDSM, IHFWDSM, IHFGDSM, and IHFWGDSM.
3. Based on the investigated measures, certain special cases were also evaluated. Furthermore, by using the discovered measures, medical diagnosis and pattern recognition problems were determined.
4. Finally, we determined the supremacy of the explored work and the sensitivity analysis and advantages of the explored measures. Their geometrical expressions are also discussed.

Our recent work focused on the prevailing information computed based on complex q-rung orthopair FSs [32], spherical FSs (SFSs) [33], Aczel-Alsina operational laws [34], different types of measures [35,36], Aczel-Alsina aggregation operators [37], Maclaurin operators [38], Complex SFSs [39,40], linguistic group decision-making techniques [41], and unbalanced linguistic information [42], and we aim to employ it in the field of computer science, road signals, software engineering, and decision-making.

Author Contributions: Conceptualization, M.A.; Formal analysis, M.A.; Funding acquisition, M.A.; Investigation, M.A. and T.M.; Methodology, T.M.; Project administration, M.A. and T.M.; Supervision, T.M.; Validation, T.M. All authors contributed equally. All authors have read and agreed to the published version of the manuscript.

Funding: The Deanship of Scientific Research (DSR) at King Abdulaziz University (KAU), Jeddah, Saudi Arabia has funded this Project under grant No. (G: 445-130-1443).

Institutional Review Board Statement: Not applicable.

Informed Consent Statement: Not applicable.

Data Availability Statement: The data utilized in this study are hypothetical and artificial, and one can use these data before prior permission by citing this paper.

Acknowledgments: This project was funded by the Deanship of Scientific Research (DSR) at King Abdulaziz University (KAU), Jeddah, Saudi Arabia under grant No. (G:445-130-1443). The authors, therefore, acknowledge with thanks DSR for technical and financial support.

Conflicts of Interest: The authors declare no conflict of interest.

Ethics Declaration Statement: The authors state that this is their original work, and it is neither submitted nor under consideration in any other journal simultaneously.

References

1. Zadeh, L.A. Fuzzy sets. *Inf. Control* **1965**, *8*, 338–353. [CrossRef]
2. Aydın, S. A fuzzy MCDM method based on new Fermatean fuzzy theories. *Inter. J. Infor. Tech. Decis. Mak.* **2021**, *20*, 881–902. [CrossRef]
3. John, R. Type 2 fuzzy sets: An appraisal of theory and applications. *Inter. J. Uncertain. Fuzziness Knowl. Based Syst.* **1998**, *6*, 563–576. [CrossRef]
4. Mendel, J.M.; John, R.B. Type-2 fuzzy sets made simple. *IEEE Trans. Fuzzy Syst* **2002**, *10*, 117–127. [CrossRef]
5. Mahmood, T. A Novel Approach towards Bipolar Soft Sets and Their Applications. *J. Math.* **2020**, *2020*, 4690808. [CrossRef]
6. Torra, V. Hesitant fuzzy sets. *Inter. J. Intel. Syst.* **2010**, *25*, 529–539. [CrossRef]
7. Meng, F.; Chen, X. Correlation coefficients of hesitant fuzzy sets and their application based on fuzzy measures. *Cogn. Comput.* **2015**, *7*, 445–463. [CrossRef]
8. Li, D.; Zeng, W.; Li, J. New distance and similarity measures on hesitant fuzzy sets and their applications in multiple criteria decision making. *Eng. Appl. Artif. Intel.* **2015**, *40*, 11–16. [CrossRef]
9. Su, Z.; Xu, Z.; Liu, H.; Liu, S. Distance and similarity measures for dual hesitant fuzzy sets and their applications in pattern recognition. *J. Intel Fuzzy Syst.* **2015**, *29*, 731–745. [CrossRef]
10. Wei, C.; Yan, F.; Rodriguez, R.M. Entropy measures for hesitant fuzzy sets and their application in multi-criteria decision-making. *J. Intel Fuzzy Syst.* **2016**, *31*, 673–685. [CrossRef]
11. Atanassov, K. Intuitionistic fuzzy sets. *Fuzzy Sets Syst.* **1986**, *20*, 87–96. [CrossRef]
12. Ye, J. Cosine similarity measures for intuitionistic fuzzy sets and their applications. *Math. Comput. Model.* **2011**, *53*, 91–97. [CrossRef]
13. Rani, D.; Garg, H. Distance measures between the complex intuitionistic fuzzy sets and their applications to the decision-making process. *Inter. J. Uncertain. Quantif.* **2017**, *7*, 211–227. [CrossRef]
14. Liang, Z.; Shi, P. Similarity measures on intuitionistic fuzzy sets. *Pattern Recognit. Lett.* **2003**, *24*, 2687–2693. [CrossRef]
15. Xu, Z.S.; Chen, J. An overview of distance and similarity measures of intuitionistic fuzzy sets. *Inter. J. Uncertain. Fuzziness Knowl. -Based Syst.* **2008**, *16*, 529–555. [CrossRef]
16. Xu, Z. Some similarity measures of intuitionistic fuzzy sets and their applications to multiple attribute decision making. *Fuzzy Optim. Decis. Mak.* **2007**, *6*, 109–121. [CrossRef]
17. Garg, H.; Rani, D. A robust correlation coefficient measure of complex intuitionistic fuzzy sets and their applications in decision-making. *Appl. Intel.* **2019**, *49*, 496–512. [CrossRef]
18. Zeshui, X. On similarity measures of interval-valued intuitionistic fuzzy sets and their application to pattern recognitions. *J. Southeast. Univ.* **2007**, *3*, 27–41.
19. Wei, C.P.; Wang, P.; Zhang, Y.Z. Entropy, similarity measure of interval-valued intuitionistic fuzzy sets and their applications. *Infor. Sci.* **2011**, *181*, 4273–4286. [CrossRef]
20. Wang, W.; Xin, X. Distance measure between intuitionistic fuzzy sets. *Pattern Recognit. Lett.* **2005**, *26*, 2063–2069. [CrossRef]

21. Beg, I.; Rashid, T. Group decision making using intuitionistic hesitant fuzzy sets. *Inter. J. Fuzzy Logic. Intel Syst.* **2014**, *14*, 181–187. [CrossRef]
22. Peng, J.J.; Wang, J.Q.; Wu, X.H.; Zhang, H.Y.; Chen, X.H. The fuzzy cross-entropy for intuitionistic hesitant fuzzy sets and their application in multi-criteria decision-making. *Inter. J. Syst. Sci.* **2015**, *46*, 2335–2350. [CrossRef]
23. Zhai, Y.; Xu, Z.; Liao, H. Measures of probabilistic interval-valued intuitionistic hesitant fuzzy sets and the application in reducing excessive medical examinations. *IEEE Trans. Fuzzy Syst.* **2017**, *26*, 1651–1670. [CrossRef]
24. Tang, Y.; Wen, L.L.; Wei, G.W. Approaches to multiple attribute group decision making based on the generalized Dice similarity measures with intuitionistic fuzzy information. *Inter. J. Knowl. Based Intel Eng. Syst.* **2017**, *21*, 85–95. [CrossRef]
25. Ye, S.; Ye, J. Dice similarity measure between single valued neutrosophic multisets and its application in medical diagnosis. *Neutrosophic Sets Syst.* **2014**, *6*, 9–23.
26. Ye, J. *Vector Similarity Measures of Simplified Neutrosophic Sets and Their Application in Multicriteria Decision Making*; Infinite Study: West Conshohocken, PA, USA, 2014.
27. Ulucay, V.; Deli, I.; Şahin, M. Similarity measures of bipolar neutrosophic sets and their application to multiple criteria decision making. *Neural Comput. Appl.* **2018**, *29*, 739–748. [CrossRef]
28. Chen, J.; Ye, J.; Du, S. Vector similarity measures between refined simplified neutrosophic sets and their multiple attribute decision-making method. *Symmetry* **2017**, *9*, 153. [CrossRef]
29. Chatterjee, R.; Majumdar, P.; Samanta, S.K. Similarity measures in neutrosophic sets-I. *Fuzzy Multi-Criteria Decis. Mak. Using Neutrosophic Sets* **2019**, *369*, 249–294.
30. Pramanik, S.; Biswas, P.; Giri, B.C. Hybrid vector similarity measures and their applications to multi-attribute decision making under neutrosophic environment. *Neural Comput. Appl.* **2017**, *28*, 1163–1176. [CrossRef]
31. Mondal, K.; Pramanik, S. Decision making based on some similarity measures under interval rough neutrosophic environment. *Neutrosophic Sets Syst.* **2015**, *10*, 3–19.
32. Ali, Z.; Mahmood, T. Maclaurin symmetric mean operators and their applications in the environment of complex q-rung orthopair fuzzy sets. *Comput. Appl. Math.* **2020**, *39*, 161. [CrossRef]
33. Mahmood, T.; Ullah, K.; Khan, Q.; Jan, N. An Approach Towards Decision Making and Medical Diagnosis Problems Using the Concept of Spherical Fuzzy Sets. *Neural Comput. Appl.* **2019**, *31*, 7041–7053. [CrossRef]
34. Hussain, A.; Ullah, K.; Yang, M.S.; Pamucar, D. Aczel-Alsina Aggregation Operators on T-Spherical Fuzzy (TSF) Information with Application to TSF Multi-Attribute Decision Making. *IEEE Access* **2022**, *10*, 26011–26023. [CrossRef]
35. Abid, M.N.; Yang, M.S.; Karamti, H.; Ullah, K.; Pamucar, D. Similarity Measures Based on T-Spherical Fuzzy Information with Applications to Pattern Recognition and Decision Making. *Symmetry* **2022**, *14*, 410. [CrossRef]
36. Khan, R.; Ullah, K.; Pamucar, D.; Bari, M. Performance measure using a multi-attribute decision making approach based on Complex T-spherical fuzzy power aggregation operators. *J. Computat. Cogn. Eng.* **2022**. [CrossRef]
37. Hussain, A.; Ullah, K.; Alshahrani, M.N.; Yang, M.S.; Pamucar, D. Novel Aczel–Alsina Operators for Pythagorean Fuzzy Sets with Application in Multi-Attribute Decision Making. *Symmetry* **2022**, *14*, 940. [CrossRef]
38. Ullah, K. Picture fuzzy maclaurin symmetric mean operators and their applications in solving multiattribute decision-making problems. *Math. Probl. Eng.* **2021**, *2021*, 1098631. [CrossRef]
39. Ali, Z.; Mahmood, T.; Yang, M.S. TOPSIS method based on complex spherical fuzzy sets with Bonferroni mean operators. *Mathematics* **2020**, *8*, 1739. [CrossRef]
40. Ali, Z.; Mahmood, T.; Yang, M.S. Complex T-spherical fuzzy aggregation operators with application to multi-attribute decision making. *Symmetry* **2020**, *12*, 1311. [CrossRef]
41. Zhang, Z.; Li, Z. Personalized individual semantics-based consistency control and consensus reaching in linguistic group decision making. *IEEE Trans. Syst Man Cybern. Syst.* **2021**. [CrossRef]
42. Zhang, Z.; Li, Z.; Gao, Y. Consensus reaching for group decision making with multi-granular unbalanced linguistic information: A bounded confidence and minimum adjustment-based approach. *Inf. Fusion* **2021**, *74*, 96–110. [CrossRef]

mathematics

MDPI

Article

A Novel Driving-Strategy Generating Method of Collision Avoidance for Unmanned Ships Based on Extensive-Form Game Model with Fuzzy Credibility Numbers

Haotian Cui [1], Fangwei Zhang [1,2,*], Mingjie Li [1], Yang Cui [1] and Rui Wang [1]

1 School of Navigation and Shipping, Shandong Jiaotong University, Weihai 264209, China
2 College of Transport and Communications, Shanghai Maritime University, Shanghai 201306, China
* Correspondence: 210103@sdjtu.edu.cn

Abstract: This study aims to solve the problem of intelligent collision avoidance of unmanned ships at sea, and it proposes a novel driving strategy generating method of collision avoidance based on an extensive-form game mode with fuzzy credibility numbers. The innovation of this study is to propose an extensive-form game model of unmanned ships under the situation of two-sides clamping and verify the validity by fuzzy credibility. Firstly, this study divides the head-on situation of the ship at sea quantitatively to help the unmanned ship take targeted measures when making collision avoidance decisions. Secondly, this study adopts an extensive-form game model to model the problem of collision avoidance of an unmanned ship in the case of clamping on two sides. Thirdly, the extensive-form game model is organically combined with the fuzzy credibility degree to judge whether the collision avoidance game of unmanned ship achieves the optimal collision avoidance result. The effectiveness of the introduced game model is verified by case analysis and simulation. Finally, an illustrative example shows that the proposed mathematical model can better help unmanned ships make real-time game decisions at sea in the scenario of two-sides clamping effectively.

Keywords: collision avoidance; encounter situation; fuzzy credibility numbers; intelligent unmanned ships; extensive-form game model

MSC: 90C70

Citation: Cui, H.; Zhang, F.; Li, M.; Cui, Y.; Wang, R. A Novel Driving-Strategy Generating Method of Collision Avoidance for Unmanned Ships Based on Extensive-Form Game Model with Fuzzy Credibility Numbers. *Mathematics* **2022**, *10*, 3316. https://doi.org/10.3390/math10183316

Academic Editor: Mariano Luque

Received: 2 August 2022
Accepted: 5 September 2022
Published: 13 September 2022

1. Introduction

In actual maritime navigation, the entire collision avoidance operation of unmanned ships revolves around the three stages of "observation, judgment and decision making" [1]. At the same time, the specific water environment and different encounter states will also affect the collision avoidance decision-making process of the unmanned ship. Under the above background, to help unmanned ships take targeted measures to avoid collision decisions, this study analyzes the situation of ships under the condition of both sides.

The two-sides clamping scenario is a condition in which a ship sails between two ships while at sea. Investigation illustrates that it is dangerous when a ship is trapped in this two-sides clamping situation. Considering that the collision avoidance operations of unmanned ships is a game process, this study proposes an anti-collision decision model for unmanned ships based on extensive-form game model [2].

1.1. Literature Review

According to investigation, the issues of unmanned ship collision avoidance in a two-side scenario is focused on. At present, scholars' research on ship collision avoidance mostly focuses on three aspects: strategies for avoiding ship collisions, application of game theory to ships, and practical application of fuzzy credibility numbers.

In the past five years, avoiding collision problems have been mainly studied from the viewpoints of risk assessment, variable distribution, safety domain, etc. Scholars have performed research on strategies for ships avoiding collisions. In the study by Li et al. [3], by balancing the safety and economy of ship collision avoidance, the avoidance angle and the time to the action point are used as the variables encoded by the algorithm, and the fuzzy ship domain is used to calculate the collision avoidance risk to achieve collision avoidance. Thereafter, Lee et al. [4] proposed a heuristic search technology for collision avoidance operations for autonomous ships. Based on the multi-vessel collision avoidance problem, Wang et al. [5] researched the decision-making for obstacle avoidance based on deep reinforcement learning to solve the problem of intelligent collision avoidance for unmanned ships in unknown environments. Based on the mathematical model group's ship motion mathematical model, Xing et al. [6] proposed an open sea ship collision prevention approach to enhance the prediction of ship collision risk and the real-time and dependability of collision avoidance method.

At present, the application fields of extensive-form game mode are concluding containing transportation; Lisowski [7] introduced the application of the game control processes in marine navigation. The control goal has been defined first. Then, the approximated models of multi-stage positional game and multi-step matrix game of the safe ship steering in a collision situation has been presented. Subsequently, Lisowski et al. [8] described six methods of optimal and game theory and artificial neural network for synthesis of safe control in collision situations at sea. The optimal control algorithm and game control algorithm were used to determine the safe track. Afterwards, Zou et al. [9] identified the safety evaluation indicator system and evaluation standards and established an after-collisions safety evaluation model of maritime ships based on the extension cloud theory. Considering the defects of the classic extensive game method in ship collision avoidance decision-making, Tu et al. [10] proposed the improved extensive game method based on the velocity obstacle method.

Up to now, fuzzy credibility numbers were mainly used to solve decision making problem, project scheduling problem, multi-objective fuzzy-interval credibility-constrained non-linear programming, etc. Ran et al. [11] aimed at the problems of inaccurate evaluation results caused by experts in the process of simulation credibility evaluation based on traditional fuzzy comprehensive evaluation according to personal preferences or expectations, and unreasonable selection of fuzzy synthetic calculations, and a simulation credibility evaluation method based on improved fuzzy comprehensive evaluation was proposed. Moreover, Ye et al. [12] proposed the concept of a fuzzy credibility number as a new extension of the fuzzy concept. Thereafter, Vercher et al. [13] presented a new forecasting scheme based on the credibility distribution of fuzzy events. In the same year, Zhou et al. [14] proposed a decision support model for USVs to improve the accuracy of collision avoidance decision-making.

Based on the aforementioned analysis, the collision avoidance of unmanned ships is studied. The main innovation of this study is combining extensive-form game model and FCN together. Specifically, by using the extensive-form game model, the collision avoidance strategy of unmanned ships is studied for the special situation between the two-sides. By using FCN, the danger of collision is quantified.

1.2. Goals and Contributions

The purpose of this study is to explore the decision-making problem of collision avoidance for unmanned ships at sea under the two-sides clamp scenario. In response to the aforementioned problems, this study establishes an extensive-form game model based on the two-sides clamping scenario and applies it to solve the specific collision avoidance problem.

The contribution of this study is as follows. Firstly, based on the extensive-form game model, this study establishes a description of the ship collision avoidance structure under the situation of two-sides clamping. Secondly, this study chooses driving strategy following a priority principle on ship collision avoidance and introduces a utility function to describe it. By combining this utility function and the extensive-form game model, a set of utilities of the own ship and the target ship are collected and compared to find the optimal collision avoidance decision. Thirdly, this study establishes a ship collision risk fuzzy credibility operator to judge whether the ship has escaped from collision danger.

The rest of this study is organized as follows. Section 2 clarifies the research basis. Section 3 proposes the driving-strategy generating method for collision avoidance. Section 4 carries out simulation verification for the proposed method. Section 5 summarizes and points out possible future work. The structure of this study is shown in Figure 1.

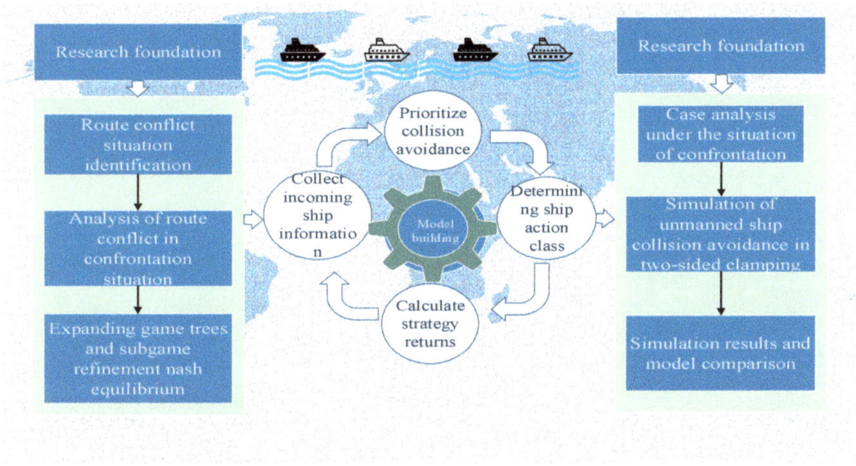

Figure 1. Research process.

2. Research Basis

This part mainly introduces the conflict identification of the ships' encounter situation at sea, which quantitatively analyzes the ship's encounter situation, and introduces the relevant knowledge of extensive-form game tree and sub-game refinement Nash equilibrium.

2.1. Route Conflict Situation Identification

The identification of the conflict situation on the route and the division of ship responsibilities are based on the 1972 International Regulations for Preventing Collisions at Sea, namely COLREGS. In actual navigation, the collision avoidance measures taken by unmanned ships are based on the collision avoidance rules listed in COLREGS combined with various ship identification devices for automatic collision avoidance [15]. The uncoordinated collision avoidance measures may lead to the uncoordinated collision avoidance process of the entire ship so that the best avoidance opportunity is missed [16]. According to the different angles of encounter of ships, the encounter situation will be divided into three types: head-on situation, overtaking situation, and cross encounter situation. Head-on situation is the situation that ships often encounter at sea, and it is also the main situation that causes the ship to be in imminent danger or to collide. Therefore, this study researches the collision avoidance strategy of ships in the confrontation situation.

2.2. Judgment of Head-On Ship Situation

The "International Regulations for Preventing Collisions at Sea" has the following four points to judge the head-on situation of ship [17]. Firstly, both ships must be motorized ships. Secondly, the sailing directions of the two ships are in an opposite or almost opposite confrontation on the route. Thirdly, one motorized ship is sailing directly in front of or nearly in front of the other. Finally, the two ships are seeing each other and constitute a collision hazard. Therein, the heading angle of B the two ships in the confrontation situation ΔC is the relative azimuth. The heading opposite or close to the opposite means that the heading difference between the two ships is within $174^{\circ} \leq \Delta C \leq 186^{\circ}$ the range. From the point of view of encountering the relative orientation of the two ships, the heading of the two ships is close to the opposite, which means that one ship is located within 6° on the left and right in front of the other ship. Therefore, the relative orientation of the confronting situation should satisfy $B \leq 005^{\circ}$ or $B \geq 351^{\circ}$; the specific details are shown in Figure 2.

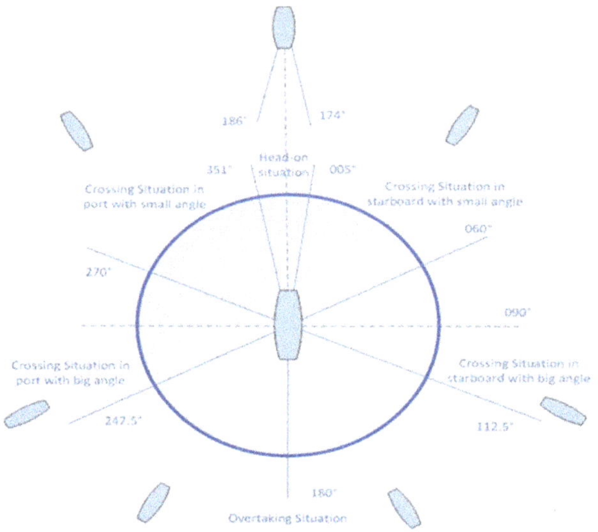

Figure 2. Schematic diagram of Head-on situation.

2.3. *Extensive-Form Game Model Tree*

The extensive-form game is dynamic. The difference between it and the static game is that the extensive-form dynamic game needs to determine the order of actions [18]. Each knot on the "game tree" represents a player's decision point, and this point is said to belong to the player acting at that point [19]. The branches represent the possible actions of the players, and each branch connects two knots, which has a direction from one knot to the other. Each branch of the game tree may or may not be expanded. Meanwhile, each branch in the game tree can be regarded as a new game tree, called a sub-tree, as shown in Figure 3. The part of A is the sub-game of B, and A is also the sub-game of the whole game. The nodes are expanded outward, as shown in Figure 4.

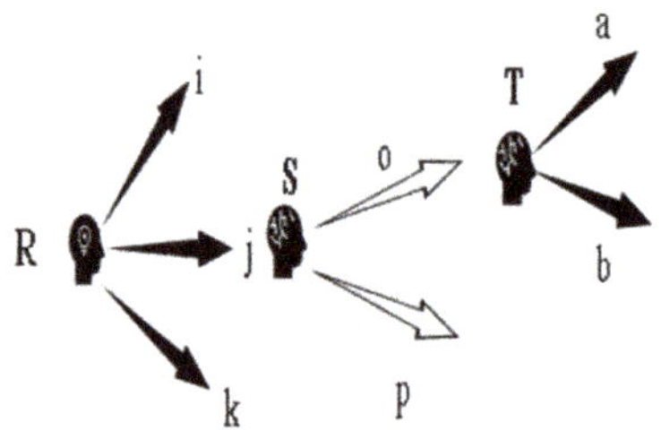

Figure 3. Extensive-form game model tree.

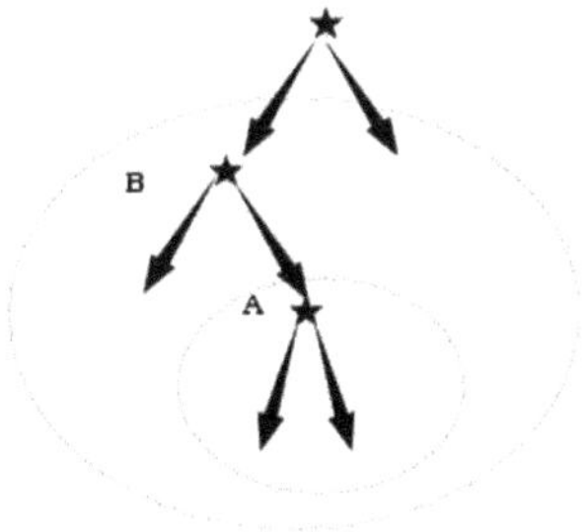

Figure 4. The sub-tree of the game tree.

2.4. Sub-Game Refinement Nash Equilibrium

The Nash equilibrium strategy is that all players in the game adopt the best strategy that is beneficial to them [20]. In the whole process of the game, the players of each game are rational and intelligent. The combination of actions taken in each game is the optimal strategy, and the sub-game developed by the game tree is also the optimal solution. The combination of action strategies taken in the game process conforms to the Nash equilibrium strategy. Sub-game refined Nash equilibrium is the most effective tool for analyzing perfect information dynamic games in the game theory [21].

3. Unmanned Ship Collision Avoidance Model in Two-Sides Clamp Scenario

This subsection adopts the fuzzy mathematics method, which organically combines the extensive-form game with the collision risk fuzzy credibility numbers. This study analyzes the collision avoidance game problem of route conflict in the situation where the unmanned ship is under two-sides clamping situation particularly. In this model, the fuzzy confidence degree of collision risk is used to calculate whether the ship escapes from the collision risk after the collision avoidance game, so as to judge whether the ship adopts the optimal collision avoidance strategy.

3.1. A Novel Ship Collision Avoidance Model

In this subsection, a ship collision avoidance model is proposed in two-sides clamp scenario. Specific steps are as follows.

Step 1: Determination of priority. When ships encounter emergency and dangerous situations in the course of navigation, if they want to recognize each other's game information through various ship identification equipment on unmanned ships, they also need to play sequential dynamic game on ships. To determine the action sequence of the players in the game process, this study proposes a ship priority function. This study makes two assumptions about the gross tonnage of the ship and the sailing speed of the ship regarding the actual sailing experience. The larger the gross tonnage of the ship in the voyage, the higher the priority in the game situation. Then, it is assumed that the higher the speed of the ship during the voyage, the higher the priority in the ship game. The following formula is given for the aforementioned two assumptions:

$$p_i = w_1(G_i/\sum_{i=1}^{n} G_i) + w_2(V_i/\sum_{i=1}^{n} V_i). \quad (1)$$

In Equation (1), it is noteworthy that p_i represents the priority index of player i in the game, G_i represents the total tonnage of player i in the game, and V_i represents i the speed of the player in the game. Among them w_1 and w_2 represents the weight of the gross tonnage of the ship and the speed, at the same time $w_1 + w_2 = 1$. After p_i has been determined, players p_i alternately make action decisions based on the magnitude of the index.

Step 2: Action space (Action set). After obtaining the corresponding action sequence based on the ship collision avoidance priority in step 1, it is assumed that ship i start to act. The set of game decision it makes in the current situation is called the action set of the ship i. In this set of action strategies, the number of action strategies made by ship i is related to the complexity of the game situation; the number of action strategies made by the ship is related to the complexity of the game situation. The more complex the game model, the more actions can be made, the more combinations of actions, and the longer the solution process will take. This study only adopts steering avoidance as a collision avoidance measure to simplify the development space of the game and reduce the time required for the game-solving process. In sailing practice, the steering angle is too large, which will cause inconvenience in resuming the voyage. Therefore, the upper and lower limits of steering are 30° in this study, and each turn is 10° as an action strategy, then the action set of the ship i can be represented as $A_i = \{-30^{\circ}, -20^{\circ}, -10^{\circ}, 0^{\circ}, 10^{\circ}, 20^{\circ}, 30^{\circ}\}$.

Step 3: Profit function. After determining the ships' collision avoidance priority and the ships' decision-making action set, this study only considers the ships' offset as a profit on the premise of ensuring that the ship can sail safely and establishes a profit function. In collision avoidance, the lower the ships' drift, the lower the ships' cost, and the more the ship benefits throughout the game process. Set the initial position of the ship as , x_0, y_0,, the speed of the ship as v, the heading angle as ψ, and the time interval of the ship game as t. This study only studies a series of games between our ship and the other two ships under the special situation of the two-sides. It is assumed that one of the ships is an environmental variable, that is, the ship does not take any steering measures to maintain direction and speed. If the planned course is sailing at a constant speed, t is the displacement increments of the abscissa, and the increment of the x_l ordinate of the ship in time y_l are:

$$x_l = \begin{cases} vt\sin(\psi), & 0^{\circ} \leq \psi \leq 90^{\circ}; \\ vt\cos(\psi - 90^{\circ}), & 90^{\circ} < \psi \leq 180^{\circ}; \\ -vt\sin(\psi - 180^{\circ}), & 180^{\circ} < \psi \leq 270^{\circ}; \\ -vt\cos(\psi - 270^{\circ}), & 270^{\circ} < \psi < 360^{\circ}. \end{cases} \tag{2}$$

$$y_l = \begin{cases} vt\cos(\psi), & 0^{\circ} \leq \psi \leq 90^{\circ}; \\ -vt\sin(\psi - 90^{\circ}), & 90^{\circ} < \psi \leq 180^{\circ}; \\ -vt\cos(\psi - 180^{\circ}), & 180^{\circ} < \psi \leq 270^{\circ}; \\ vt\sin(\psi - 270^{\circ}), & 270^{\circ} < \psi < 360^{\circ}. \end{cases} \tag{3}$$

After the i-th decision is made, the coordinates where the ship arrives (x_p, y_p) according to the planned course and constant speed, it gets:

$$x_p = x_0 + ix, y_p = y_0 + iy_l. \tag{4}$$

During the actual ship's action, the ship's expected position (x_i, y_i) will be affected by the last decision. If the ship's position after making a decision is (x_{i-1}, y_{i-1}), then:

$$x_i = x_{i-1} + x_m, y_i = y_{i-1} + y_m \tag{5}$$

where ψ_i represents the new heading angle of the ship after the i-th decision is executed:

$$\psi_i = \begin{cases} \psi_i & 0^{\circ} \leq \psi_i < 360^{\circ} \\ \psi_i - 360^{\circ} & \psi_i \geq 360^{\circ} \\ \psi_i + 360^{\circ} & \psi_i < 0^{\circ} \end{cases}. \tag{6}$$

However, environmental variables should be taken into account when considering collision avoidance strategies. Therefore, the relevant distance variable is introduced in combination with the collision risk μ. The unmanned ship will take measures to avoid the ship when it encounters the nearest distance. The influence of bump measure on revenue function is as follows:

$$\mu = \frac{1}{2} - \frac{1}{2}\sin\left[\frac{\pi}{d_2 - d_1}\left(\omega - \frac{d_1 + d_2}{2}\right)\right]. \tag{7}$$

Among them, d_1 and d_2 are the safety field value of the ship and the safe passing distance of the ship, respectively, and the distance ω between our ship and the environmental variable ship.

To sum up, it can be extracted that the ships' offset S in the i-th decision of the player S is shown in Equation (8):

$$S=\begin{cases}\sqrt{\begin{matrix}[x_0+vt\sin(\psi)(i-1)+vt\sin(\psi_i)-(x_0+vt\sin(\psi)i)]^2\\+\\ [y_0+vt\cos(\psi)(i-1)+vt\cos(\psi_i)-(y_0+vt\cos(\psi)i)]^2\end{matrix}}+\mu,0^{\circ}\le\psi\le 90^{\circ};\\ \sqrt{\begin{matrix}\left[x_0+vt\cos(\psi-\frac{\pi}{2}),i-1,+vt\cos(\psi_i-\frac{\pi}{2})-(x_0+vt\cos(\psi-\frac{\pi}{2})i)\right]^2\\+\\ \left[y_0-vt\sin(\psi-90^{\circ})(i-1)-vt\sin(\psi_i-\frac{\pi}{2})-(y_0-vt\sin(\psi-\frac{\pi}{2})i)\right]^2\end{matrix}}+\mu,90^{\circ}<\psi\le 180\\ \sqrt{\begin{matrix}[x_0-vt\sin(\psi-\pi),i-1,-vt\sin(\psi_i-\pi)-(x_0-vt\sin(\psi-\pi)i)]^2\\+\\ [y_0-vt\cos(\psi-\pi)(i-1)-vt\cos(\psi_i-\pi)-(y_0-vt\cos(\psi-\pi)i)]^2\end{matrix}}+\mu,180^{\circ}<\psi\le 270\\ \sqrt{\begin{matrix}\left[x_0-vt\cos(\psi-\frac{3\pi}{2}),i-1,-vt\cos(\psi_i-\frac{3\pi}{2})-(x_0-vt\cos(\psi-\frac{3\pi}{2})i)\right]^2\\+\\ \left[y_0+vt\sin(\psi-270^{\circ})(i-1)+vt\sin(\psi_i-\frac{3\pi}{2})-(y_0+vt\sin(\psi-\frac{3\pi}{2})i)\right]^2\end{matrix}}+\mu,270^{\circ}<\psi<36\end{cases} \quad (8)$$

Step 4: Collision avoidance decision. In the dynamic game with complete information, the reverse solution from the final decision position is the most effective method to solve Nash equilibrium [22]. In order to facilitate understanding, the following complete information dynamic game is taken as an example to analyze.

Suppose there are two ships No. 1 and No. 2, in which ship No. 1 can choose an action a_1 from the action set A_1 and ship No. 2 can choose an action a_2 from the action set A_2. At the same time, $U_1(a_1, a_2)$ and $U_2(a_1, a_2)$ represent the value of the ship's profit of No. 1 and the ship's profit of No. 2, respectively. Based on the principle of the inverse solution method, it is assumed that ship No.1 in this example makes an action decision first, so the analysis starts from ship No. 2. Assuming that ship No. 1 is selected from the action set first a_1, then ship No. 2 needs to choose an action from its own action set that is the most profitable for itself in the environment affected by the decision of ship No. 1. Therefore, ship No. 2 faces that the decision problem is denoted as $\max U_2(a_1, a_2), a_2 \in A_2, \forall a_1 \in A_1$, the optimal strategy made by ship No. 2 after ship No. 1 makes the action decision is denoted by $F_2(a_1)$, and there is one and only one optimal strategy.

When inferring the decision made by ship No. 1 in the process of reverse solving, ship No. 1 predicts that ship No. 2 will take the next action according to its decision. Therefore, ship No. 1 only needs to arbitrarily find an action that can maximize its benefits in its own action set. So, the decision-making problem of ship No. 1 is written as follows. At this time, (a_1, a_2) represents the maximum value of ship No. 1 and ship No. 2 which are the best combination of actions.

In summary, the choice of ship collision avoidance strategy based on perfect information game is mainly divided into four steps. Firstly, the surrounding environment of the ship is checked during the voyage. Secondly, the occurrence of collision risk is judged in the encounter situation. Thirdly, priority action sequence is taken into account. Finally, the optimal strategy to play the game is calculated according to the action sequence. Specifically, the ship collision avoidance strategy and flow chart of the perfect information game are shown in Figure 5.

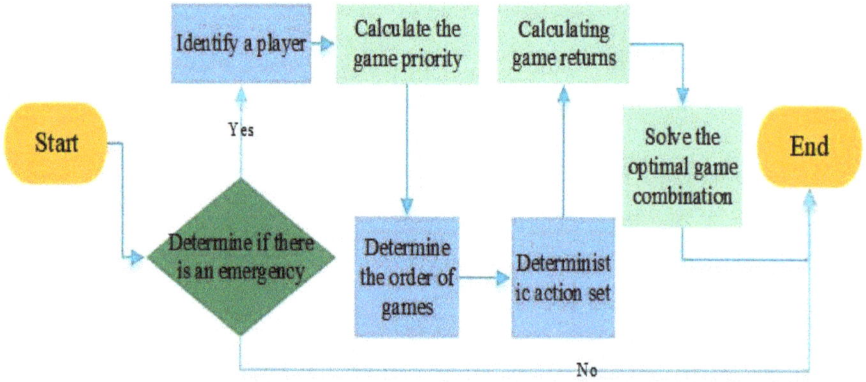

Figure 5. Flow chart of ship collision avoidance strategy in the perfect information game.

3.2. Expansion of Unmanned Ship Collision Avoidance Game Tree

The unmanned ship collision avoidance game model constructed in the previous section is the process of game tree expansion. The game tree designed in this study is a breadth search tree [23]. The nodes in the state space of the whole game tree can be divided into three categories: UNSEARCH nodes, OPEN node sets, and CLOSE node sets. Taking the game expansion tree with game round 3 as an example, node 1 is the head node, which contains the heading angle, offset, and collision risk of the unmanned ship in the current encounter situation. Node 1 is expanded to generate sub-nodes 2, 3, and 4. The three sub-nodes are respected in the new ship state formed by the combination of different actions taken by the ship in the situation. The aforementioned three nodes (including all the information in the new state) are initialized, listed in sequence after the head node, and pointed the parent pointer to node 1. After node 1 is expanded, the next node is sequentially expanded in the queue, namely node 2. Then, node 2 becomes the current node, and then it expands based on the space state of node 2. If the collision risk degree in the space state of node 3 is greater than 0.5, there is a possibility of collision risk if the node in this space state is expanded. So, node 3 is skipped and node 4 becomes the current node [24]. By analogy, until the end of the game round, the schematic diagram of the game algorithm is shown in Figure 6.

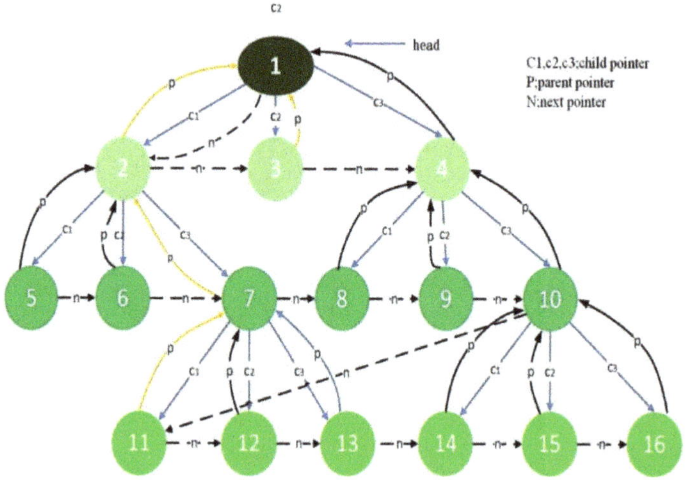

Figure 6. Game expansion tree of ship A and ship B.

The process of solving is to find the node with the largest profit in the last layer of nodes, that is, the node with the smallest sum of the offset of the two ships, in which the value of the collision risk of the node members must be less than 0.5. After finding the node with the greatest profit, it can follow its parent pointer for a reverse solution until the root node of the entire extended game tree is found, and the final optimal solution is the action combination information contained in the game strategy combination sequence node of the two ships.

3.3. Collision Risk Fuzzy Credibility Number

After the collision avoidance game, the ship collision risk can be determined by using the fuzzy credibility number of the ship collision risk [25]. There are many methods to calculate ship collision risk: fuzzy mathematical calculation method, BP neural network method, hazard mode immune control algorithm, bacterial foraging algorithm, and so on. The fuzzy mathematical method has high calculation accuracy. BP neural network method has strong self-learning ability, small calculation error, but high failure probability and long calculation time. Therefore, this study uses the fuzzy mathematics method to measure the ship collision risk.

In the introduced encounter situation, the judgment of whether there is a danger of collision between ships mainly depends on the distance to the closest point of approach D_{CPA}, the time to the closest point of approach value between the ships T_{CPA}, the ship speed ratio between the ships K, the distance between the ships D, the azimuth angle of the target ship relative to the own ship θ, and other related factors. In this study, the method of fuzzy mathematics is used to calculate the collision risk index (*CRI*) [26]. When $CRI = 0$, it means that there is no danger of collision between two ships. When $CRI = 1$, it means that the collision cannot be avoided and $CRI = 1$. Let U_{DCPA}, U_{TCPA}, U_θ, U_D, U_K be the D_{CPA}, the T_{CPA}, the azimuth angle between two ships, D between the two ships, and the risk membership degree of the shipping speed ratio K, respectively, and its belong to $[0,1]$. Then, it gets:

$$
\begin{aligned}
CRI = \ & a\left\{\frac{1}{2} - \frac{1}{2}\sin\left[\frac{\pi}{d_2-d_1}\left(D_{CPA} - \frac{d_1+d_2}{2}\right)\right]\right\} \\
& +b\left[\left(\frac{t_2-|T_{CPA}|}{t_2-t_1}\right)^2\right] \\
& +c\left\{\frac{1}{2}\left[\cos(\theta - 19^\circ) + \sqrt{\frac{440}{289} + \cos^2(\theta - 19^\circ)}\right] - \frac{5}{17}\right\} \\
& +d\left[\left(\frac{H_1 \cdot H_2 \cdot 1.7\cos(\theta-19^\circ)+\sqrt{4.4+2.89\cos^2(\theta-19^\circ)}-D}{\left[H_1 \cdot H_2 \cdot 1.7\cos(\theta-19^\circ)+\sqrt{4.4+2.89\cos^2(\theta-19^\circ)}\right]-(H_1 \cdot H_2 \cdot DLA)}\right)^2\right] \\
& +\frac{e}{1+\frac{2}{K\sqrt{K^2+1+2K|\sin(|\psi_t-\psi_o|)|}}}.
\end{aligned} \tag{9}
$$

Among the d_1 and d_2 are the value of the safety field of the ship safety threshold and the safe passing distance of the ship, respectively. At the same time $a + b + c + d + e = 1$. Then, the aforementioned ship collision time t_1 and ship attention time t_2 are obtained as:

$$
t_1 = \begin{cases} \dfrac{\sqrt{D_1{}^2 - D_{CPA}{}^2}}{V_r}, D_{CPA} \le D_1, \\ \dfrac{D_1 - D_{CPA}}{V_r}, D_{CPA} > D_1, \end{cases} \tag{10}
$$

and:

$$
t_2 = \frac{\sqrt{12^2 - D_{CPA}{}^2}}{V_r}. \tag{11}
$$

It is noteworthy that, in Equations (10) and (11), D_1 represents the closest avoidance removal and D_2 represents the distance at which the approaching ship should take avoidance actions. V_r is defined as the velocity vector of the incoming ship relative to the present ship. Meanwhile, the schematic diagram of the latest avoidance distance D_1 is shown in Figure 7, where D_{LA} is defined as the latest distance to turn the rudder. Here, the value of D_{LA} is valued as 12 times the length of the ship for convenience [27]. Especially, on the conditions that $D_{CPA} \leq d_1$, $0 \leq |T_{CPA}| \leq t_1$ and $D \leq D_1$, the value of U_{DCPA}, U_D, U_{TCPA}, and CRI are all obtained as 1. In this situation, the ship is collided. Meanwhile, on the conditions that $d_2 < D_{CPA}$ and $D_2 \leq D$, it gets the value of U_{DCPA}, U_D, and U_{TCPA} which are all 0, which means there is no danger of collision between the two ships.

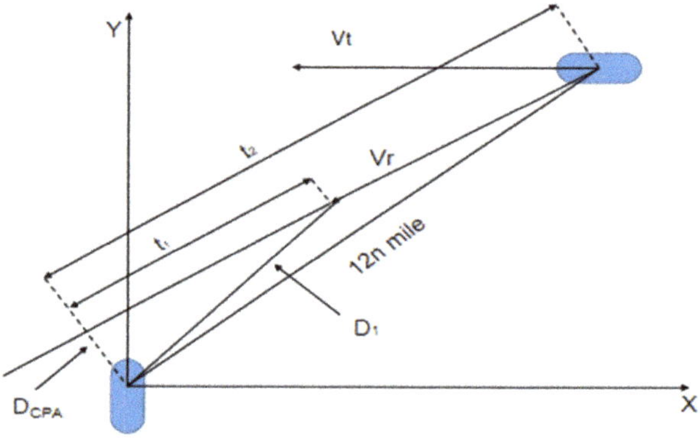

Figure 7. Geometric diagram of the latest avoidance distance.

4. Illustrative Example

To explain and verify the aforementioned extensive-form game model, an illustrative example is given as follows.

4.1. Problem Introduction

On 26 March 2019, the Xinde Maritime Network released news that on the 24th local time in the port of Fujairah, the United Arab Emirates, a tragic and incredible ship collision accident occurred. An exceptionally large tanker collided with another LNG carrier. The accident is a typical conflict scenario where the two ships sail down, as shown in Figure 8.

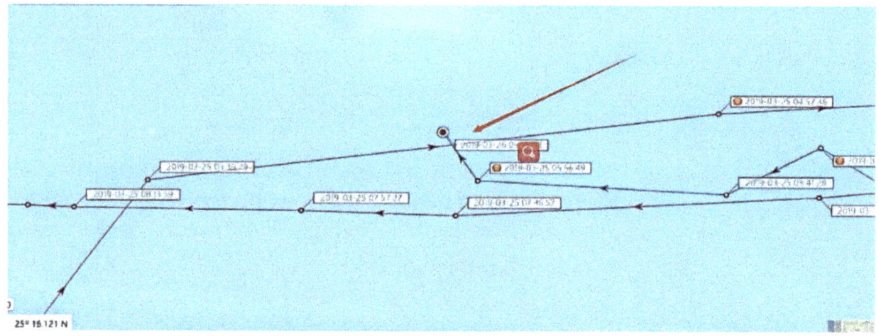

Figure 8. Course conflict scenario between two ships.

In such a situation where the two-sides are clamped, the ships can judge by the conflict of the routes during the encounter: the ships in the blue route are the right-give-way vessels, the purple-route vessels under the right-give-way vessels are the left-give-way vessels, and the vessels located in the right-give-way vessels are the left-give-way vessels. The pink route ships aforementioned are treated as environmental parameter variables in the whole game situation. In this collision avoidance game, the action set of the ship in the green route is $\{10^{\circ},20^{\circ},30^{\circ}\}$, the action set of the ship in the purple route is $\{-30^{\circ},-20^{\circ},-10^{\circ}\}$, and then the action combination of the two ships is $\{(10^{\circ},-30^{\circ}),(10^{\circ},-20^{\circ}),\ldots,(30^{\circ},-20^{\circ}),(30^{\circ},-10^{\circ})\}$. The next action combination will change, and the action set will change. Otherwise, the ship will be greatly offset, which is not in line with the benefits.

Consider the two-sides clamping scenario combine with the head-on situation, the target ships on two-sides of my ship approached at the $174^{\circ} \leq \Delta C \leq 186^{\circ}$ relative course of my ship. At this point, the ship is in the head-on sides of the two-sides clamping scenario. The schematic diagram of the head on scenario analysis is shown in Figure 9.

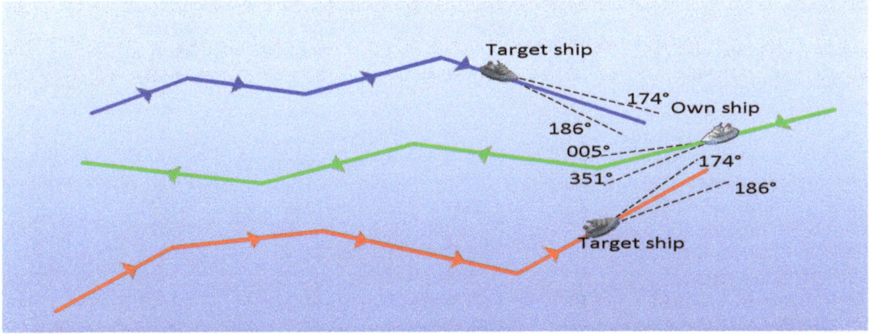

Figure 9. Head-on situation of two-sides clipping scenario schematic diagram.

4.2. Simulation Process and Analysis

In this section, two ships, i.e., "Own Ship" and the "Target Ship" are taken into account in this simulation sample, where the ship length of "Own Ship" is 105 m, the maximum speed of "Own Ship" is 18 kn, and the gross tonnage of "Own Ship" is 6000 tons, whereas the ship length of "Target Ship" is 139.8 m, the maximum speed of "Target Ship" is 13.5 kn, and the gross tonnage of "Target Ship "is 6000 tons.

For convenience, the game round is valued as 3. According to the relevant parameters of the two ships, the position of them is initialized. According to the 1972 International Collision Avoidance Regulations, "the two ships should each take a right turn to avoid collision" in encounter situation, which makes ship A as its own ship. In this case, the relevant parameter variables are obtained as in Table 1. By using Equation (9), the original collision risk between the two ships is 0.5911. Then, each ship starts to make a collision avoidance decision at 0 s [28]. In the first round, the own ship takes a 10° right turn to avoid collision. The target ship takes a 20° right turn to avoid collision. At the time node of 300 s in the second round, the own ship takes a 10° right turn to avoid collision, the target ship takes a 20° left turn to avoid collision, and the collision risk is 0.4635. At the time node of 600 s in the third round, the own ship takes a 10° left turn to avoid collision, the target ship takes a 10° left turn to avoid collision, and the collision risk is 0.4329. In the third round, because all ships completed the collision avoidance operation and there is no risk of subsequent collision, the course is readjusted, and the original course is restored. The simulation results of the confrontation situation based on the aforementioned are shown in Table 2. The optimal collision avoidance sequence combination composed of the obtained sub-game Nash equilibrium is $\{(10^{\circ},20^{\circ}),(0^{\circ},0^{\circ}),(0^{\circ},0^{\circ})\}$. All collision avoidance behaviors are consistent with COLREGS.

Table 1. Related parameter variables.

Ship Parameters							
V_r	24 kn	t_2	1800 s	D	6 n mile	U_θ	0.9558
φ_r	180°	D_{CPA}	0 n mile	D_1	0.9057 n mile	U_D	0
d_1	1.12 n mile	T_{CPA}	900 s	D_2	4.278 n mile	U_K	0.4143
d_2	2.21 n mile	U_{DCPA}	1	t_1	135.874 s	CRI	0.5912
θ	0°	U_{TCPA}	0.2926				

Table 2. Simulation results of encounter situation.

Time/s	Vessel	Decision	Course Angle	The Abscissa (Nautical Miles)	Y-Coordinate (Nautical Miles)	Offset (Nautical Miles)	Sum of Offset	Risk Collision Index
initial time	A		0°	5	1	0	0	0.5911
	B		180°	5	7	0		
0	A	Turn right 10 °	10°	5	1	0	0	0.5911
	B	Turn right 20°	200°	5	7	0		
300	A	Turn right 10°	20°	5.1736	1.9848	0.1743	0.5216	0.4635
	B	Turn left 10°	190°	4.658	6.0603	0.3473		
600	A	Turn left 10°	10°	5.5157	2.9245	0.5212	1.0423	0.4329
	B	Turn left 10°	190°	4.4843	5.755	0.5212		
900	A	Restore the course	10°	5.6893	3.9093	0.6953	1.2164	0.4803
	B	Restore the course	180°	4.4843	4.0755	0.5212		

5. Conclusions

This study proposes a decision-making problem on the collision avoidance of unmanned ships at sea in the situation of two-sides clamping. This study introduces the decision process of collision avoidance of unmanned ships at sea based on the extensive-form game model and verifies the effectiveness of collision avoidance by using fuzzy credibility numbers. Specifically, the main innovations of this study are concluded as follows.

Firstly, this study proposes a two-sides clamping intelligent collision avoidance strategy for unmanned ships. This strategy can provide real-time collision avoidance measures for unmanned ships at sea. The example analysis shows that this strategy can effectively improve the efficiency of collision avoidance of unmanned ships.

Secondly, the simulation experiment is carried out with the navigation simulator to realize the ship's extended game collision avoidance decision-making system. The simulation of two unmanned ships is carried out in the situation where two unmanned ships in the case of clamping on two-sides. Aiming at the intelligent collision avoidance problem of unmanned ships in the situation of two-sides, this study establishes a dynamic collision avoidance game model for ships based on the extensive-form game model. The unmanned ship can make the optimal collision avoidance action in the situation of being clamped on two sides.

Thirdly, a novel collision risk fuzzy credibility number is used to calculate the ship collision risk at the comprehensive fuzzy assessment based on the same time. The evaluation indicators include D_{CPA}, T_{CPA}, the distance between the two ships, the relative orientation between the two ships, the speed ratio of the two ships, and other factors.

Moreover, by using fuzzy credibility numbers, the decision-making efficiency on collision avoidance of ships in uncertain environments can be improved. In future work, the fuzzy credibility numbers can be considered in more decision-making situations in shipping management. Furthermore, the Fermatean fuzzy sets [29] is applied to the collision avoidance process of unmanned ships. Considering the multiple fuzzy factors that affect the collision avoidance of unmanned ships at sea, this study combines Fermatean fuzzy sets and links it with extensive-form game to provide support for the intelligent collision avoidance of unmanned ships at sea.

Author Contributions: Conceptualization, F.Z.; Data curation, M.L.; Formal analysis, R.W.; Investigation, Y.C.; Writing—original draft, H.C. All authors have read and agreed to the published version of the manuscript.

Funding: The work is partially supported by Shanghai Pujiang Program (2019PJC062), the Natural Science Foundation of Shandong Province (ZR2021MG003), the Research Project on Undergraduate Teaching Reform of Higher Education in Shandong Province (No. Z2021046), the National Natural Science Foundation of China (51508319), and the Nature and Science Fund from Zhejiang Province Ministry of Education (Y201327642).

Institutional Review Board Statement: Not applicable.

Informed Consent Statement: Not applicable.

Data Availability Statement: Not applicable.

Conflicts of Interest: The authors declare no conflict of interest.

References

1. Dinh, G.H.; Im, N.K. The combination of analytical and statistical method to define polygonal ship domain and reflect human experiences in estimating dangerous area. *Int. J. e-Navig. Marit. Econ.* **2016**, *4*, 97–108. [CrossRef]
2. Zhao, Y.; Li, W.; Shi, P. A real-time collision avoidance learning system for Unmanned Surface Vessels. *Neurocomputing* **2016**, *182*, 255–266. [CrossRef]
3. Li, J.; Wang, H.; Zhao, W.; Xue, Y. Ship's trajectory planning based on improved multiobjective algorithm for collision avoidance. *J. Adv. Transp.* **2019**, *2019*, 4068783. [CrossRef]
4. Lee, Y.I.; Kim, S.G.; Kim, Y.G. Fuzzy relational product for collision avoidance of autonomous ships. *Intell. Autom. Soft Comput.* **2015**, *21*, 21–38. [CrossRef]
5. Wang, C.; Zhang, X.; Cong, L.; Li, J.; Zhang, J. Research on intelligent collision avoidance decision-making of unmanned ship in unknown environments. *Evol. Syst. Ger.* **2019**, *10*, 649–658. [CrossRef]
6. Xing, S.; Xie, H.; Zhang, W. A method for unmanned vessel autonomous collision avoidance based on model predictive control. *Syst. Sci. Control. Eng.* **2022**, *10*, 255–263. [CrossRef]
7. Lisowski, J. The safety of marine navigation based on a game theory. *Risk Anal.* **2014**, *47*, 467–478.
8. Lisowski, J. Analysis of methods of determining the safe ship trajectory. *TransNav* **2016**, *10*, 2. [CrossRef]
9. Zou, Y.; Zhang, Y.; Ma, Z. Emergency Situation Safety Evaluation of Marine Ship Collision Accident Based on Extension Cloud Model. *J. Mar. Sci. Eng.* **2021**, *9*, 1370. [CrossRef]
10. Tu, Y.; Xiong, Y.; Mou, J. Decision-making method for multi-ship collision avoidance based on improved extensive game model. *Marit. Technol. Res.* **2022**, *4*, 4. [CrossRef]
11. Ran, P.; Li, W.; Bao, R.; Ma, P. A simulation credibility assessment method based on improved fuzzy comprehensive evaluation. *J. Syst. Simul.* **2020**, *32*, 2469.
12. Ye, J.; Song, J.; Du, S.; Rui, Y. Weighted aggregation operators of fuzzy credibility numbers and their decision-making approach for slope design schemes. *Comput. Appl. Math.* **2021**, *40*, 155. [CrossRef]
13. Vercher, E.; Rubio, A.; Bermúdez, J.D. Fuzzy prediction intervals using credibility distributions. *Eng. Proc.* **2021**, *5*, 51.
14. Zhou, J.; Ding, F.; Yang, J. Navigation safety domain and collision risk index for decision support of collision avoidance of USVs. *Int. J. Nav. Arch. Ocean.* **2021**, *13*, 340–350. [CrossRef]
15. Zhang, J.; Zhang, D.; Yan, X. A distributed anti-collision decision support formulation in multi-ship encounter situations under COLREGs. *Ocean. Eng.* **2015**, *105*, 336–348.
16. Liu, R.W.; Liang, M.; Nie, J. Deep learning-powered vessel trajectory prediction for improving smart traffic services in maritime Internet of Things. *IEEE Trans. Netw. Sci. Eng.* **2022**, *9*, 3080–3094. [CrossRef]
17. Li, Q.; Fan, H.S.L. A simulation model for detecting vessel conflicts within a seaport. *Trans. Nav.* **2012**, *6*, 11–17.
18. Li, G.; Yang, Y.; Zhang, T.; Qu, X.; Cao, D.; Cheng, B.; Li, K. Risk assessment based collision avoidance decision-making for autonomous vehicles in multi-scenarios. *Transp. Res. Part C Emerg. Technol.* **2021**, *122*, 102820. [CrossRef]
19. De, R.; Mahapatra, R.P.; Chakraborty, P.S. Exploring and Expanding the World of Artificial Intelligence. *Int. J. Comput. Appl.* **2016**, *975*, 8887. [CrossRef]
20. Bueno, L.F.; Haeser, G.; Rojas, F.N. Optimality conditions and constraint qualifications for generalized Nash equilibrium problems and their practical implications. *SIAM J. Optimiz.* **2019**, *29*, 31–54. [CrossRef]
21. Etessami, K. The complexity of computing a (quasi-) perfect equilibrium for an n-player extensive form game. *Game Econ. Behav.* **2021**, *125*, 107–140. [CrossRef]
22. Fei, L.; Feng, Y. A dynamic framework of multi-attribute decision making under Pythagorean fuzzy environment by using Dempster–Shafer theory. *Eng. Appl. Artif. Intell.* **2021**, *101*, 104213. [CrossRef]
23. Zilu, O.; Hongdong, W.; Jianyao, W. Automatic collision avoidance algorithm for unmanned surface vessel based on improved Bi-RRT algorithm. *Chin. J. Ship Res.* **2019**, *14*, 9.

24. Kishimoto, A.; Winands, M.H.M.; Müller, M. Game-tree search using proof numbers: The first twenty years. *ICGA J.* **2012**, *35*, 131–156. [CrossRef]
25. Goerlandt, F.; Kujala, P. On the reliability and validity of ship–ship collision risk analysis in light of different perspectives on risk. *Saf. Sci.* **2014**, *62*, 348–365. [CrossRef]
26. Hu, Y.; Zhang, A.; Tian, W. Multi-ship collision avoidance decision-making based on a collision risk index. *J. Mar. Sci. Eng.* **2020**, *8*, 640. [CrossRef]
27. Li, S.; Zheng, Z.; Mi, J. The latest minute action of ship. *ICTETS SPIE* **2021**, *12058*, 644–649.
28. Liu, H.; Fang, Z.; Li, R. Credibility-based chance-constrained multimode resource-constrained project scheduling problem under fuzzy uncertainty. *Comput. Ind. Eng.* **2022**, *171*, 108402. [CrossRef]
29. Senapati, T.; Yager, R.R. Fermatean fuzzy sets. *J. Amb. Intel. Hum. Comp.* **2020**, *11*, 663–674. [CrossRef]

mathematics

Article

Study on Chaotic Multi-Attribute Group Decision Making Based on Weighted Neutrosophic Fuzzy Soft Rough Sets

Fu Zhang [1,2] and Weimin Ma [1,*]

1 School of Economics and Management, Tongji University, Shanghai 200092, China
2 Business School, Shanghaidianji University, Shanghai 201306, China
* Correspondence: mawm@tongji.edu.cn

Abstract: In this article, we have proposed a multi-attribute group decision making (MAGDM) with a new scenario or new condition named Chaotic MAGDM, in which not only the weights of the decision makers (DMs) and the weights of the decision attributes are considered, but also the familiarity of the DMs with the attributes are considered. Then we applied the weighted neutrosophic fuzzy soft rough set theory to Chaotic MAGDM and proposed a new algorithm for MAGDM. Moreover, we provide a case study to demonstrate the application of the algorithm. Our contributions to the literature are as follows: (1) familiarity is rubbed into MAGDM for the first time in the context of neutrosophic fuzzy soft rough sets; (2) a new MAGDM model based on neutrosophic fuzzy soft rough sets has been designed; (3) a sorting/ranking algorithm based on a neutrosophic fuzzy soft rough set is constructed.

Keywords: multi-attribute group decision making; fuzzy soft rough sets; neutrosophic fuzzy soft rough sets

MSC: 90B50

Citation: Zhang, F.; Ma, W. Study on Chaotic Multi-Attribute Group Decision Making Based on Weighted Neutrosophic Fuzzy Soft Rough Sets. *Mathematics* **2023**, *11*, 1034. https://doi.org/10.3390/math11041034

Academic Editors: Yanhui Guo and Jun Ye

Received: 29 December 2022
Revised: 25 January 2023
Accepted: 14 February 2023
Published: 18 February 2023

1. Introduction

Multi-attribute decision making (MADM) is an important branch of modern decision theory and methodology with a wide range of practical contexts, such as human resource performance assessment, economic performance assessment, political election assessment, military performance assessment, etc. However, due to the limitations of human knowledge and the specialization of professions, as well as the diversity and complexity of real-world decision making, a single decision maker (DM) cannot make the optimal option. As a result, in most MADMs, decision makers (DMs) from diverse sectors, areas of expertise, or knowledge backgrounds are frequently required to collaborate in order to make more scientifically sound conclusions. That is, multi-attribute group decision making (MAGDM). In addition, there is a lot of uncertainty and ambiguity in practical MAGDM. Therefore, the study of MAGDM under fuzzy scenarios has become a popular research direction in recent years. Considering that different DMs have different professional backgrounds, areas of knowledge, expertise, etc. Therefore, in a MAGDM, how to engage DMs to evaluate the attributes of the alternatives in their areas of expertise and familiarity is an issue that must be considered in the decision making. In the existing literature or research results, there are two common approaches to address this concern. One is to assign weights to DMs, the other is to group DMs according to certain rules.

- Assigning weights to DMs.

Liu et al. [1] proposed a variable weighting approach by considering DMs and attributes weights together for MAGDM problems under interval-valued intuitionistic fuzzy sets. Yu et al. [2] developed a novel consensus reaching process for MAGDM based on hesitant fuzzy linguistic term sets (HFLTSs) which not only can deal with multi-granular

HFLTSs, but also considers the weight vectors of DMs and attributes in the proposed consensus model. Liu et al. [1] presented a hybrid approach based on variable weights for multi-attribute group decision making, and so on [3–12].

- Grouping DMs according to certain rules.

For example, Su et al. [13] proposed a MAGDM approach for evaluation and self-confidence in online learning platforms based on probabilistic linguistic term sets. Sun et al. [14] provided diverse fuzzy multi-branch rough sets based on binary relations for MAGDM. Sun et al. [15] analyzed the diversified MAGDM problem with the personal preference parameters, etc. [16–19].

2. Comparison and Motivation

To date, research on MAGDM in fuzzy scenarios has produced a very large number of theoretical and practical results. All these studies have investigated the relationship between DM and attributes from different perspectives either by assigning weights or by grouping; however, the following issues still need to be further explored.

- Most of the existing studies are empowered weight according to the level of DMs. In other words, DMs with high weight will have a higher proportion of the evaluation of all attributes, even if it is an attribute that he or she is not familiar with.
- The scope of DMs' expertise is ambiguous, so in practice it is difficult to group DMs according to their expertise only. For example, if each DM only examines attributes that relate to their own expertise, who should evaluate these combined attributes that encompass a variety of expertise?
- Grouping DMs based only on their professional background would inevitably lose many very valuable evaluations. Because most DMs have a great deal of practical experience, they are well-placed to evaluate attributes even in non-specialist areas.
- Most studies did not consider the relationship between DMs and decision attributes. That is, the familiarity of DMs with the decision attributes.

The following situations are often encountered in decision making. For example, in a large-scale fire rescue, an important attribute of the rescue plan is the ability to quickly rescue trapped people. This attribute is not only related to the cause of the fire, but also to the structure of the building, construction materials, etc. Therefore, both fire experts and building experts should be important evaluators of this attribute. Hence, it is obviously unreasonable to have only one of the expert group (e.g., the fire expert group) to evaluate it. In other words, simply grouping experts would result in a lack of many valuable evaluations. Furthermore, fires are related to weather conditions, and a meteorologist is a very well-known expert whose evaluation of this property will be less important if he or she is not familiar with the fire scene. On the contrary, the person who built the building, even if he is just a worker, will have a very important evaluation because he is more familiar with the structure of the building at the fire site.

This suggests that the weights of DM's evaluations are related to familiarity with the attributes, rather than just considering the weights of the DM. In other words, in MAGDM, not only the weight of the DM and the weight of the attribute are considered, but also the familiarity of the DM with the attributes.

There will be a lot of uncertainty and fuzziness in the actual MAGDM problem, and also the familiarity of the DM with the decision attributes will change in different scenarios. Therefore, in the absence of an explicit method to determine the familiarity between DM and decision attributes, using fuzzy theory to describe the familiarity between DM and attributes is a good choice. In order to allow DMs to focus their attention on evaluation scoring without considering the limitations of scoring values, we chose NFN (Neutrosophic Fuzzy Number) for evaluation scoring.

In summary, we have proposed a MAGDM with a new scenario or new condition in which not only the weights of the DMs and the weights of the decision attributes are considered, but also the familiarity of the DMs with the attributes needs to be considered.

That is Chaotic MAGDM (CMAGDM) which was proposed by Zhang et al. [20]. Then we applied neutrosophic fuzzy soft rough set theory to CMAGDM and proposed a new approach for CMAGDM. Our contributions are mainly as follows.

- Familiarity is rubbed into MAGDM for the first time in the context of neutrosophic fuzzy soft rough sets;
- A new Chaotic MAGDM model based on neutrosophic fuzzy soft rough sets has been designed;
- A sorting/ranking algorithm based on a neutrosophic fuzzy soft rough set is constructed.

The remainder of this paper is structured as follows: Section 3 briefly introduces the basic concepts and framework of MAGDM, provides a brief overview of fuzzy theory and several concepts in fuzzy theory. In Section 4, combining the neutrosophic fuzzy soft rough set and CMAGDM, we provide a new Chaotic MAGDM model based on neutrosophic fuzzy soft rough sets. A case study and the numerical analysis of the proposed model is illustrated in Section 5. At last, conclusions are given in Section 6.

3. Theoretical Background

In this section, first, we will review the basic concepts and framework of MAGDM. Second, we will provide a brief overview of fuzzy theory. Finally, we will review several important concepts in fuzzy theory, as well as their basic rules and properties.

3.1. The Basic Concepts and Framework of MAGDM

The problem of selecting the best answer from a list of potential solutions, based on a set of attributes or criteria, can be summarized as a decision problem. In real-world decision making, there will always be a group of DMs. The corresponding MADM translates into multi-attribute group decision making (MAGDM). The problem of MAGDM is represented by the following notation [20]:

- $A = \{a_1, a_2, \dots, a_m\}$ is the set of considered alternatives;
- $C = \{c_1, c_2, \dots, c_n\}$ is the set of attribute which are used for evaluating of alternatives;
- $E = \{e_1, e_2, \dots, e_l\}$ is the set of the DMs involved in the decision process;
- $w = (w_1, w_2, \cdots, w_n)^T$ is the weight vector of the attribute ($w \geqslant 0, \sum_{j=1}^{n} w_j = 1$);
- $\tau = (\tau_1, \tau_2, \cdots, \tau_l)^T$ is the weight vector of the DMs ($\tau \geqslant 0, \sum_{k=1}^{l} \tau_k = 1$);
- x_{ij}^k is the kth DM evaluation of the ith alternative against to jth attribute ($k = 1, 2, \cdots, l, i = 1, 2, \cdots, m, j = 1, 2, \cdots, n$);
- $X = \{X(e_k) | e_k \in E\}$ is the decision matrix set, and $X(e_k)$ is the decision matrix of the kth DM:

$$X(e_k) = \begin{array}{c} \\ a_1 \\ a_2 \\ \vdots \\ a_m \end{array} \begin{array}{c} \begin{array}{cccc} c_1 & c_2 & \cdots & c_n \end{array} \\ \begin{pmatrix} x_{11}^k & x_{12}^k & \cdots & x_{1n}^k \\ x_{21}^k & x_{22}^k & \cdots & x_{2n}^k \\ \vdots & \vdots & \ddots & \vdots \\ x_{m1}^k & x_{m2}^k & \cdots & x_{mn}^k \end{pmatrix} \end{array} \quad k = 1, 2, \cdots, l. \tag{1}$$

In order to describe MAGDM more clearly and concisely, a MAGDM problem can usually be represented by a sextuple $\langle A, C, E, w, \tau, X \rangle$, that is:

$$MAGDM = \langle A, C, E, w, \tau, X \rangle. \tag{2}$$

3.2. A Brief Overview of Fuzzy Theory

In order to better combine uncertainty in MAGDM with fuzzy theory, we will make a brief review of the development of fuzzy set (FS).

Zadeh [21] (1965) first proposed fuzzy theory. It breaks through the limitations of classical set theory by introducing a membership function to represent uncertainty. Atanassov [22] (1983) suggested a generalization of fuzzy sets making the degrees of membership (μ) and non-membership (ν) intervene to describe the vinculation of an element to a set, and the sum of these degrees is less or equal to 1 ($\mu + \nu \leq 1$), that is, an intuitionistic fuzzy set (IFS). However, an IFS fails when the sum of these degrees is more then 1. So, Yager [23] (2016) developed the concept of q-rung orthopair fuzzy sets (q-ROFS) and considered an efficient method to explain the vagueness of MADM problems. In q-ROFS, the sum of two degrees can be more then 1, it just needs to satisfy the condition $\mu^q + \eta^q \leq 1, (q \geq 1)$. It is clear that, for $q = 1$, it is an IFS, it is a Pythagorean fuzzy set (Yager and Abbasov [24] 2013) (PyFS) if $q = 2$, and it is a Fermatean fuzzy set (Senapati and Yager [25] 2020) (FFS) if $q = 3$, thus, q-ROFS generalize the IFS, PyFS, FFS. When we face human opinions involving more types of answers: yes, abstain, no, or refusal. Voting can be a good example of such a situation as the human voters may be divided into four group of those who : vote for, abstain, vote against, or refuse to vote. In order to solve this problem, Cuong and Kreinovich [26,27] in 2013 introduced a new notion of picture fuzzy set (PFS), which are directly extensions of FS and of intuitionistic fuzzy set (IFS). In PFS, the following three dimensions are considered simultaneously: degree of positive membership (μ), degree of neutral membership (η), and degree of negative membership (ν), and satisfy the following condition $\mu + \eta + \nu \leq 1$. The structure of PFS is of great importance as it has the ability to deal with human opinion efficiently. It is observed that the constraint on PFS makes us unable to assign values by own choice. In simple words, one can say that the domain of PFS is restricted. The concept of spherical fuzzy set (SFS) and T-spherical fuzzy set (T-SFS) is introduced as a generalization of FS, IFS, and PFS by Mahmood et al. [28] in 2019. In T-SFS, the sum of three degrees can more then 1, instead, need to satisfy condition $\mu^t + \eta^t + \nu^t \leq 1, (t \geq 1)$. Obviously, when $t = 1$, T-SFS degenerates to PFS, and if $t = 2$, T-SFS is a SFS. Neutrosophic set (NS) was introduced by Smarandache [29]. In NS, there is a need to satisfy the condition $\mu_A(x) + \eta_A(x) + \nu_A(x) \leq 3$. Although there are many other fuzzy sets, such as fuzzy multi-set (FMS) [30], interval-valued fuzzy set (IVFS) [31], hesitant fuzzy set (HFS) [32], hybrid fuzzy set, and so on. These theories play a very important role in practice and theory. However, due to limited space and the focus of our article, we will not repeat them here.

According to the above review and analysis of FSs, we can clearly draw the relationship between different FSs, as shown in Figure 1. There are two ideas for the promotion of the FS, one is from the dimension of the variable, the other is from the domain of the variable.

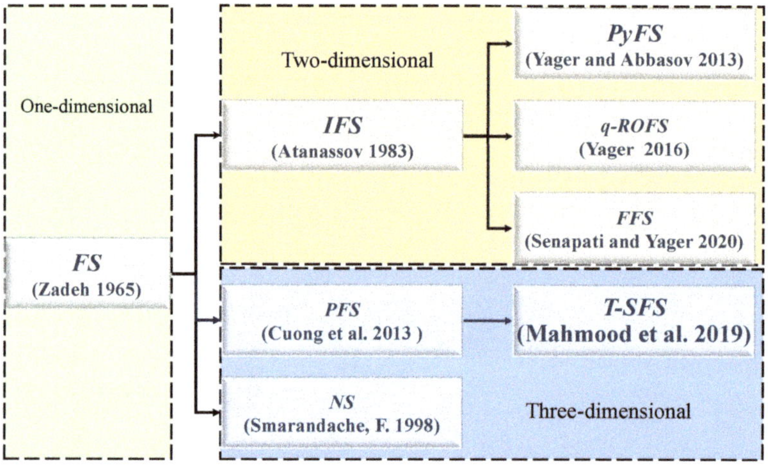

Figure 1. Dimension-Based Fuzzy Set Classification [21–26,28,29].

Researchers have expanded the fuzzy set from the perspective of variable dimension and its domain. These works not only played a very important role in the development of fuzzy set theory, but also played a very important role in practical applications. However, these ideas for generalizing fuzzy sets only consider the dimension and domain of fuzzy variables and do not consider the ambiguity of the attribute. Here we still use [26]'s voting example to illustrate our opinions. Suppose there are two candidates p_1 and p_2 participating in the campaign. A voter is very familiar with p_1, but only knows about p_2 based on his campaign speech. Now, the voter evaluates the two candidates using the PFS method and receives the same score. Clearly, it is unreasonable to consider these two evaluations as identical due to the difference in familiarity with the candidates. Phenomena such as the above will be frequently encountered in MAGDM problems. Fortunately, in addition to fuzzy sets, there are soft sets [33] and rough sets [34] that can describe unclear and fuzzy relations. Especially, the combination of these uncertainty theories can describe more details in MAGDM. These include fuzzy soft set (FSS) [35], fuzzy soft rough set (FSRS), picture fuzzy soft rough set (PFSRF), spherical fuzzy soft rough set (SFSRS), T-spherical fuzzy soft rough set (T-SFSRS) [36] and so on.

3.3. Concepts of Fuzzy Set

Classes and sets in the traditional mathematical sense do not include things such as "the class of all real numbers which are significantly bigger than 1," "the class of attractive ladies," or "the class of tall men." However, it is still true that such loosely defined "classes" are crucial to human thought. In essence, rather than the existence of random variables, the source of imprecision is the absence of well specified criteria for class membership. Zadeh [21] explored a concept which may be of use in dealing with "classes" of the type cited above. That is the fuzzy set (FS), a "class" with a continuum of grades of membership. It is defined as follows:

Definition 1 (Zadeh [21])**.** *A fuzzy set A on a universe X is an object of the form*

$$A = \{(x, \mu_A(x)) | x \in X\}, \tag{3}$$

where $\mu_A(x) \in [0,1]$ is called the "degree of membership of x in A". The variable $\mu_A(x)$ is called a Fuzzy Number (FN).

Intuitionistic fuzzy set (IFS) was developed by Atanassov [22] and is suitable for situations in which there is uncertainty about the degree of membership of an element in a defined set: each element in an IFS has a membership degree and a nonmembership degree between 0 and 1 [37].

Definition 2 (Atanassov [22])**.** *A intuitionistic fuzzy set A on a universe X is an object of the form*

$$A = \{(x, \mu_A(x), \nu_A(x)) | x \in X\}, \tag{4}$$

where $\mu_A(x) \in [0,1]$ is called the "degree of membership of x in A", $\nu_A(x) \in [0,1]$ is called the "degree of non-membership of x in A", and where $\mu_A(x)$ and $\nu_A(x)$ satisfy the following condition:

$$\forall x \in X, \mu_A(x) + \nu_A(x) \leq 1.$$

The pair $(\mu_A(x),\ \nu_A(x))$ is called an Intuitionistic Fuzzy Number (IFN).

When we face human opinions involving more types of answers such as yes, abstain, and refusal. Cuong and Kreinovich [26] introduced the concept of picture fuzzy set (PFS), and Mahmood et al. [28] provided the concept of T-spherical fuzzy set (T-SFS), both of which are direct extensions of the fuzzy set (FS) and the intuitonistic fuzzy set (IFS).

Definition 3 (Cuong and Kreinovich [26])**.** *A picture fuzzy set A on a universe X is an object of the form*

$$A = \{(x, \mu_A(x), \eta_A(x), \nu_A(x)) | x \in X\} \tag{5}$$

where $\mu_A(x) \in [0,1]$ is called the "degree of positive membership of x in A", $\eta_A(x) \in [0,1]$ is called the "degree of neutral membership of x in A" and $v_A(x) \in [0,1]$ is called the "degree of negative membership of x in A", and where $\mu_A(x)$, $\eta_A(x)$ and $v_A(x)$ satisfy the following condition:

$$\forall x \in X, \quad \mu_A(x) + \eta_A(x) + v_A(x) \leq 1.$$

Then for $x \in X$, let $\pi_A(x) = 1 - (\mu_A(x) + \eta_A(x) + v_A(x))$, $\pi_A(x)$ could be called the "degree of refusal membership of x in A". Let $PFS(X)$ denote the set of all the picture fuzzy sets on a universe X. A triplet $(\mu_A(x), \eta_A(x), v_A(x))$ can be referred to as a Picture Fuzzy Number (PFN).

Definition 4 (Mahmood et al. [28]). *A T-spherical fuzzy set A on a universe X is an object of the form*

$$A = \{(x, \mu_A(x), \eta_A(x), v_A(x)) | x \in X\} \tag{6}$$

where $\mu_A(x) \in [0,1]$ is called the "degree of positive membership of x in A", $\eta_A(x) \in [0,1]$ is called the "degree of neutral membership of x in A", and $v_A(x) \in [0,1]$ is called the "degree of negative membership of x in A", and where $\mu_A(x)$, $\eta_A(x)$ and $v_A(x)$ satisfy the following condition:

$$\forall x \in X, \quad \mu^t_A(x) + \eta^t_A(x) + v^t_A(x) \leq 1.$$

Then for $x \in X$, let $\pi_A(x) = 1 - (\mu^t_A(x) + \eta^t_A(x) + v^t_A(x))$, $\pi_A(x)$ could be called the "degree of refusal membership of x in A". Let T-$SFS(X)$ denote the set of all the T-spherical fuzzy sets (T-SFS)on a universe X. A triplet $(\mu_A(x), \eta_A(x), v_A(x))$ can be identified as a spherical fuzzy number (T-SFN). If $t = 2$ the T-spherical fuzzy set is called spherical fuzzy set (SFS), and corresponding $SFS(X)$ denote the set of all the spherical fuzzy sets on a universe X. A triplet $(\mu_A(x), \eta_A(x), v_A(x))$ can be referred to as a spherical fuzzy number (SFN).

Smarandache [29] generalized intuitionistic fuzzy sets (IFSs) to neutrosophic sets (NSs). A neutrosophic set (NS) contains three parameters: truth membership function, indeterminacy membership function, and falsity membership function. Unlike the PFS and T-SFS, the NS has a broader definition domain, giving DMs more options for evaluating scores in MAGDM.

Definition 5 (Smarandache [29]). *A neutrosophic set A on a universe X is an object of the form*

$$A = \{(x, \mu_A(x), \eta_A(x), v_A(x)) | x \in X\} \tag{7}$$

where $\mu_A(x) \in [0,1]$ is called the "truth membership function of x in A", $\eta_A(x) \in [0,1]$ is called the "indeterminacy membership function of x in A" and $v_A(x) \in [0,1]$ is called the "falsity membership function of x in A", and where $\mu_A(x)$, $\eta_A(x)$ and $v_A(x)$ satisfy the following condition:

$$\forall x \in X, \quad \mu_A(x) + \eta_A(x) + v_A(x) \leq 3.$$

A triplet $(\mu_A(x), \eta_A(x), v_A(x))$ can be referred to as a Neutrosophic Fuzzy Number (NFN).

By comparing the above concepts, it is easy to conclude that NS has a broader field of definition, thus allowing the DM to focus more on scoring the options without having to think too much about the constraints that need to be met for the evaluation scores. It is with this in mind that the Neutrosophic Fuzzy Number (NSN) will be chosen for scoring in this paper.

3.4. The Concept of Rough Set

Pawlak [34] introduced the concept of rough sets (RS) in 1982, which can handle uncertainty, imprecision, and ambiguity in sets. It is defined as follows.

Definition 6 (Pawlak [34]). *Let R be an equivalence relation on the universe X ($X \neq \varnothing$), (X, R) be a Pawlak approximation space. A subset $A \subseteq X$ is called definable if $\underline{R}(x) = \overline{R}(x)$; in the opposite case, i.e., if $\overline{R}(x) - \underline{R}(x) \neq \emptyset$, A is said to be a rough set, where the two operations are defined as:*

$$\underline{R}(x) = \{x \in X | [x]_R \subseteq A\} \tag{8}$$
$$\overline{R}(x) = \{x \in X | [x]_R \cap A \neq \varnothing\} \tag{9}$$

As an illustration, let us consider the following example (Example 1).

Example 1. *Table 1 is a information system of RS. The universe $X = \{a_i | i = 1, 2, \cdots, 8\}$ and $A = \{a_2, a_3, a_4, a_5, a_7\}$. Suppose the equivalence relation R is that the attributes c_2 and c_4 have the same value, then $[X]_R = \{\{a_1\}, \{a_2, a_4, a_6\}, \{a_3, a_7\}, \{a_5\}, \{a_8\}\}$. Thus, we can obtain the following results: $\underline{R}(A) = \{a_3, a_5, a_7\}, \overline{R}(A) = \{a_2, a_3, a_4, a_5, a_6, a_7\}$. Since $\overline{R}(A) - \underline{R}(A) \neq \emptyset$, therefore, A is a rough set.*

Table 1. A information system of rough set.

	c_1	c_2	c_3	c_4	c_5
a_1	0.50	0.75	0.25	0.25	0.50
a_2	0.50	0.25	0.25	0.75	0.50
a_3	0.75	0.25	0.25	0.50	0.25
a_4	0.25	0.25	0.50	0.75	0.50
a_5	0.75	0.50	0.25	0.75	0.50
a_6	0.25	0.25	0.50	0.75	0.50
a_7	0.75	0.25	0.25	0.50	0.25
a_8	0.25	0.50	0.50	0.50	0.50

3.5. The Concept of Soft Set

Molodtsov [33] proposed in 1999 a mathematical approach to dealing with uncertain information with the core idea of emphasizing the study of uncertainty and ambiguity of information from a parametric perspective, which is known as soft set (SS) theory. The concept of SS is as follows.

Definition 7 (Molodtsov [33]). *Let X be the universe. C is a set of parameters (attributes) about objects in X, (X, C) is called a soft space, and $\mathcal{C} \subseteq C$, φ is a mapping given by $\varphi : \mathcal{C} \to 2^X$; here 2^X is the power set of X, then a pair $(\varphi, \mathcal{C})$ is named a soft set (SS) over the universe X.*

To illustrate the point, let us consider the following example (Example 2).

Example 2. *Suppose Table 2 is a information system of soft set. $A = \{a_1, a_2, a_3, a_4\}$ is a universe of soft set, $C = \{c_1, c_2, c_3, c_4, c_5\}$ is the set of parameters.*

Table 2. A information system of soft set.

	c_1	c_2	c_3	c_4	c_5
a_1	1	0	1	0	0
a_2	0	1	1	1	1
a_3	1	0	1	1	0
a_4	0	1	0	1	1

According to the Definition of soft set, we could easily come to the results as follows:

$$\varphi(c_1) = \{a_1, a_3\}\ ;\ \varphi(c_2) = \{a_2, a_4\}\ ;\ \varphi(c_3) = \{a_1, a_2, a_3\};$$
$$\varphi(c_4) = \{a_2, a_3, a_4\}\ ;\ \varphi(c_5) = \{a_2, a_4\};$$

Take $\varphi(c_3)$ as an example, it means that the objects with attribute c_3 are a_1, a_2 and a_3.

3.6. The Concept of Fuzzy Soft Set

Maji et al. [35] combined the theory of fuzzy sets and soft sets in 2001 and proposed the definition of fuzzy soft sets. A fuzzy soft set can essentially be seen as a parametric fuzzy set for a given universe, which is a representation model that combines parameters and fuzzy information together. It is no longer restricted to the 0 and 1 values of the parameters in the soft set, but is a more flexible form of parameter selection that can be used in a wide range of uncertainty areas [15]. The concept is as follows.

Definition 8 (Maji et al. [35]). *Let X be a universal set, C be a collection of parameters regarding X, and $FS(X)$ represents the collection of all FSs over the universe X. A pair $(\varphi, \mathcal{C})$ is said to ba a Fuzzy Soft Set (FSS) over X, where $\mathcal{C} \subseteq C$ and $\varphi : \mathcal{C} \to FS(X)$, where $FS(X)$ represents the collection of all the fuzzy sets (FSs) on a universe X. For every $x \in X$, the FSS can be defined as follows:*

$$S = \{(x, \varphi(x)) | x \in \mathcal{C}, \varphi(x) \in FS(X)\} \quad (10)$$

Particularly, when $|\mathcal{C}| = 1$, the fuzzy soft set degenerates to a fuzzy set.

Specially, when the fuzzy set is PFS, the corresponding concept has the following form:

Definition 9 (Khan et al. [38]). *Let X be a universal set, C be a collection of parameters (attributes) regarding to X and $PFS(X)$ represents the collection of all picture fuzzy set over the universe X. A pair $(\varphi, \mathcal{C})$ is said to ba a Picture Fuzzy Soft Set (PFSS) over X, where $\mathcal{C} \subseteq C$ and $\varphi : \mathcal{C} \to PFS(X)$. For every $x \in X$, the PFSS can be defined as follows:*

$$S = \{(x, \varphi(x)) | x \in \mathcal{C}, \varphi(x) \in PFS(X)\} \quad (11)$$

Obviously, when $|\mathcal{C}| = 1$, the picture fuzzy soft set degenerates to a picture fuzzy set.

3.7. The Concept of Fuzzy Soft Rough Set

Combining fuzzy sets, soft sets, and rough sets can lead to a more flexible method of describing parameters, namely fuzzy soft rough sets. If the fuzzy set is a picture fuzzy set, the corresponding fuzzy soft rough set is called a picture fuzzy soft rough set and is defined as follows.

Definition 10 (Muahmmad and Martino [36]). *Let X be the universe. C be a set of parameters (attributes) about objects in X, $PFS\{X\}$ be the collection of all picture fuzzy soft sets over the universe X, $\mathbb{R}$ be a picture fuzzy soft set relation from universe X to $\mathcal{C}$ (That is $\forall c \in \mathcal{C} \subseteq C, \mathbb{R}(c) \in PFSS(X)$), and ψ be a mapping given by $\psi : \mathcal{C} \to PFSS\{X\}$. Then $(\psi, \mathcal{C}, \mathbb{R})$ is known as a Picture Fuzzy Soft Rough Approximation Space. For every $\mathcal{F} \in PFS(\mathcal{C})$, the lower and upper approximation of $\mathcal{F}$ can be defined as follows:*

$$\underline{\mathbb{R}}(\mathcal{F}) = \{(x, \underline{\mu}(x), \underline{\eta}(x), \underline{\nu}(x)) | x \in X\} \quad (12)$$

$$\overline{\mathbb{R}}(\mathcal{F}) = \{(x, \overline{\mu}(x), \overline{\eta}(x), \overline{\nu}(x)) | x \in X\} \quad (13)$$

where

$$\underline{\mu}(x) = \wedge_{c \in \mathcal{C}}(\mu_{\mathbb{R}}(x,c) \wedge \mu_{\mathcal{F}}(c)), \quad (14)$$

$$\underline{\eta}(x) = \vee_{c \in \mathcal{C}}(\eta_{\mathbb{R}}(x,c) \vee \eta_{\mathcal{F}}(c)), \quad (15)$$

$$\underline{\nu}(x) = \vee_{c \in \mathcal{C}}(\nu_{\mathbb{R}}(x,c) \vee \nu_{\mathcal{F}}(c)). \quad (16)$$

and

$$\overline{\mu}(x) = \vee_{c \in \mathcal{C}}(\mu_{\mathbb{R}}(x,c) \vee \mu_{\mathcal{F}}(c)), \quad (17)$$

$$\overline{\eta}(x) = \wedge_{c \in \mathcal{C}}(\eta_{\mathbb{R}}(x,c) \wedge \eta_{\mathcal{F}}(c)), \quad (18)$$

$$\overline{\nu}(x) = \wedge_{c \in \mathcal{C}}(\nu_{\mathbb{R}}(x,c) \wedge \nu_{\mathcal{F}}(c)). \quad (19)$$

Here, $0 \leq \underline{\mu}(x) + \underline{\eta}(x) + \underline{\nu}(x) \leq 1$, $0 \leq \overline{\mu}(x) + \overline{\eta}(x) + \overline{\nu}(x) \leq 1$
Then

$$\mathbb{R}(\mathcal{F}) = (\underline{\mathbb{R}}(\mathcal{F}), \overline{\mathbb{R}}(\mathcal{F})) = (x, (\underline{\mu}(x), \overline{\mu}(x)), (\underline{\eta}(x), \overline{\eta}(x)), (\underline{\nu}(x), \overline{\nu}(x))). \quad (20)$$

The score function can be defined as:

$$\mathcal{S}(\mathbb{R}(\mathcal{F})) = \underline{\mu}(x) + \overline{\mu}(x) - \underline{\eta}(x) - \overline{\eta}(x) - \underline{\nu}(x) - \overline{\nu}(x). \quad (21)$$

Example 3. *In a (MADM), suppose the alternatives set is $A = \{a_1, a_2, a_3\}$ and the attributes set is $C = \{c_1, c_2, c_3, c_4\}$, as presented in Table 3. Let $\mathcal{F}$ be a PFS over C as follows.*

$$\mathcal{F} = \{(c_1, 0.30, 0.30, 0.20), (c_2, 0.50, 0.30, 0.10), (c_3, 0.70, 0.20, 0.10), (c_4, 0.20, 0.60, 0.10)\}$$

Table 3. A information system of PFS.

	c_1	c_2	c_3	c_4
	(0.30, 0.30, 0.20)	(0.50, 0.30, 0.10)	(0.70, 0.20, 0.10)	(0.20, 0.60, 0.10)
a_1	(0.30, 0.20, 0.40)	(0.10, 0.20, 0.50)	(0.10, 0.40, 0.20)	(0.20, 0.50, 0.10)
a_2	(0.50, 0.20, 0.10)	(0.80, 0.10, 0.00)	(0.30, 0.40, 0.20)	(0.20, 0.60, 0.10)
a_3	(0.60, 0.10, 0.20)	(0.50, 0.10, 0.30)	(0.50, 0.20, 0.20)	(0.40, 0.00, 0.10)

Then, we can calculate the corresponding lower and upper approximation of $\mathcal{F}$ as follows.

$$\underline{\mathbb{R}}(\mathcal{F}) = \{(a_1, 0.10, 0.60, 0.50), (a_2, 0.20, 0.60, 0.20), (a_3, 0.20, 0.60, 0.30)\};$$

$$\overline{\mathbb{R}}(\mathcal{F}) = \{(a_1, 0.70, 0.20, 0.10), (a_2, 0.80, 0.10, 0.00), (a_3, 0.70, 0.00, 0.10)\}.$$

Finally, according to Equation (21) we obtain:

$$\mathcal{S}(a_1) = -0.6, \mathcal{S}(a_2) = 0.1, \mathcal{S}(a_3) = -0.1.$$

Obviously, $\mathcal{S}(a_2) > \mathcal{S}(a_3) > \mathcal{S}(a_1)$. Therefore, it follows that: $a_2 \succ a_3 \succ a_1$.

4. A Novel Neutrosophic FSRS-Based Method for Chaotic MAGDM

4.1. Chaotic Multi-Attribute Group Decision Making

Zhang et al. [20] proposed the concept of Chaotic MAGDM, in which not only the weights of DMs and decision attributes are considered, but also the familiarity of DMs with the decision attributes. With the crossover factor of familiarity, Chaotic MAGDM is brought closer to the real decision problem. The relevant concepts are as follows.

Definition 11 (Zhang et al. [20])**.** *A MAGDM is called Chaotic MAGDM if there exists at least one decision attribute such that at least two DMs have the different familiarity with it.*

For convenience, the symbols of the variables used in Chaotic MAGDM are summarized as follows: [20]:

- $A = \{a_1, a_2, \ldots, a_m\}$ is the set of considered alternatives;
- $C = \{c_1, c_2, \ldots, c_n\}$ is the set of attribute which are used for evaluating of alternatives;
- $E = \{e_1, e_2, \ldots, e_l\}$ is the set of the DMs involved in the decision process;
- $w = (w_1, w_2, \cdots, w_n)^T$ is the weight vector of the attribute ($w \geqslant 0, \sum_{j=1}^{n} w_j = 1$);
- $\tau = (\tau_1, \tau_2, \cdots, \tau_l)^T$ is the weight vector of the DMs ($\tau \geqslant 0, \sum_{k=1}^{l} \tau_k = 1$);
- x_{ij}^k is the kth DM evaluation of the ith alternative against to jth attribute ($k = 1, 2, \cdots, l; i = 1, 2, \cdots, m; j = 1, 2, \cdots, n$);
- $X = \{X^k | k = 1, 2, \cdots, l\}$ is the decision matrix set, and X^k is the decision matrix of the kth DM. As shown in Equation (1);
- f_j^k is the familiarity of the kth DM against to the jth attribute ($k = 1, 2, \cdots, l; j = 1, 2, \cdots, n$);
- $\mathcal{F} = \{\mathcal{F}_1, \mathcal{F}_2, \cdots, \mathcal{F}_l\}$ is the vector of the familiarity of DMs with attributes, where $\mathcal{F}_k$ is the familiarity of kth DM.

Since we consider the relationship between DMs and decision attributes in Chaotic MAGDM, we added the familiarity variable $\mathcal{F}$ to the MAGDM. That is using septuple $\langle A, C, E, w, \tau, X, \mathcal{F} \rangle$ to represent the Chaotic MAGDM. As shown in Equation (22)

$$ChaoticMAGDM = \langle A, C, E, w, \tau, X, \mathcal{F} \rangle. \tag{22}$$

Clearly, the diversified multi-attribute group decision making proposed by Sun et al. [14] is a special case of chaotic MAGDM. One of the core ideas of diversified MAGDM is that by establishing a pluralistic binary fuzzy relationship between the set of evaluation attribute indicators and different decision makers.

In order to describe a Chaotic MAGDM more visually, it can be represented by a information form. As shown in Table 4.

Table 4. The Chaotic MAGDM Information Form.

		$c_1(w_1)$	$c_2(w_2)$	$\cdots$	$c_n(w_n)$
$e_1(\tau_1)$		f_1^1	f_2^1	$\cdots$	f_n^1
	a_1	x_{11}^1	x_{12}^1	$\cdots$	x_{1n}^1
	a_2	x_{21}^1	x_{22}^1	$\cdots$	x_{2n}^1
	$\vdots$	$\vdots$	$\vdots$	$\ddots$	$\vdots$
	a_m	x_{m1}^1	x_{m2}^1	$\cdots$	x_{mn}^1
	$\vdots$	$\vdots$	$\vdots$	$\vdots$	$\vdots$
$e_i(\tau_i)$		f_1^i	f_2^i	$\cdots$	f_n^i
	a_1	x_{11}^i	x_{12}^i	$\cdots$	x_{1n}^i
	a_2	x_{21}^i	x_{22}^i	$\cdots$	x_{2n}^i
	$\vdots$	$\vdots$	$\vdots$	$\ddots$	$\vdots$
	a_m	x_{m1}^i	x_{m2}^i	$\cdots$	x_{mn}^i
	$\vdots$	$\vdots$	$\vdots$	$\vdots$	$\vdots$
$e_l(\tau_l)$		f_1^l	f_2^l	$\cdots$	f_n^l
	a_1	x_{11}^l	x_{12}^l	$\cdots$	x_{1n}^l
	a_2	x_{21}^l	x_{22}^l	$\cdots$	x_{2n}^l
	$\vdots$	$\vdots$	$\vdots$	$\ddots$	$\vdots$
	a_m	x_{m1}^l	x_{m2}^l	$\cdots$	x_{mn}^l

4.2. Weighted Neutrosophic Fuzzy Soft Rough Sets

In the Definition 10 proposed by Muahmmad and Martino. If, for example, in MADM, $\mathcal{F}$ denotes the familiarity of the DMs against to the attributes rather then the values of evaluation, then the defining functions of the upper and lower bounds of $\mathbb{R}(\mathcal{F})$ must be changed accordingly. So, we give the new definition as follows:

Definition 12. *Let X be the universe. C be a set of parameters (attributes) about objects in X, $NFS\{X\}$ be the collection of all neutrosophic fuzzy soft sets over the universe X, $\mathbb{R}$ be a neutrosophic fuzzy soft set relation from universe X to $\mathcal{C}$ (That is $\forall c \in \mathcal{C} \subseteq C, \mathbb{R}(c) \in NFSS(X)$), ψ be a mapping given by $\psi : \mathcal{C} \to NFSS\{X\}$. Then $(\psi, \mathcal{C}, \mathbb{R})$ is known as a Neutrosophic Fuzzy Soft Rough Approximation Space. For every $\mathcal{F} \in NFS(\mathcal{C})$, the lower and upper approximation of $\mathcal{F}$ can be defined as follows:*

$$\underline{\mathbb{R}}(\mathcal{F}) = \{(x, \underline{\mu}(x), \underline{\eta}(x), \underline{\nu}(x)) | x \in X\} \tag{23}$$

$$\overline{\mathbb{R}}(\mathcal{F}) = \{(x, \overline{\mu}(x), \overline{\eta}(x), \overline{\nu}(x)) | x \in X\} \tag{24}$$

where

$$\underline{\mu}(x) = \min_{c \in \mathcal{C}}(\mu_{\mathbb{R}}(x,c) \cdot \min(\mu_{\mathcal{F}}(c), (2 - \eta_{\mathcal{F}}(c) - \nu_{\mathcal{F}}(c)))), \tag{25}$$

$$\underline{\eta}(x) = \max_{c \in \mathcal{C}}(\eta_{\mathbb{R}}(x,c) \cdot \max(\eta_{\mathcal{F}}(c), (2 - \mu_{\mathcal{F}}(c) - \nu_{\mathcal{F}}(c)))), \tag{26}$$

$$\underline{\nu}(x) = \max_{c \in \mathcal{C}}(\nu_{\mathbb{R}}(x,c) \cdot \max(\nu_{\mathcal{F}}(c), (2 - \mu_{\mathcal{F}}(c) - \eta_{\mathcal{F}}(c)))). \tag{27}$$

and

$$\overline{\mu}(x) = \max_{c \in C}(\mu_{\mathbb{R}}(x,c) \cdot \max(\mu_{\mathcal{F}}(c), (2 - \eta_{\mathcal{F}}(c) - \nu_{\mathcal{F}}(c)))), \tag{28}$$

$$\overline{\eta}(x) = \min_{c \in C}(\eta_{\mathbb{R}}(x,c) \cdot \min(\eta_{\mathcal{F}}(c), (2 - \mu_{\mathcal{F}}(c) - \nu_{\mathcal{F}}(c)))), \tag{29}$$

$$\overline{\nu}(x) = \min_{c \in C}(\nu_{\mathbb{R}}(x,c) \cdot \min(\nu_{\mathcal{F}}(c), (2 - \mu_{\mathcal{F}}(c) - \eta_{\mathcal{F}}(c)))). \tag{30}$$

where, $0 \leq \underline{\mu}(x) + \underline{\eta}(x) + \underline{\nu}(x) \leq 3, 0 \leq \overline{\mu}(x) + \overline{\eta}(x) + \overline{\nu}(x) \leq 3$

Then,

$$\mathbb{R}(\mathcal{F}) = (\underline{\mathbb{R}}(\mathcal{F}), \overline{\mathbb{R}}(\mathcal{F})) \tag{31}$$

$$= (x, (\underline{\mu}(x), \overline{\mu}(x)), (\underline{\eta}(x), \overline{\eta}(x)), (\underline{\nu}(x), \overline{\nu}(x))) \tag{32}$$

The score function is as following:

$$\mathcal{S}(\mathbb{R}(\mathcal{F})) = \underline{\mu}(x) + \overline{\mu}(x) - \underline{\eta}(x) - \overline{\eta}(x) - \underline{\nu}(x) - \overline{\nu}(x). \tag{33}$$

Example 4. *Still analyzing the data in Example 3, under the new Definition 12, the corresponding results are:*

$$\underline{\mathbb{R}}(\mathcal{F}) = \{(a_1, 0.04, 0.85, 0.60), (a_2, 0.04, 1.02, 0.22), (a_3, 0.08, 0.24, 0.36)\};$$
$$\overline{\mathbb{R}}(\mathcal{F}) = \{(a_1, 0.45, 0.06, 0.01), (a_2, 1.28, 0.03, 0.00), (a_3, 0.09, 0.00, 0.01)\}.$$

Furthermore, $\mathcal{S}(a_1) = -1.03, \mathcal{S}(a_2) = 0.05, \mathcal{S}(a_3) = 0.37$, *then* $\mathcal{S}(a_3) > \mathcal{S}(a_2) > \mathcal{S}(a_1)$. *So, the final sorting is* $a_3 \succ a_2 \succ a_1$.

If different attributes c ($c \in C$) have different weights, then the neutrosophic fuzzy soft rough set will become a weighted neutrosophic fuzzy soft rough set.

Definition 13. *Let X be the universe. C be a set of parameters (attributes) with the weight w_c about objects in X, $NFS\{X\}$ be the collection of all neutrosophic fuzzy soft sets over the universe X, $\mathbb{R}$ be a neutrosophic fuzzy soft set relation from universe X to C (That is $\forall c \in \mathcal{C} \subseteq C, \mathbb{R}(c) \in NFSS(X)$), and ψ be a mapping given by $\psi : \mathcal{C} \to NFSS\{X\}$. Then $(\psi, C, \mathbb{R}, w_c)$ is known as a weighted neutrosophic Fuzzy Soft Rough Approximation Space. For every $\mathcal{F} \in NFS(\mathcal{C})$, the lower and upper approximation of $\mathcal{F}$ can be defined as follows:*

$$\underline{\mathbb{R}}(\mathcal{F}) = \{(x, \underline{\mu}(x), \underline{\eta}(x), \underline{\nu}(x)) | x \in X\} \tag{34}$$

$$\overline{\mathbb{R}}(\mathcal{F}) = \{(x, \overline{\mu}(x), \overline{\eta}(x), \overline{\nu}(x)) | x \in X\} \tag{35}$$

where

$$\underline{\mu}(x) = \min_{c \in C}(w_c \cdot \mu_{\mathbb{R}}(x,c) \cdot \min(\mu_{\mathcal{F}}(c), (2 - \eta_{\mathcal{F}}(c) - \nu_{\mathcal{F}}(c)))), \tag{36}$$

$$\underline{\eta}(x) = \max_{c \in C}(w_c \cdot \eta_{\mathbb{R}}(x,c) \cdot \max(\eta_{\mathcal{F}}(c), (2 - \mu_{\mathcal{F}}(c) - \nu_{\mathcal{F}}(c)))), \tag{37}$$

$$\underline{\nu}(x) = \max_{c \in C}(w_c \cdot \nu_{\mathbb{R}}(x,c) \cdot \max(\nu_{\mathcal{F}}(c), (2 - \mu_{\mathcal{F}}(c) - \eta_{\mathcal{F}}(c)))). \tag{38}$$

and

$$\overline{\mu}(x) = \max_{c \in C}(w_c \cdot \mu_{\mathbb{R}}(x,c) \cdot \max(\mu_{\mathcal{F}}(c), (2 - \eta_{\mathcal{F}}(c) - \nu_{\mathcal{F}}(c)))), \quad (39)$$

$$\overline{\eta}(x) = \min_{c \in C}(w_c \cdot \eta_{\mathbb{R}}(x,c) \cdot \min(\eta_{\mathcal{F}}(c), (2 - \mu_{\mathcal{F}}(c) - \nu_{\mathcal{F}}(c)))), \quad (40)$$

$$\overline{\nu}(x) = \min_{c \in C}(w_c \cdot \nu_{\mathbb{R}}(x,c) \cdot \min(\nu_{\mathcal{F}}(c), (2 - \mu_{\mathcal{F}}(c) - \eta_{\mathcal{F}}(c)))). \quad (41)$$

here, $0 \leq \underline{\mu}(x) + \underline{\eta}(x) + \underline{\nu}(x) \leq 3$, $0 \leq \overline{\mu}(x) + \overline{\eta}(x) + \overline{\nu}(x) \leq 3$

Then,

$$\mathbb{R}(\mathcal{F}) = (\underline{\mathbb{R}}(\mathcal{F}), \overline{\mathbb{R}}(\mathcal{F})) \quad (42)$$

$$= (x, (\underline{\mu}(x), \overline{\mu}(x)), (\underline{\eta}(x), \overline{\eta}(x)), (\underline{\nu}(x), \overline{\nu}(x))). \quad (43)$$

The evaluation function is

$$\mathcal{S}(\mathbb{R}(\mathcal{F})) = \underline{\mu}(x) + \overline{\mu}(x) - \underline{\eta}(x) - \overline{\eta}(x) - \underline{\nu}(x) - \overline{\nu}(x). \quad (44)$$

Considering that as the parameters $\underline{\eta}, \overline{\eta}, \underline{\nu}$ and $\overline{\nu}$ increase, the value of the evaluation function becomes very close to zero or even negative, this does not facilitate numerical calculations and comparisons. Therefore, the evaluation function needs to be improved accordingly. The new evaluation function is as following:

Definition 14. *The score function is as following:*

$$\mathcal{S}(\mathbb{R}(\mathcal{F})) = \underline{\mu}(x) + \overline{\mu}(x) + (2 - \underline{\eta}(x) - \underline{\nu}(x)) + (2 - \overline{\eta}(x) - \overline{\nu}(x)) \quad (45)$$

$$= 4 + \underline{\mu}(x) + \overline{\mu}(x) - \underline{\eta}(x) - \overline{\eta}(x) - \underline{\nu}(x) - \overline{\nu}(x). \quad (46)$$

Example 5. *Still analyzing the data in Example 3 with the weight vector* $w = \{0.50, 0.25, 0.15, 0.10\}^T$, *under the new Definition 13 and Equation (14), the corresponding upper and lower bounds and score functions are as follows.*

$\underline{\mathbb{R}}(\mathcal{F}) = \{(a_1, 0.0040, 0.1500, 0.2800), (a_2, 0.0040, 0.1500, 0.0700), (a_3, 0.0080, 0.0750, 0.1400)\}$;
$\overline{\mathbb{R}}(\mathcal{F}) = \{(a_1, 0.2250, 0.0120, 0.0010), (a_2, 0.3750, 0.0075, 0.0000), (a_3, 0.4500, 0.0000, 0.0010)\}$;
$\mathcal{S}(a_1) = 3.7860$;
$\mathcal{S}(a_2) = 4.1515$;
$\mathcal{S}(a_3) = 4.2420$,

Obviously, $\mathcal{S}(a_3) > \mathcal{S}(a_2) > \mathcal{S}(a_1)$, *so the final ordering is* $a_3 \succ a_2 \succ a_1$.

In general, in chaotic multi-attribute group decision making (CMAGDM), different DMs have different decision weights, so the total evaluation function of the corresponding CMAGDM is as follows.

Definition 15 (Zhang et al. [20]). *The total evaluation score function for the CMAGDM is* $\mathbb{S}(a_i)$:

$$\mathbb{S}(a_i) = \sum_{k=1}^{l} \tau_k \mathcal{S}_k(a_i) \quad (47)$$

where $\mathcal{S}_k(a_i)$ *is the kth DM's score for the ith alternative* $(i = 1, ..., m; k = 1, 2, ..., l)$; τ_k *is the weight of the kth DM* $(\tau_k \geqslant 0, \sum_{k=1}^{l} \tau_k = 1)$.

4.3. The Algorithm for CMAGDM

Through the analysis in the previous subsection, we summarize the algorithm for solving CMAGDM as Algorithm 1.

Algorithm 1 The Algorithm for CMAGDM.

Input:

$A, C, E, w, \tau, X, \mathcal{F}$.

Output:

Optimal sorting: $a_1^* \succ a_2^* \succ \cdots \succ a_m^*$

Step 1: Calculate the NFSRSs vector $\mathbb{R}(\mathcal{F})$:
for $k = 1$ to l **do**
  Calculate $\mathbb{R}(\mathcal{F}_k)$ according to the Definition (13)
end for
Step 2: Calculate the score vector $\mathcal{S}(A)$:
for $k = 1$ to l **do**
  for $i = 1$ to m **do**
    Calculate $\mathcal{S}_k(a_i)$ according to the Definition (14)
  end for
end for
Step 3: Calculate the total score vector $\mathbb{S}(A)$:
for $i = 1$ to m **do**
  Calculate $\mathbb{S}(a_i)$ according to the Definition (15)
end for
Step 4: Ranking of the total evaluation function for all alternatives $\mathbb{S}(a_1^*) \geq \mathbb{S}(a_2^*) \geq \cdots \geq \mathbb{S}(a_m^*)$.
Return $a_1^* \succ a_2^* \succ \cdots \succ a_m^*$

5. Numerical Analysis

In this section, we will analyze a real-life home purchase problem to explain the specific application of our proposed method.

5.1. Problem Statement

A family with three members (husband, wife, and daughter) are planning to buy one of four houses. Suppose the family considers the following factors in purchasing: price, construction materials, decoration, convenience for shopping (e.g., availability of supermarkets, food markets, shops, etc.), and convenience of transportation (e.g., availability of public parking, bus stops, metro stations, etc.). Assume that the purchase decision weight of the husband is 40%, the wife is 35%, and the daughter is 25%. The weights of the purchasing factors are as follows: 35% for price, 20% for construction materials, 15% for decoration, 10% for convenience of shopping, and 20% for convenience of transportation.

Obviously, this is a MAGDM problem, it can be represented by a sextuple $\langle A, C, E, w, \tau, X \rangle$, that is:

$$MAGDM = \langle A, C, E, w, \tau, X \rangle. \tag{48}$$

where $A = \{a_1, a_2, a_3, a_4\}$, a_1, a_2, a_3, a_4 denote the first house, the second house, the third house, and the fourth house, respectively. $C = \{c_1, c_2, c_3, c_4, c_5\}$, c_1, c_2, c_3, c_4, c_5 denote price, construction materials, decoration, convenience for shopping, and convenience of transportation, respectively, and the corresponding weight vector is w, where $w = \{0.35, 0.20, 0.15, 0.10, 0.20\}$. $E = \{e_1, e_2, e_3\}$, e_1, e_2, e_3 denote husband, wife, and daughter, respectively, the corresponding weight vector is τ, where $\tau = \{0.40, 0.35, 0.25\}$. $X = \{X(e_k) | e_k \in E\}$ is the decision matrix set, and $X(e_k)$ is the decision matrix of the kth DM.

Usually, however, the wife and daughter are not familiar with attribute (indicator) construction materials, while the husband is not particularly familiar with attribute (indicator) decoration. Since DMs have different levels of familiarity with attributes, this is not a regular MAGDM problem, but a CMAGDM problem. It can be represented by a septuple $\langle A, C, E, w, \tau, X, \mathcal{F} \rangle$, that is:

$$CMAGDM = \langle A, C, E, w, \tau, X, \mathcal{F} \rangle. \tag{49}$$

Assuming that the DMs select NFSS for scoring evaluation, the evaluation form is shown in Table 5.

Table 5. CMAGDM information of house purchase.

	c_1 **(0.35)**	c_2 **(0.20)**	c_3 **(0.15)**	c_4 **(0.10)**	c_5 **(0.20)**
e_1 (0.40)	(0.92, 0.21, 0.07)	(0.87, 0.10, 0.16)	(0.83, 0.48, 0.43)	(0.86, 0.29, 0.32)	(0.77, 0.48, 0.35)
a_1	(0.97, 0.55, 0.24)	(0.91, 0.04, 0.38)	(0.91, 0.07, 0.01)	(0.71, 0.45, 0.47)	(0.76, 0.03, 0.30)
a_2	(0.75, 0.23, 0.20)	(0.97, 0.41, 0.09)	(0.85, 0.46, 0.47)	(0.83, 0.32, 0.09)	(0.66, 0.27, 0.34)
a_3	(0.93, 0.42, 0.42)	(0.66, 0.58, 0.29)	(0.97, 0.48, 0.24)	(0.94, 0.56, 0.01)	(0.68, 0.16, 0.17)
a_4	(0.83, 0.04, 0.16)	(0.80, 0.11, 0.23)	(0.80, 0.22, 0.26)	(0.68, 0.47, 0.39)	(0.70, 0.28, 0.35)
e_2 (0.35)	(0.98, 0.43, 0.08)	(0.92, 0.54, 0.14)	(0.98, 0.56, 0.39)	(0.96, 0.50, 0.47)	(0.99, 0.56, 0.33)
a_1	(0.80, 0.24, 0.23)	(0.96, 0.30, 0.01)	(0.83, 0.27, 0.27)	(0.70, 0.30, 0.18)	(0.88, 0.07, 0.43)
a_2	(0.74, 0.14, 0.28)	(0.77, 0.23, 0.30)	(0.89, 0.33, 0.37)	(0.82, 0.39, 0.43)	(0.98, 0.19, 0.40)
a_3	(0.83, 0.58, 0.29)	(0.99, 0.22, 0.08)	(0.88, 0.14, 0.44)	(0.78, 0.35, 0.44)	(0.77, 0.23, 0.30)
a_4	(0.83, 0.08, 0.31)	(0.90, 0.42, 0.33)	(0.90, 0.02, 0.38)	(0.84, 0.22, 0.49)	(0.66, 0.53, 0.16)
e_3 (0.25)	(0.73, 0.60, 0.03)	(0.75, 0.42, 0.47)	(0.83, 0.42, 0.32)	(0.88, 0.51, 0.02)	(0.92, 0.47, 0.38)
a_1	(0.71, 0.22, 0.10)	(0.93, 0.15, 0.15)	(0.94, 0.12, 0.01)	(0.86, 0.37, 0.34)	(0.91, 0.59, 0.26)
a_2	(0.68, 0.26, 0.11)	(0.73, 0.41, 0.04)	(0.91, 0.11, 0.24)	(0.72, 0.41, 0.05)	(0.93, 0.19, 0.11)
a_3	(0.70, 0.28, 0.40)	(0.85, 0.51, 0.05)	(0.95, 0.43, 0.30)	(0.85, 0.51, 0.36)	(0.67, 0.10, 0.12)
a_4	(0.99, 0.11, 0.01)	(0.86, 0.08, 0.25)	(0.93, 0.07, 0.01)	(0.76, 0.03, 0.02)	(0.99, 0.03, 0.16)

5.2. Numerical Computations

According to the Algorithm 1, we can obtain following results:

Step 1: By Equation (42), the NFSRSs vector $\mathbb{R}(\mathcal{F})$ can be found. To illustrate the exact process of calculation, we take the husband's evaluation of alternative a_1 as an example.

$$\begin{aligned}
\underline{\mu}(a_1) &= \min_{c \in C}(w_c \cdot \mu_{\mathbb{R}}(a_1, c) \cdot \min(\mu_{\mathcal{F}}(c), (2 - \eta_{\mathcal{F}}(c) - \nu_{\mathcal{F}}(c)))) \\
&= \min(0.35 \times 0.97 \times \min(0.92, (2 - 0.21 - 0.07)), \\
&\qquad 0.20 \times 0.91 \times \min(0.87, (2 - 0.10 - 0.16)), \\
&\qquad 0.15 \times 0.91 \times \min(0.83, (2 - 0.48 - 0.43)), \\
&\qquad 0.10 \times 0.71 \times \min(0.86, (2 - 0.29 - 0.32)), \\
&\qquad 0.20 \times 0.76 \times \min(0.77, (2 - 0.48 - 0.35))) \\
&= 0.06106,
\end{aligned}$$

$$\begin{aligned}\overline{\mu}(a_1) &= \max_{c\in C}(w_c \cdot \mu_{\mathbb{R}}(a_1,c) \cdot \max(\mu_{\mathcal{F}}(c),(2-\eta_{\mathcal{F}}(c)-\nu_{\mathcal{F}}(c))))\\ &= \max(0.35\times 0.97\times \max(0.92,(2-0.21-0.07)),\\ &\quad 0.20\times 0.91\times \max(0.87,(2-0.10-0.16)),\\ &\quad 0.15\times 0.91\times \max(0.83,(2-0.48-0.43)),\\ &\quad 0.10\times 0.71\times \max(0.86,(2-0.29-0.32)),\\ &\quad 0.20\times 0.76\times \max(0.77,(2-0.48-0.35)))\\ &= 0.58394.\end{aligned}$$

Similarly, the values of $\underline{\eta}(a_1), \overline{\eta}(a_1), \underline{\nu}(a_1), \overline{\nu}(a_1)$ can be calculated. The same approach could be used to obtain the husband's evaluation of alternative a_2, a_3, a_4. Using the same method, we can obtain all the evaluation of wife and daughter. Finally, the NFSRSs vector $\mathbb{R}(\mathcal{F})$ can be found. As shown in Table 6.

Table 6. Information of the NFSRSs vector $\mathbb{R}(\mathcal{F})$.

		$\underline{\mu}$	$\underline{\eta}$	$\underline{\nu}$	$\overline{\mu}$	$\overline{\eta}$	$\overline{\nu}$
Husband	a1	0.06106	0.33110	0.14448	0.58394	0.00462	0.00125
	a2	0.07138	0.14268	0.12040	0.45150	0.02752	0.00774
	a3	0.08084	0.25284	0.25284	0.55986	0.02464	0.00086
	a4	0.05848	0.06552	0.09632	0.49966	0.01288	0.03237
Wife	a1	0.06720	0.12516	0.11995	0.41720	0.01386	0.00184
	a2	0.07872	0.07301	0.14602	0.38591	0.03744	0.04128
	a3	0.07488	0.30247	0.15124	0.43285	0.02058	0.01472
	a4	0.08064	0.11766	0.16167	0.43285	0.00294	0.03168
Daughter	a1	0.07568	0.13570	0.05980	0.34045	0.01494	0.00125
	a2	0.06336	0.12467	0.05275	0.32606	0.01370	0.00440
	a3	0.07480	0.13426	0.19180	0.33565	0.01840	0.00750
	a4	0.06688	0.05275	0.05550	0.47471	0.00264	0.00125

Step 2: By Definition (14), the score vector $\mathcal{S}$ can be obtained. Here is an example of the calculation process using the husband's scoring of the alternative a_1.

$$\begin{aligned}\mathcal{S}_1(a_1) &= 4+\underline{\mu}(a_1)+\overline{\mu}(a_1)-\underline{\eta}(a_1)-\overline{\eta}(a_1)-\underline{\nu}(a_1)-\overline{\nu}(a_1)\\ &= 4+0.06106+0.58394-0.33110-0.14448-0.00462-0.00125\\ &= 4.16356.\end{aligned}$$

Using the same method, all evaluation scores can be derived, then the information of the score vector $\mathcal{S}$ can be found, as shown in Table 7.

Table 7. Information of the score vector $\mathcal{S}$.

		$\mathcal{S}$
Husband	a_1	4.16356
	a_2	4.22454
	a_3	4.10952
	a_4	4.35105
Wife	a_1	4.22360
	a_2	4.16688
	a_3	4.01872
	a_4	4.19954
Daughter	a_1	4.20444
	a_2	4.19391
	a_3	4.05849
	a_4	4.42946

Step 3: By Definition 15, we can obtain the corresponding total evaluation score for four houses as shown in Table 8. Using alternative a_1 as an example, the process for calculating its overall evaluation score is as follows.

Table 8. Final evaluation score for all the alternative.

A	$\mathbb{S}$
a_1	4.19479
a_2	4.19670
a_3	4.06498
a_4	4.31762

$$\begin{aligned}\mathbb{S}(a_1) &= 4.16356 \times 0.40 + 4.22360 \times 0.35 + 4.20444 \times 0.25 \\ &= 4.19479.\end{aligned}$$

Step 4: Obtain the final ranking. By the calculation in the previous step, obviously, we can obtain: $\mathbb{S}(a_4) > \mathbb{S}(a_2) > \mathbb{S}(a_1) > \mathbb{S}(a_3)$. That is, $a_4 \succ a_2 \succ a_1 \succ a_3$. The optimal alternative $a^* = a_4$, so the $4th$ house is the optimal choice.

6. Conclusions

Most of the current studies on multi-attribute group decision problems mainly give the corresponding solutions in different practical applications or when DMs use different fuzzy sets [1,9–11,39–41]. Typically, the study of traditional group decision models and methods consists of two main aspects: consensus building and optimal choice. The former refers to how to make the opinions of all experts as consensual as possible among all candidate alternatives, while the latter focuses on how to select the optimal decision alternative from all candidates based on group preference opinions [14,15]. However, there are few research results that consider the structure of the multi-attribute group decision problem itself, such as the relationship between DMs and attributes. Based on such considerations, we propose a chaotic multi-attribute group decision model that considers the familiarity of DMs with attributes, which can well avoid the drawbacks arising from grouping or weighting of decision makers by introducing familiarity in multi-attribute group decision making. At the same time, we combine neutrosophic set with a wider definition domain with soft set and rough set to give the concept of weighted neutrosophic fuzzy soft rough set and apply it to chaotic multi-attributes group decision making to obtain the corresponding algorithm. The validity of the model and the flexibility of the algorithm are well illustrated by practical case studies.

Despite our attempts to solve more realistic problems, however, there are still many shortcomings in our work. We have only considered the evaluation scoring of decision makers using neutrosophic fuzzy sets, whereas in practical decision making, decision makers can choose different evaluation methods, such as using different fuzzy sets or precise numbers or linguistic variables, etc. This is a drawback of our work and is certainly a direction for future research. In addition, this paper does not give a scheme for determining familiarity. However, how to determine the familiarity, just like how to assign weights to decision makers or decision attributes, is still the key to multi-attribute cluster decision making, so this will be another popular direction for future research.

Author Contributions: Conceptualization, F.Z. and W.M.; methodology, F.Z. and W.M.; software, F.Z.; writing—original draft preparation, F.Z.; writing—review and editing, F.Z.; funding acquisition, W.M. All authors have read and agreed to the published version of the manuscript.

Funding: This research was funded by National Social Science Foundation of China grant number 20BGL115.

Acknowledgments: The authors would like to thank the editors and the anonymous reviewers for their constructive comments and suggestions, which have helped to improve the paper.

Conflicts of Interest: The authors declare no conflicts of interest.

Abbreviations

The following abbreviations are used in this manuscript:

MAGDM	Multi-attribute Group Decision Making
DM	Decision Maker
FS	Fuzzy set
PFS	Picture Fuzzy Set
NFS	Neutrosophic Fuzzy Set
RS	Rough Set
SS	Soft Set
FSS	Fuzzy Soft Set
FSRS	Fuzzy Soft Rough Set
PFSRS	Picture Fuzzy Soft Rough Set
NFSRS	Neutrosophic Fuzzy Soft Rough Set

References

1. Liu, S.; Yu, W.; Chan, F.T.S.; Niu, B. A variable weight-based hybrid approach for multi-attribute group decision making under interval-valued intuitionistic fuzzy sets. *Int. J. Intell. Syst.* **2021**, *36*, 1015–1052. [CrossRef]
2. Yu, W.; Zhang, Z.; Zhong, Q. Consensus reaching for MAGDM with multi-granular hesitant fuzzy linguistic term sets: A minimum adjustment-based approach. *Ann. Oper. Res.* **2021**, *300*, 443–466. [CrossRef]
3. Ashraf, S.; Mahmood, T.; Abdullah, S.; Khan, Q. Different Approaches to Multi-Criteria Group Decision Making Problems for Picture Fuzzy Environment. *Bull. Braz. Math. Soc.* **2019**, *50*, 373–397. [CrossRef]
4. Jin, H.; Ashraf, S.; Abdullah, S.; Qiyas, M.; Bano, M.; Zeng, S. Linguistic Spherical Fuzzy Aggregation Operators and Their Applications in Multi-Attribute Decision Making Problems. *Mathematics* **2019**, *7*, 413. [CrossRef]
5. Lin, M.; Xu, Z.; Zhai, Y.; Yao, Z. Multi-attribute group decision-making under probabilistic uncertain linguistic environment. *J. Oper. Res. Soc.* **2018**, *69*, 157–170. [CrossRef]
6. Liu, P.; Liu, J.; Chen, S.M. Some intuitionistic fuzzy Dombi Bonferroni mean operators and their application to multi-attribute group decision making. *J. Oper. Res. Soc.* **2018**, *69*, 1–24. [CrossRef]
7. Liu, P.; Liu, J.; Merigo, J.M. Partitioned Heronian means based on linguistic intuitionistic fuzzy numbers for dealing with multi-attribute group decision making. *Appl. Soft Comput.* **2018**, *62*, 395–422. [CrossRef]
8. Liu, P.; Liu, J. Some q-Rung Orthopai Fuzzy Bonferroni Mean Operators and Their Application to Multi-Attribute Group Decision Making. *Int. J. Intell. Syst.* **2018**, *33*, 315–347. [CrossRef]
9. Mu, Z.; Zeng, S.; Wang, P. Novel approach to multi-attribute group decision-making based on interval-valued Pythagorean fuzzy power Maclaurin symmetric mean operator. *Comput. Ind. Eng.* **2021**, *155*, 107049. [CrossRef]
10. Zhan, J.; Sun, B.; Alcantud, J.C.R. Covering based multigranulation (I, T)-fuzzy rough set models and applications in multi-attribute group decision-making. *Inf. Sci.* **2019**, *476*, 290–318. [CrossRef]
11. Zhao, M.; Wei, G.; Wei, C.; Wu, J.; Wei, Y. Extended CPT-TODIM method for interval-valued intuitionistic fuzzy MAGDM and its application to urban ecological risk assessment. *J. Intell. Fuzzy Syst.* **2021**, *40*, 4091–4106. [CrossRef]
12. Xu, Y.; Wen, X.; Zhang, W. A two-stage consensus method for large-scale multi-attribute group decision making with an application to earthquake shelter selection. *Comput. Ind. Eng.* **2018**, *116*, 113–129. [CrossRef]
13. Su, W.; Luo, D.; Zhang, C.; Zeng, S. Evaluation of online learning platforms based on probabilistic linguistic term sets with self-confidence multiple attribute group decision making method. *Expert Syst. Appl.* **2022**, *208*, 118153. [CrossRef]
14. Sun, B.; Zhou, X.; Lin, N. Diversified binary relation-based fuzzy multigranulation rough set over two universes and application to multiple attribute group decision making. *Inf. Fusion* **2020**, *55*, 91–104. [CrossRef]
15. Sun, B.; Zhang, M.; Wang, T.; Zhang, X. Diversified multiple attribute group decision-making based on multigranulation soft fuzzy rough set and TODIM method. *Comput. Appl. Math.* **2020**, *39*, 1–30. [CrossRef]
16. Chakhar, S.; Saad, I. Incorporating stakeholders' knowledge in group decision-making. *J. Decis. Syst.* **2014**, *23*, 113–126. [CrossRef]
17. Li, N.; Sun, M.; Bi, Z.; Su, Z.; Wang, C. A new methodology to support group decision-making for IoT-based emergency response systems. *Inf. Syst. Front.* **2014**, *16*, 953–977. [CrossRef]
18. Du, Y.; Liu, D. A novel approach to relative importance ratings of customer requirements in QFD based on probabilistic linguistic preferences. *Fuzzy Optim. Decis. Mak.* **2021**, *20*, 365–395. [CrossRef]
19. Wagatsuma, K.; Sato, R.; Yamazaki, S.; Iwaya, M.; Takahashi, Y.; Nojima, A.; Oseki, M.; Abe, T.; Phyu, W.W.; Tamura, T.; et al. Genomic Epidemiology Reveals Multiple Introductions of Severe Acute Respiratory Syndrome Coronavirus 2 in Niigata City, Japan, Between February and May 2020. *Front. Microbiol.* **2021**, *12*, 749149. [CrossRef]
20. Zhang, F.; Ma, W.; Ma, H. Dynamic Chaotic Multi-Attribute Group Decision Making under Weighted T-Spherical Fuzzy Soft Rough Sets. *Symmetry* **2023**, *15*, 307. [CrossRef]
21. Zadeh, L.A. Information and control. *Fuzzy Sets* **1965**, *8*, 338–353.

22. Atanassov, K. Intuitionistic fuzzy sets VII ITKR's Session. *Sofia June* **1983**, *1*, 983.
23. Yager, R.R. Generalized orthopair fuzzy sets. *IEEE Trans. Fuzzy Syst.* **2016**, *25*, 1222–1230. [CrossRef]
24. Yager, R.R.; Abbasov, A.M. Pythagorean membership grades, complex numbers, and decision making. *Int. J. Intell. Syst.* **2013**, *28*, 436–452. [CrossRef]
25. Senapati, T.; Yager, R.R. Fermatean fuzzy sets. *J. Ambient. Intell. Humaniz. Comput.* **2020**, *11*, 663–674. [CrossRef]
26. Cuong, B.C.; Kreinovich, V. Picture Fuzzy Sets—A New Concept for Computational Intelligence Problems. In Proceedings of the 2013 Third World Congress on Information and Communication Technologies (WICT 2013), Hanoi, Vietnam, 15–18 December 2013; IEEE: Piscataway, NJ, USA, 2013; pp. 1–6.
27. Cuong, B.C.; Kreinovich, V. Picture fuzzy sets. *J. Comput. Sci. Cybern.* **2014**, *30*, 409–420.
28. Mahmood, T.; Ullah, K.; Khan, Q.; Jan, N. An approach toward decision-making and medical diagnosis problems using the concept of spherical fuzzy sets. *Neural Comput. Appl.* **2019**, *31*, 7041–7053. [CrossRef]
29. Smarandache, F. Neutrosophic set—A generalization of the intuitionistic fuzzy set. In Proceedings of the 2006 IEEE International Conference on Granular Computing, Atlanta, GA, USA, 10–12 May 2006; pp. 38–42. [CrossRef]
30. Yager, R.R. On the theory of bags. *Int. J. Gen. Syst.* **1986**, *13*, 23–37. [CrossRef]
31. Zadeh, L.A. The concept of a linguistic variable and its application to approximate reasoning—I. *Inf. Sci.* **1975**, *8*, 199–249. [CrossRef]
32. Torra, V.; Narukawa, Y. On hesitant fuzzy sets and decision. In Proceedings of the 2009 IEEE International Conference on Fuzzy Systems, Jeju Island, Republic of Korea, 20–24 August 2009; IEEE: Piscataway, NJ, USA, 2009; pp. 1378–1382.
33. Molodtsov, D. Soft set theory—first results. *Comput. Math. Appl.* **1999**, *37*, 19–31. [CrossRef]
34. Pawlak, Z. Rough sets. *Int. J. Comput. Inf. Sci.* **1982**, *11*, 341–356. [CrossRef]
35. Maji, P.K.; Biswas, R.; Roy, A. Fuzzy soft sets. *J. Fuzzy Math.* **2001**, *9*, 589–602.
36. Muahmmad, A.; Martino, A. Multi-attribute group decision making based on T-spherical fuzzy soft rough average aggregation operators. *Granul. Comput.* **2023**, *8*, 171–207. [CrossRef]
37. Pelissari, R.; Oliveira, M.C.; Abackerli, A.J.; Ben-Amor, S.; Assumpção, M.R.P. Techniques to model uncertain input data of multi-criteria decision-making problems: A literature review. *Int. Trans. Oper. Res.* **2021**, *28*, 523–559. [CrossRef]
38. Khan, M.J.; Kumam, P.; Liu, P.; Kumam, W. An adjustable weighted soft discernibility matrix based on generalized picture fuzzy soft set and its applications in decision making. *J. Intell. Fuzzy Syst.* **2020**, *38*, 2103–2118. [CrossRef]
39. Liu, P. Some Hamacher Aggregation Operators Based on the Interval-Valued Intuitionistic Fuzzy Numbers and Their Application to Group Decision Making. *IEEE Trans. Fuzzy Syst.* **2014**, *22*, 83–97. [CrossRef]
40. Liu, Y.; Dong, Y.; Liang, H.; Chiclana, F.; Herrera-Viedma, E. Multiple Attribute Strategic Weight Manipulation With Minimum Cost in a Group Decision Making Context With Interval Attribute Weights Information. *IEEE Trans. Syst. Man, Cybern. Syst.* **2019**, *49*, 1981–1992. [CrossRef]
41. Liu, P.; Jin, F. Methods for aggregating intuitionistic uncertain linguistic variables and their application to group decision making. *Inf. Sci.* **2012**, *205*, 58–71. [CrossRef]

mathematics

MDPI

Article

Some Operations and Properties of the Cubic Intuitionistic Set with Application in Multi-Criteria Decision-Making

Shahzad Faizi [1], Heorhii Svitenko [2,3], Tabasam Rashid [4], Sohail Zafar [4] and Wojciech Sałabun [3,*]

1 Department of Mathematics, Virtual University of Pakistan, Lahore 54000, Pakistan
2 Department of Software Engineering, Kharkiv National University of Radio Electronics, Nauky Ave. 14, 61166 Kharkiv, Ukraine
3 Research Team on Intelligent Decision Support Systems, Department of Artificial Intelligence and Applied Mathematics, Faculty of Computer Science and Information Technology, West Pomeranian University of Technology in Szczecin, ul. Żołnierska 49, 71-210 Szczecin, Poland
4 Department of Science and Humanities, University of Management and Technology, Lahore 54770, Pakistan
* Correspondence: wojciech.salabun@zut.edu.pl; Tel.: +48-91-449-5580

Abstract: This paper proposes some operations on the cubic intuitionistic set along with useful properties. We propose the internal cubic intuitionistic set (ICIS), the external cubic intuitionistic set (ECIS), P-order, R-order order (P-(R-) order), P-union, R-union (P-(R-) union), P-intersection, and R-intersection (P-(R-) intersection). We further investigate several properties of the P-(R-) union and P-(R-) intersection of ICISs and ECISs, and present some examples in this context. Some important theorems related to ICISs and ECISs are also presented with proof. Finally, an application example is given to measure the effectiveness and significance of the proposed operations by solving a multi-criteria decision-making (MCDM) problem.

Keywords: fuzzy set; interval-valued fuzzy set; intuitionistic fuzzy set; interval-valued intuitionistic fuzzy set; cubic set; cubic intuitionistic set

MSC: 03E72;94D05

Citation: Faizi, S.; Svitenko, H.; Rashid, T.; Zafar, S.; Sałabun, W. Some Operations and Properties of the Cubic Intuitionistic Set with Application in Multi-Criteria Decision-Making. *Mathematics* **2023**, *11*, 1190. https://doi.org/10.3390/math11051190

Academic Editors: Jun Ye and Yanhui Guo

Received: 9 January 2023
Revised: 15 February 2023
Accepted: 21 February 2023
Published: 28 February 2023

1. Introduction

Zadeh [1] proposed the idea of fuzzy sets in 1965 and further extended this idea to an interval-valued fuzzy set (IVFS) [2]. Some complex decision-making problems in the economy, engineering, social science, environmental science, etc., exist that cannot be completely modeled by methods of classical mathematics because of the presence of various types of uncertainties. Others, on the other hand, use certain data processed by methods that are hybrid approaches, such as the INVAR method [3] or the CODAS-COMET method [4]. However, to handle the vagueness and uncertainty occurring in such decision-making problems, some well-known mathematical theories have been introduced, such as fuzzy set theory [1], intuitionistic fuzzy set (IFS) theory [5], interval-valued intuitionistic fuzzy set (IVIFS) theory [6,7], hesitant fuzzy set theory [8], hesitant fuzzy linguistic set theory [9], soft set theory [10], fuzzy soft set theory [11], etc. An example of this could be the use of triangular fuzzy numbers in a fuzzy extension of a simplified best–worst method [12].

At times, uncertainty research uses generalized approaches to better cope with the decision-making process via approaches related to the Dempster–Shafer evidence theory (DSET) [13], or quantum evidence theory (QET) [14]. Other ways are to use methods based on either entropy [15] or distance measures [16]. Most of the researchers studied IVFS [12]. For example, Zhang et al. [17] investigated the entropy of IVFSs based on distance measures. Zeng and Guo [18] discussed the similarity measure, inclusion of the measure, and entropy of IVFSs, while Grzegorzewski [19] proposed IVFSs based on the

Hausdorff metric. Furthermore, IVFSs have been widely used and applied in real-life applications. For example, Sambuc [20] and Kohout [21] used the concept of IVFSs in medical diagnoses in thyroid pathology and medicine in a CLINAID system, respectively. Gorzalczany [22] used the idea of IVFSs in approximate reasoning. Turksen [23,24] further used the same idea of IVFSs in interval-valued logic in preference modeling [25].

Jun et al. [26] proposed the idea of a cubic set and presented its two important types, called the internal cubic set and the external cubic set by using the idea of the fuzzy set and IVFS. They further introduced some operations of union and intersection regarding the cubic sets, such as the P-(R-) union and P-(R-) intersection, and studied important related properties. Jun [27] further extended the idea of the cubic set, introduced the notion of the cubic intuitionistic set, and discussed its useful applications in BCK/BCI-algebras. Recently, studies on the cubic set theory have rapidly grown. For example, Jun et al. [28] proposed the concept of cubic IVIFS and discussed its important applications in BCK/BCI-algebra. With the help of using a cubic set and a neutrosophic set, Ali et al. [29] presented the notion of a neutrosophic cubic set and studied some useful properties. Kang and Kim [30] investigated the images and inverse images of almost-stable cubic sets and discussed the complement, the P-union, and the P-intersection of inverse images of almost-stable cubic sets. Chinnadurai et al. [31] investigated several properties of the P-(R-) union and P-(R-) intersection of cubic sets and studied some properties of cubic ideals of near rings. Jun et al. [32] proposed the ideas of cubic α-ideals and cubic p-ideals and studied several useful properties.

Cubic sets are widely studied and are important in many areas, as discussed in the literature by various researchers. Motivated by the advantages of cubic sets, this paper proposes the notion of CIS based on IVFSs and intuitionistic fuzzy sets. Although Jun [27] previously introduced the idea of CIS as cubic intuitionistic sets and discussed their applications in BCK/BCI-algebras, this paper presents a completely different research work under the framework of CIS. We first propose two important types of CIS, named ICIS and ECIS. We then investigate the complement of CIS, the P-(R-) cubic intuitionistic subsets, and the P-(R-) union and the intersection of CISs. Furthermore, we prove various important theorems and results related to the proposed union and intersection operations. Finally, we present an application example to demonstrate the validity of the proposed operations by solving a MCDM problem.

The remainder of the paper can be summarized briefly as follows. Some basic concepts related to the work are presented in Section 2. The notions of CIS, ICIS, and ECIS are introduced in Section 3. We further investigate P-(R-)order, P-(R-)union, P-(R-)intersection, and related important properties with proof in the same section. A MCDM approach using CISs is presented in Section 4 along with an application example. We conclude the paper with some concluding remarks in Section 5.

2. Preliminary

This section introduces necessary notions and presents a few auxiliary results that we need in the rest of the paper. Throughout this paper, we let $[I]$, I^X, and $[I]^X$ stand for the set of all closed subintervals of $[0,1]$, the collection of all fuzzy sets in a set X, and IVFSs in X, respectively.

Definition 1. *Let X be a non-empty set. A fuzzy set in set X is defined as function $f : X \to [0,1]$. the relation $\leq$, join ($\vee$), meet ($\wedge$), and complement of I^X for all $x \in X$ can be defined, respectively, as follows:*

$$f_1 \leq f_2 \Leftrightarrow f_1(x) \leq f_2(x) \text{ for all} f_1, f_2 \in I^X,$$

$$(f_1 \vee f_2)(x) = f_1(x) \vee f_2(x) = \max\{f_1(x), f_2(x)\},$$

$$(f_1 \wedge f_2)(x) = f_1(x) \wedge f_2(x) = \min\{f_1(x), f_2(x)\},$$

$$f_1^c(x) = 1 - f_1(x),$$

where f_1^c represents the complement of f_1.

Definition 2. *By an interval number, we mean a closed sub-interval $a = [a^-, a^+]$ of Iwhere $0 \leq a^- \leq a^+ \leq 1$. The complement a^c of $a \in [I]$ is defined as follows:*

$$a^c = [1 - a^+, 1 - a^-].$$

The refined minimum and refined maximum (briefly, rmin and rmax) and the symbols $\succeq$, $\preceq$, $=$ of the elements $a_1 = [a_1^-, a_1^+]$ and $a_2 = [a_2^-, a_2^+]$ of $[I]$ is defined as follows:

$$\text{rmin}\{a_1, a_2\} = [\min\{a_1^-, a_2^-\}, \min\{a_1^+, a_2^+\}],$$
$$\text{rmax}\{a_1, a_2\} = [\max\{a_1^-, a_2^-\}, \max\{a_1^+, a_2^+\}],$$
$$a_1 \succeq a_2 \text{if and only if } a_1^- \geq a_2^- \text{ and } a_1^+ \geq a_2^+.$$

Similarly, we can define $a_1 \preceq a_2$ and $a_1 = a_2$.

Definition 3. *For a non-empty set X, a function $A : X \to [I]$ is called an IVFS in X. The element $A = [A^-(x), A^+(x)]$ for every $A \in [I]^X$ and $x \in X$, is called the membership degree of an element x to the set A. The IVFS is simply denoted as $A = [A^-, A^+]$. The complement A^c of A can be defined as $A^c = [1 - A^+, 1 - A^-]$.*

For every $A_1, A_2 \in [I]^X$, the following are true:

$$A_1 \subseteq A_2 \text{ if and only if} A_1 \preceq A_2,$$

$$A_1 = A_2 \text{ if and only if} A_1 = A_2.$$

Definition 4 ([5]). *Let E be a crisp set. An IFS $\tilde{A}$ can be defined as*

$$\tilde{A} = \{\langle x, \mu_{\tilde{A}}(x), \nu_{\tilde{A}}(x)\rangle : x \in E\}.$$

where $\mu_{\tilde{A}} : E \to [0,1]$ and $\nu_{\tilde{A}} : E \to [0,1]$ indicate, respectively, the membership and non-membership degrees of $x \in E$ with the condition $0 \leq \mu_{\tilde{A}}(x) + \nu_{\tilde{A}}(x) \leq 1$ for every $x \in E$.

Definition 5 ([6]). *An expression of the form given by*

$$B = \{\langle x, M_B(x), N_B(x)\rangle : x \in X\}$$

is called the IVIFS in X, where $M_B : X \to [I]$ and $N_B : X \to [I]$ are IVFSs with the condition that

$$0 \leq M_B^+(x) + N_B^+(x) \leq 1 \text{for all } x \in X.$$

The intervals M_B and N_B denote, respectively, the membership and non-membership degrees of $x \in X$.

Definition 6 ([26]). *A mathematical structure of the form*

$$A = \{\langle x, A(x), \lambda(x)\rangle : x \in X\},$$

is called the cubic set in X, where A and λ are, respectively, the IVFS and a fuzzy set in X. Jun [27] introduced the notion of the cubic intuitionistic set as follows:

Definition 7 ([27]). *A mathematical structure of the form*

$$A = \{\langle x, A(x), \lambda(x)\rangle : x \in X\},$$

is called the cubic intuitionistic set where A is an IVIFS in X and λ is an IFS in X.

3. Some Operations on the Cubic Intuitionistic Set

This section introduces the concept of CIS with some modifications as proposed by Jun in [27] as follows:

Definition 8. *By CIS in a non-empty set X, we mean a mathematical structure of the form*

$$\mathbf{A} = \{\langle x, M_A(x)/\alpha_A(x), N_A(x)/\beta_A(x)\rangle | x \in X\}$$

where $M_A : X \to [I]$ *and* $N_A : X \to [I]$ *are IVFSs of the form* $M_A(x) = [M^-(x), M^+(x)]$, $N_A(x) = [N^-(x), N^+(x)]$ *with the conditions that*

$$0 \leq M_A^+(x) + N_A^+(x) \leq 1 \, and \, 0 \leq \alpha_A(x) + \beta_A(x) \leq 1 \, for \, all \, x \in X.$$

$M_A(x)$ *and* $N_A(x)$ *denote, respectively, the membership and non-membership degrees of x and* $\alpha_A : X \to [0,1]$, $\beta_A : X \to [0,1]$ *are fuzzy sets in X. For simplicity, we denote* $CIS(X)$ *as the collection of all CISs* $\mathbf{A} = \langle M_A/\alpha_A, N_A/\beta_A\rangle$ *in X. In the rest of the paper, we will use the same notations with symbols for CIS as presented in the above definition.*

Remark 1. *For any non-empty set X, let* $1(x) = 1$ *and* $0(x) = 0$ *for all* $x \in X$. *Then,* $\mathbf{A} = \langle M_A/1, N_A/0(x)\rangle$, $\mathbf{B} = \langle M_B/0, N_B/1\rangle$ *and* $\mathbf{C} = \langle M_C/\frac{M_C^- + M_C^+}{2}, N_C/\frac{N_C^- + N_C^+}{2}\rangle$ *are all CISs in X.*

Definition 9. *For* $\mathbf{A} = \langle M_A/\alpha_A, N_A/\beta_A\rangle \in CIS(X)$, *the score value of* $\mathbf{A}$ *is defined as*

$$Sc(\mathbf{A}) = \frac{1}{3}[(M_A^- + M_A^+ + \alpha_A) - (N^-(x) + N^+(x) + \beta_A)]$$

where $Sc(\mathbf{A}) \in [-1, 1]$.

Definition 10. *Let* $\mathbf{A} = \langle M_A/\alpha_A, N_A/\beta_A\rangle$ *and* $\mathbf{B} = \langle M_B/\alpha_B, N_B/\beta_B\rangle \in CIS(X)$, *then*

(i) $\mathbf{A} = \mathbf{B} \Leftrightarrow M_A = M_B, \alpha_A = \alpha_B; N_A = N_B, \beta_A = \beta_B$ *(Equality)*

(ii) $\mathbf{A} \subseteq_P \mathbf{B} \Leftrightarrow M_A \subseteq M_B, \alpha_A \leq \alpha_B; N_A \supseteq N_B, \beta_A \geq \beta_B$ *(P-order)*

(iii) $\mathbf{A} \subseteq_R \mathbf{B} \Leftrightarrow M_A \subseteq M_B, \alpha_A \geq \alpha_B; N_A \supseteq N_B, \beta_A \leq \beta_B$ *(R-order)*

Definition 11. *Let* $\mathbf{0} = [0,0]$ *and* $\mathbf{1} = [1,1]$. *Then, a CIS* $\mathbf{A} = \langle M_A/\alpha_A, N_A/\beta_A\rangle$ *in which* $M_A = \mathbf{0}$, $\alpha_A = 1$, $N_A = \mathbf{1}$ *and* $\beta_A = 0$ *(respectively,* $M_A = \mathbf{1}$, $\alpha_A = 0$, $N_A = \mathbf{0}$ *and* $\beta_A = 1$*) is denoted by* $\ddot{0}$ *(respectively* $\ddot{1}$*).*

A CIS $\mathbf{B} = \langle M_B/\alpha_B, N_B/\beta_B\rangle$ *in which* $M_B = \mathbf{0}$, $\alpha_B = 0$, $N_B = \mathbf{1}$, $\beta_B = 1$ *(respectively* $M_B = \mathbf{1}$, $\alpha_B = 1$, $N_B = \mathbf{0}$ *and* $\beta_A = 0$*) is denoted by* $\hat{0}$ *(respectively,* $\hat{1}$*).*

We can see that the score values of $\ddot{0}, \ddot{1}, \hat{0}$ and $\hat{1}$ can be computed, respectively, as $Sc(\ddot{0}) = -0.33$, $Sc(\ddot{1}) = 0.33$, $Sc(\hat{0}) = -1$ and $Sc(\hat{1}) = 1$.

Definition 12. *Consider the family of CISs* $\mathbf{A}_i = \langle M_i/\alpha_i, N_i/\beta_i\rangle$, $i \in \mho$ *in X, we define*

(a) *P-union*

$$\cup_{P_{i \in \mho}} \mathbf{A}_i = \langle \cup_{i \in \mho} M_i / \vee_{i \in \mho} \alpha_i, \cap_{i \in \mho} N_i / \wedge_{i \in \mho} \beta_i\rangle$$

(b) *P-intersection*

$$\cap_{P_{i \in \mho}} \mathbf{A}_i = \langle \cap_{i \in \mho} M_i / \wedge_{i \in \mho} \alpha_i, \cup_{i \in \mho} N_i / \vee_{i \in \mho} \beta_i\rangle$$

(c) *R-union*

$$\cup_{R_{i \in \mho}} \mathbf{A}_i = \langle \cup_{i \in \mho} M_i / (\wedge_{i \in \mho} \alpha_i, \cap_{i \in \mho} N_i / \vee_{i \in \mho} \beta_i\rangle$$

(d) *R-intersection*

$$\cap_{R \, i\in\mho} \mathbf{A}_i = \langle \cap_{i\in\mho} M_i / \vee_{i\in\mho} \alpha_i \rangle, \cup_{i\in\mho} N_i / (\wedge_{i\in\mho} \beta_i \rangle$$

Remark 2. *The complement of* $\mathbf{A} = \langle M_A/\alpha_A, N_A/\beta_A \rangle$ *is defined as*

$$\mathbf{A}^c = \langle M_A^c/1-\alpha_A, N_A^c/1-\beta_A \rangle.$$

Obviously, $(\mathbf{A}^c)^c = \mathbf{A}$, $\ddot{0}^c = \ddot{1}$, $\ddot{1}^c = \ddot{0}$, $\hat{0}^c = \hat{1}$, $\hat{1}^c = \hat{0}$.

Remark 3. *For the family of CISs* $\mathbf{A}_i = \langle M_i/\alpha_i, N_i/\beta_i \rangle$, $i \in \mho$ *in X, we have* $(\cup_{P \, i\in\mho} \mathbf{A}_i)^c = \cap_{P \, i\in\mho} (\mathbf{A}_i)^c$, $(\cap_{P \, i\in\mho} \mathbf{A}_i)^c = \cup_{P \, i\in\mho} (\mathbf{A}_i)^c$, $(\cup_{R \, i\in\mho} \mathbf{A}_i)^c = \cap_{R \, i\in\mho} (\mathbf{A}_i)^c$ *and* $(\cap_{R \, i\in\mho} \mathbf{A}_i)^c = \cup_{R \, i\in\mho} (\mathbf{A}_i)^c$.

Definition 13. *Let X be a non-empty set.*

1 *A CIS* $\mathbf{A} = \langle M_A/\alpha_A, N_A/\beta_A \rangle$ *is said to be ICIS if* $M_A^- \leq \alpha_A \leq M_A^+$ *and* $N_A^- \leq \beta_A \leq N_A^+$.
2 *A CIS* $\mathbf{B} = \langle M_B/\alpha_B, N_B/\beta_B \rangle$ *in X is said to be ECIS if* $\alpha_B \notin (M_B^-, M_B^+)$ *and* $\beta_B \notin (N_B^-, N_B^+)$.

Example 1. *For a non-empty set X,*

1 *Let* $\mathbf{A} = \langle M_A/\alpha_A, N_A/\beta_A \rangle$ *be a CIS with* $M_A = [0.1, 0.3]$, $\alpha_A = 0.2$, $N_A = [0.4, 0.6]$ *and* $\beta_A = 0.5$, *then* $\mathbf{A}$ *is ICIS.*
2 *Let* $\mathbf{B} = \langle M_B/\alpha_B, N_B/\beta_B \rangle$ *be a CIS with* $M_B = [0.2, 0.4]$, $\alpha_B = 0.1$, $N_B = [0.5, 0.6]$ *and* $\beta_B = 0.7$, *then* $\mathbf{B}$ *is ECIS.*

Remark 4. *Every CIS in X can be considered a Zadeh fuzzy set, IFS, IVFS, IVIFS, and cubic set according to* $(M = N = \mathbf{0}, \beta = 0)$, $(M = N = \mathbf{0})$, $(N = \mathbf{0}, \beta = 0)$, $(\beta = \alpha = 0)$ *and* $(N = \mathbf{0}, \beta = 0)$, *respectively.*

Theorem 1. *Let* $\mathbf{A} = \langle M_A/\alpha_A, N_A/\beta_A \rangle$ *be A CIS which is not an ECIS in X. Then there exist* $x \in X$ *such that* $\alpha_A(x) \in (M_A^-(x), M_A^+(x))$ *and* $\beta_A(x) \in (N_A^-(x), N_A^+(x))$.

Proof. Straightforward. □

Theorem 2. *Let* $\mathbf{A} = \langle M_A/\alpha_A, N_A/\beta_A \rangle$ *be A CIS in X. If* $\mathbf{A}$ *is both ICIS and ECIS, then* $\alpha(x) \in U(M) \cup L(M)$ *and* $\beta(x) \in U(N) \cup L(N)$ *for all* $x \in X$ *where* $U(M) = \{M^+(x) | x \in X\}$, $L(M) = \{M^-(x) | x \in X\}$, $U(N) = \{N^+(x) | x \in X\}$ *and* $L(N) = \{N^-(x) | x \in X\}$.

Proof. Assume that $\mathbf{A}$ is both ICIS and ECIS. Then, using Definition 13, we have $M^-(x) \leq \alpha(x) \leq M^+(x)$, $N^-(x) \leq \beta(x) \leq N^+(x)$ and $\alpha(x) \notin (M^-(x), M^+(x))$, $\beta(x) \notin (N^-(x), N^+(x))$ for all $x \in X$. Thus $\alpha(x) = M^-(x)$ or $\alpha(x) = M^+(x)$ and $\beta(x) = N^-(x)$ or $\beta(x) = N^+(x)$. Hence $\alpha(x) \in U(M) \cup L(M)$ and $\beta(x) \in U(N) \cup L(N)$ for all $x \in X$. □

Theorem 3. *Let* $\mathbf{A} = \langle M_A/\alpha_A, N_A/\beta_A \rangle$ *be A CIS in X. If* $\mathbf{A}$ *is ICIS (respectively, ECIS), then* A^c *is ICIS (respectively ECIS).*

Proof. Since $\mathbf{A} = \langle M_A/\alpha_A, N_A/\beta_A \rangle$ is ICIS in X, we have

$$M_A^- \leq \alpha_A \leq M_A^+ and N_A^- \leq \beta_A \leq N_A^+$$

$$(\text{respectively,} \alpha_A \notin (M_A^-, M_A^+) \, and \, \beta_A \notin (N_A^-, N_A^+)).$$

This implies that

$$1 - M_A^+ \leq 1 - \alpha_A \leq 1 - M_A^- \, and \, 1 - N_A^+ \leq 1 - \beta_A \leq 1 - N_A^-$$

$$(\text{respectively,}1-\alpha_A \notin (1-M_A^+, 1-M_A^-)\, and\, 1-\beta_A \notin (1-N_A^+, 1-N_A^-)).$$

Hence $\mathbf{A}^c = \langle M_A^c/1-\alpha_A, N_A^c/1-\beta_A\rangle$ is ICIS (respectively, ECIS) □

We will show (through the following example) that the P-union and P-intersections of ECISs are not necessarily ECISs.

Example 2. *Let* $\mathbf{A} = \langle M_A/\alpha_A, N_A/\beta_A\rangle$ *and* $\mathbf{B} = \langle M_B/\alpha_B, N_B/\beta_B\rangle$ *be two ECISs in X. Let* $M_A = [0.1, 0.3]$, $\alpha_A = 0.5$, $N_A = [0.4, 0.6]$, $\beta_A = 0.2$, $M_B = [0.4, 0.6]$, $\alpha_B = 0.2$, $N_B = [0.1, 0.3]$ *and* $\beta_B = 0.5$ *for all* $x \in X$. *Then* $\mathbf{A} \cup_p \mathbf{B} = \langle M_B/\alpha_A, N_B/\beta_A\rangle$ *and* $\mathbf{A} \cap_p \mathbf{B} = \langle M_A/\alpha_B, N_A/\beta_B\rangle$. *Hence,* $\mathbf{A} \cup_p \mathbf{B}$ *and* $\mathbf{A} \cap_p \mathbf{B}$ *are not ECISs.*

From the following example, it can be easily seen that the R-union and R-intersection of ICIS need not be ICISs.

Example 3. *Let* $\mathbf{A} = \langle M_A/\alpha_A, N_A/\beta_A\rangle$ *and* $\mathbf{B} = \langle M_B/\alpha_B, N_B/\beta_B\rangle$ *be two ICISs in X. Let* $M_A = [0.1, 0.3]$, $\alpha_A = 0.2$, $N_A = [0.5, 0.7]$, $\beta_A = 0.6$, $M_B = [0.5, 0.7]$, $\alpha_B = 0.6$, $N_B = [0.1, 0.3]$ *and* $\beta_B = 0.2$ *for all* $x \in X$. *Then* $\mathbf{A} \cup_R \mathbf{B} = \langle M_B/\alpha_A, N_B/\beta_A\rangle$ *and* $\mathbf{A} \cap_R \mathbf{B} = \langle M_A/\alpha_B, N_A/\beta_B\rangle$. *Hence,* $\mathbf{A} \cup_R \mathbf{B}$ *and* $\mathbf{A} \cap_p \mathbf{B}$ *are not ICISs.*

In the following examples, we will show that the R-union and R-intersection of ECIS may not be ECIS.

Example 4.

1 *Let* $\mathbf{A} = \langle M_A/\alpha_A, N_A/\beta_A\rangle$ *and* $\mathbf{B} = \langle M_B/\alpha_B, N_B/\beta_B\rangle$ *be two ECISs in X. Let* $M_A = [0.1, 0.3]$, $\alpha_A = 0.5$, $N_A = [0.35, 0.4]$, $\beta_A = 0.2$, $M_B = [0.4, 0.6]$, $\alpha_B = 0.7$, $N_B = [0.2, 0.3]$ *and* $\beta_B = 0.1$ *for all* $x \in X$. *Then* $\mathbf{A} \cup_R \mathbf{B} = \langle M_B/\alpha_A, N_B/\beta_A\rangle$ *and note that* $\alpha_A \in (M_B^-, M_B^+)$; *therefore,* $\mathbf{A} \cup_R \mathbf{B}$ *is not ECIS.*

2 *Let* $\mathbf{A} = \langle M_A/\alpha_A, N_A/\beta_A\rangle$ *and* $\mathbf{B} = \langle M_B/\alpha_B, N_B/\beta_B\rangle$ *be two ECISs in X. Let* $M_A = [0.2, 0.4]$, $\alpha_A = 0.1$, $N_A = [0.4, 0.6]$, $\beta_A = 0.5$, $M_B = [0.5, 0.7]$, $\alpha_B = 0.3$, $N_B = [0.1, 0.3]$ *and* $\beta_B = 0.6$ *for all* $x \in X$. *Then* $\mathbf{A} \cap_R \mathbf{B} = \langle M_A/\alpha_B, N_A/\beta_A\rangle$ *and, hence,* $\mathbf{A} \cap_R \mathbf{B}$ *is not ECIS.*

Theorem 4. *Let* $\mathbf{A} = \langle M_A/\alpha_A, N_A/\beta_A\rangle$ *and* $\mathbf{B} = \langle M_B/\alpha_B, N_B/\beta_B\rangle$ *be two ICISs in X, such that* $\max\{M_A^-, M_B^-\} \leq (\alpha_A \wedge \alpha_B)$ *and* $\min\{N_A^+, N_B^+\} \geq (\beta_A \vee \beta_B)$. *Then the R-union and R-intersection of* **A** *and* **B** *are ICISs.*

Proof. **A** and **B** are ICISs; therefore,

$$M_A^- \leq \alpha_A \leq M_A^+, N_A^- \leq \beta_A \leq N_A^+$$

$$M_B^- \leq \alpha_B \leq M_B^+ \, and \, N_B^- \leq \beta_B \leq N_B^+$$

which implies that

$$(\alpha_A \wedge \alpha_B) \leq (M_A \cup M_B)^+ \, and \, (\beta_A \vee \beta_B) \geq (N_A \cap N_B)^-.$$

It follows that

$$(M_A \cup M_B)^- = \max\{M_A^-, M_B^-\} \leq (\alpha_A \wedge \alpha_B) \leq (M_A \cup M_B)^+$$

and

$$(N_A \cap N_B)^- \leq (\beta_A \vee \beta_B) \leq \min\{N_A^+, N_B^+\} = (N_A \cap N_B)^+.$$

Hence, $\mathbf{A} \cup_R \mathbf{B}$ is ICIS. Similar arguments work in the case of $\mathbf{A} \cap_R \mathbf{B}$. □

Given two CISs $\mathbf{A} = \langle M_A/\alpha_A, N_A/\beta_A \rangle$ and $\mathbf{B} = \langle M_B/\alpha_B, N_B/\beta_B \rangle$ in X. If we exchange α_A for α_B and β_A for β_B, we denote these CISs by $\mathbf{A}^* = \langle M_A/\alpha_B, N_A/\beta_B \rangle$ and $\mathbf{B}^* = \langle M_B/\alpha_A, N_B/\beta_A \rangle$, respectively.
The next example shows that, for any two ECISs in X, $\mathbf{A}^*$ and $\mathbf{B}^*$ need not be ICISs in X.

Example 5.

1. *Let* $\mathbf{A} = \langle M_A/\alpha_A, N_A/\beta_A \rangle$ *and* $\mathbf{B} = \langle M_B/\alpha_B, N_B/\beta_B \rangle$ *be ICIS in* X*. Let* $M_A = [0.1, 0.3]$, $\alpha_A = 0.7$, $N_A = [0.5, 0.7]$, $\beta_A = 0.15$, $M_B = [0.4, 0.6]$, $\alpha_B = 0.35$, $N_B = [0.2, 0.3]$ *and* $\beta_B = 0.1$ *for all* $x \in X$*. Then it is easy to see that* $\mathbf{A}^* = \langle M_A/\alpha_B, N_A/\beta_B \rangle$ *and* $\mathbf{B}^* = \langle M_B/\alpha_A, N_B/\beta_A \rangle$ *are not ICISs in X.*
2. *Let* $X = \{a, b\}$*. Let* $\mathbf{A} = \langle M_A/\alpha_A, N_A/\beta_A \rangle$ *and* $\mathbf{B} = \langle M_B/\alpha_B, N_B/\beta_B \rangle$ *be two ECISs in X defined in Table 1. Moreover,* $\mathbf{A}^* = \langle M_A/\alpha_B, N_A/\beta_B \rangle$ *and* $\mathbf{B}^* = \langle M_B/\alpha_A, N_B/\beta_A \rangle$ *are not ICISs in X because* $\alpha_B(a) = 0.35 \notin [0.4, 0.6] = M_A(a)$, $\beta_B(a) = 0.25 \notin [0.1, 0.2] = N_A(a)$*. Moreover,* $\alpha_B(b) = 0.15 \notin [0.2, 0.4] = M_A(b)$ *and* $\beta_B(b) = 0.35 \notin [0.4, 0.6] = N_A(b)$*.*

Table 1. CISs **A** and **B**.

X	M_A/α_A	N_A/β_A	M_B/α_B	N_B/β_B
a	$[0.4, 0.6]/0.65$	$[0.1, 0.2]/0.35$	$[0.1, 0.3]/0.35$	$[0.4, 0.5]/0.25$
b	$[0.2, 0.4]/0.1$	$[0.4, 0.6]/0.7$	$[0.4, 0.5]/0.15$	$[0.1, 0.3]/0.35$

We will show through the following example that the P-union of two ECISs in X may not be an ICIS in X.

Example 6. *Consider again two ECISs,* **A** *and* **B***, as shown in Table 1. In this case,* $\mathbf{A} \cup_P \mathbf{B}$ *is not ICIS in X because* $(\alpha_A \vee \alpha_B)(a) = 0.65 \notin [0.4, 0.6] = M_A \cup M_B$, $(\beta_A \wedge \beta_B)(a) = 0.25 \notin [0.1, 0.2] = N_A \cap N_B$.

In the following result, we will find a condition for the P-union of two ECISs to be an ICIS.

Theorem 5. *For two ECISs* $\mathbf{A} = \langle M_A/\alpha_A, N_A/\beta_A \rangle$ *and* $\mathbf{B} = \langle M_B/\alpha_B, N_B/\beta_B \rangle$ *in X. If* $\mathbf{A}^* = \langle M_A/\alpha_B, N_A/\beta_B \rangle$ *and* $\mathbf{B}^* = \langle M_B/\alpha_A, N_B/\beta_A \rangle$ *are ICISs in X. Then* $\mathbf{A} \cup_P \mathbf{B}$ *and* $\mathbf{A} \cap_P \mathbf{B}$ *are ICISs in X.*

Proof. Since **A** and **B** are ECISs in X, then

$$\alpha_A \notin (M_A^-, M_A^+), \beta_A \notin (N_A^-, N_A^+),$$

$$\alpha_B \notin (M_B^-, M_B^+) \text{ and } \beta_B \notin (N_B^-, N_B^+).$$

For all $x \in X$. Since $\mathbf{A}^*$ and $\mathbf{B}^*$ are ICISs in X, then

$$M_A^- \leq \alpha_B \leq M_A^+, N_A^- \leq \beta_B \leq N_A^+$$

$$M_B^- \leq \alpha_A \leq M_B^+ \text{ and } N_B^- \leq \beta_A \leq N_B^+$$

for all $x \in X$. Thus, we can consider the following cases for any $x \in X$.

Case 1

$$\alpha_A \leq M_A^- \leq \alpha_B \leq M_A^+, \beta_A \leq N_A^- \leq \beta_B \leq N_A^+,$$
$$\alpha_B \leq M_B^- \leq \alpha_A \leq M_B^+ \text{ and } \beta_B \leq N_B^- \leq \beta_A \leq N_B^+.$$

Case 2

$$M_A^- \leq \alpha_B \leq M_A^+ \leq \alpha_A, N_A^- \leq \beta_B \leq N_A^+ \leq \beta_A,$$

$$M_B^- \leq \alpha_A \leq M_B^+ \leq \alpha_B \, and \, N_B^- \leq \beta_A \leq N_B^+ \leq \beta_B.$$

Case 3

$$\alpha_A \leq M_A^- \leq \alpha_B \leq M_A^+, \beta_A \leq N_A^- \leq \beta_B \leq N_A^+,$$
$$M_B^- \leq \alpha_A \leq M_B^+ \leq \alpha_B \, and \, N_B^- \leq \beta_A \leq N_B^+ \leq \beta_B.$$

Case 4

$$M_A^- \leq \alpha_B \leq M_A^+ \leq \alpha_A, N_A^- \leq \beta_B \leq N_A^+ \leq \beta_A,$$
$$\alpha_B \leq M_B^- \leq \alpha_A \leq M_B^+ \, and \, \beta_B \leq N_B^- \leq \beta_A \leq N_B^+.$$

The arguments in all cases are similar; therefore, we consider the first case.
We have $\alpha_A = M_A^- = M_B^- = \alpha_B$ and $\beta_A = N_A^- = N_B^- = \beta_B$.
Since $\mathbf{A}^*$ and $\mathbf{B}^*$ are ICISs in X, then

$$\alpha_B \leq M_A^+, \alpha_A \leq M_B^+, \beta_B \leq N_A^+ \, and \, \beta_A \leq N_B^+.$$

It follows that

$$\begin{aligned}(M_A \cup M_B)^- &= \max\{M_A^-, M_B^-\} = (\alpha_A \vee \alpha_B)\\ &\leq \max\{M_A^+, M_B^+\} = (M_A \cup M_B)^+ \, and\\ (N_A \cap N_B)^- &= \min\{N_A^-, N_B^-\} = (\beta_A \wedge \beta_B)\\ &\leq \min\{N_A^+, N_B^+\} = (N_A \cup N_B)^+.\end{aligned}$$

Hence, $\mathbf{A} \cup_P \mathbf{B}$ is ICIS. Similar steps can be used for $\mathbf{A} \cap_P \mathbf{B}$. □

From Example 2, it can be easily seen that the P-union and P-intersections of ECISs are not necessarily the ECISs in X. In the next result, we will show when the P-union and P-intersection of two ECISs are ECISs in X.

Theorem 6. *Let* $\mathbf{A} = \langle M_A/\alpha_A, N_A/\beta_A \rangle$ *and* $\mathbf{B} = \langle M_B/\alpha_B, N_B/\beta_B \rangle$ *be two ECISs in X, such that*

$$\begin{aligned}&\min\{\max\{M_A^+, M_B^-\}, \max\{M_A^-, M_B^+\}\} \geq (\alpha_A \wedge \alpha_B)\\ &> \max\{\min\{M_A^+, M_B^-\}, \min\{M_A^-, M_B^+\}\} and\\ &\min\{\max\{N_A^+, N_B^-\}, \max\{N_A^-, N_B^+\}\} > (\beta_A \vee \beta_B)\\ &\geq \max\{\min\{N_A^+, N_B^-\}, \min\{N_A^-, N_B^+\}\},\end{aligned}$$

then $\mathbf{A} \cap_P \mathbf{B}$ *is ECIS in X.*

Proof. Take

$$\begin{aligned}\alpha_x &= \min\{\max\{M_A^+, M_B^-\}, \max\{M_A^-, M_B^+\}\},\\ \beta_x &= \max\{\min\{M_A^+, M_B^-\}, \min\{M_A^-, M_B^+\}\},\\ \alpha_x^* &= \min\{\max\{N_A^+, N_B^-\}, \max\{N_A^-, N_B^+\}\} and\\ \beta_x^* &= \max\{\min\{N_A^+, N_B^-\}, \min\{N_A^-, N_B^+\}\}\end{aligned}$$

then α_x is one of $M_A^-, M_B^-, M_A^+, M_B^+$ and α_x^* is one of $N_A^-, N_B^-, N_A^+, N_B^+$. We will consider the case when $\alpha_x = M_A^-$ and $\alpha_x^* = N_A^-$ or $\alpha_x = M_A^+$ and $\alpha_x^* = N_A^+$. Similar arguments will work for all remaining cases.

If $\alpha_x = M_A^-$ and $\alpha_x^* = N_A^-$, then

$$M_B^- \leq M_B^+ \leq M_A^- \leq M_A^+$$

$$N_B^- \leq N_B^+ \leq N_A^- \leq N_A^+$$

and so $\beta_x = M_B^+$ and $\beta_x^* = N_B^+$. Thus,

$$M_B^- = (M_A \cap M_B)^- \leq (M_A \cap M_B)^+ = M_B^+ = \beta_x < (\alpha_A \wedge \alpha_B),$$

$$N_A^- = (N_A \cup N_B)^- = \alpha_x > (\beta_A \vee \beta_B)$$

and, hence,

$$(\alpha_A \wedge \alpha_B) \notin ((M_A \cap M_B)^-, (M_A \cap M_B)^+) and$$
$$(\beta_A \vee \beta_B) \notin ((N_A \cup N_B)^-, (N_A \cup N_B)^+).$$

If $\alpha_x = M_A^+$ and $\alpha_x^* = N_A^+$, then

$$M_B^- \leq M_A^+ \leq M_B^+ and N_B^- \leq N_A^+ \leq N_B^+$$

so

$$\beta_x = \max\{M_A^-, M_B^-\} and \beta_x^* = \max\{N_A^-, N_B^-\}.$$

Assume that $\beta_x = M_A^-$ and $\beta_x^* = N_A^-$, then

$$M_B^- \leq M_A^- < (\alpha_A \wedge \alpha_B) \leq M_A^+ \leq M_B^+ and$$

$$N_B^- \leq N_A^- \leq (\beta_A \vee \beta_B) < N_A^+ \leq N_B^+.$$

From the above inequality, we have the following cases

Case-1

$$M_B^- \leq M_A^- < (\alpha_A \wedge \alpha_B) < M_A^+ \leq M_B^+ and$$
$$N_B^- \leq N_A^- < (\beta_A \vee \beta_B) < N_A^+ \leq N_B^+$$

Case-2

$$M_B^- \leq M_A^- < (\alpha_A \wedge \alpha_B) = M_A^+ \leq M_B^+ and$$
$$N_B^- \leq N_A^- = (\beta_A \vee \beta_B) \leq N_A^+ \leq N_B^+.$$

Case-1 contradicts the fact that CISs **A** and **B** are ECISs. From Case-2, it implies that

$$(\alpha_A \wedge \alpha_B) \notin ((M_A \cap M_B)^-, (M_A \cap M_B)^+) and$$

$$(\beta_A \vee \beta_B) \notin ((N_A \cup N_B)^-, (N_A \cup N_B)^+)$$

since

$$(\alpha_A \wedge \alpha_B) = M_A^+ = (M_A \cap M_B)^+ and$$
$$(\beta_A \vee \beta_B) = N_A^- = (N_A \cup N_B)^-.$$

Assume that $\beta_x = M_B^-$ and $\beta_x^* = N_B^-$, then

$$M_A^- \leq M_B^- < (\alpha_A \wedge \alpha_B) \leq M_A^+ \leq M_B^+ and$$

$$N_A^- \leq N_B^- \leq (\beta_A \vee \beta_B) \leq N_A^+ \leq N_B^+.$$

We now have two cases.

Case-1

$$M_A^- \leq M_B^- < (\alpha_A \wedge \alpha_B) < M_A^+ \leq M_B^+ \text{ and}$$
$$N_A^- \leq N_B^- < (\beta_A \vee \beta_B) < N_A^+ \leq N_B^+.$$

Case-2

$$M_A^- \leq M_B^- < (\alpha_A \wedge \alpha_B) = M_A^+ \leq M_B^+ \text{ and}$$
$$N_A^- \leq N_B^- = (\beta_A \vee \beta_B) < N_A^+ \leq N_B^+.$$

Case-1 contradicts that **A** and **B** are ECISs. From Case-2, it implies that

$$(\alpha_A \wedge \alpha_B) \notin ((M_A \cap M_B)^-, (M_A \cap M_B)^+)\, and$$

$$(\beta_A \vee \beta_B) \notin ((N_A \cup N_B)^-, (N_A \cup N_B)^+)$$

since

$$(\alpha_A \wedge \alpha_B) = M_A^+ = (M_A \cap M_B)^+ and$$

$$(\beta_A \vee \beta_B) = N_B^- = (N_A \cup N_B)^-.$$

Similar results can be obtained if we assume

$$\beta_x = M_B^- and \beta_x^* = N_A^- or \beta_x = M_A^- \, and \, \beta_x^* = N_B^-$$

Hence, the P-intersection of **A** and **B** is ECIS in X. □

Theorem 7. *Let* $\mathbf{A} = \langle M_A/\alpha_A, N_A/\beta_A \rangle$ *and* $\mathbf{B} = \langle M_B/\alpha_B, N_B/\beta_B \rangle$ *be two ECISs in X, such that*

$$\min\{\max\{M_A^+, M_B^-\}, \max\{M_A^-, M_B^+\}\} > (\alpha_A \vee \alpha_B)$$
$$\geq \max\{\min\{M_A^+, M_B^-\}, \min\{M_A^-, M_B^+\}\} and$$
$$\min\{\max\{N_A^+, N_B^-\}, \max\{N_A^-, N_B^+\}\} \geq (\beta_A \wedge \beta_B)$$
$$> \max\{\min\{N_A^+, N_B^-\}, \min\{N_A^-, N_B^+\}\},$$

then $\mathbf{A} \cup_P \mathbf{B}$ *is ECIS in X.*

Proof. The proof is similar to Theorem 6; therefore, we omit the details. □

Example 7. *Let* $\mathbf{A} = \langle M_A/\alpha_A, N_A/\beta_A \rangle$ *and* $\mathbf{B} = \langle M_B/\alpha_B, N_B/\beta_B \rangle$ *be two ECISs in* $X = \{a, b, c\}$ *as shown in Table 2. Then,* **A** *and* **B** *always satisfy the following conditions.*

$$\min\{\max\{M_A^+, M_B^-\}, \max\{M_A^-, M_B^+\}\} = (\alpha_A \vee \alpha_B)$$
$$> \max\{\min\{M_A^+, M_B^-\}, \min\{M_A^-, M_B^+\}\} and$$
$$\min\{\max\{N_A^+, N_B^-\}, \max\{N_A^-, N_B^+\}\} > (\beta_A \wedge \beta_B)$$
$$= \max\{\min\{N_A^+, N_B^-\}, \min\{N_A^-, N_B^+\}\}.$$

However, the P-union of **A** *and* **B** *is not ECIS because* $(\alpha_A \vee \alpha_B)(a) = 0.2 \in [0.1, 0.3] = [(M_A \cup M_B)^-(a), (M_A \cup M_B)^+(a)]$ *and* $(\beta_A \wedge \beta_B)(a) = 0.45 \in [0.4, 0.5] = [(N_A \cap N_B)^-(a), (N_A \cap N_B)^+(a)]$.

Table 2. CISs **A** and **B**.

X	M_A/α_A	N_A/β_A	M_B/α_B	N_B/β_B
a	$[0.1, 0.2]/0.2$	$[0.45, 0.6]/0.45$	$[0.05, 0.3]/0.03$	$[0.4, 0.5]/0.6$
b	$[0.1, 0.4]/0.05$	$[0.5, 0.6]/0.7$	$[0.2, 0.3]/0.3$	$[0.55, 0.65]/0.55$
c	$[0.6, 0.7]/0.7$	$[0.1, 0.15]/0.1$	$[0.5, 0.8]/0.4$	$[0.05, 0.2]/0.3$

From Example 4, it can be easily observed that the R-union and R-intersection of ECISs may not be ECISs in X. In the next result, we will show that the R-union and R-intersection of two ECISs are ECISs in X.

Theorem 8. *Let* $\mathbf{A} = \langle M_A/\alpha_A, N_A/\beta_A \rangle$ *and* $\mathbf{B} = \langle M_B/\alpha_B, N_B/\beta_B \rangle$ *be two ECISs in X, such that*

$$\min\{\max\{M_A^+, M_B^-\}, \max\{M_A^-, M_B^+\}\} > (\alpha_A \wedge \alpha_B)$$

$$\geq \max\{\min\{M_A^+, M_B^-\}, \min\{M_A^-, M_B^+\}\} and$$
$$\min\{\max\{N_A^+, N_B^-\}, \max\{N_A^-, N_B^+\}\} \geq (\beta_A \vee \beta_B)$$
$$> \max\{\min\{N_A^+, N_B^-\}, \min\{N_A^-, N_B^+\}\},$$

then $\mathbf{A} \cup_R \mathbf{B}$ *is ECIS in X.*

Proof. Take

$$\alpha_x = \min\{\max\{M_A^+, M_B^-\}, \max\{M_A^-, M_B^+\}\},$$
$$\beta_x = \max\{\min\{M_A^+, M_B^-\}, \min\{M_A^-, M_B^+\}\},$$
$$\alpha_x^* = \min\{\max\{N_A^+, N_B^-\}, \max\{N_A^-, N_B^+\}\} and$$
$$\beta_x^* = \max\{\min\{N_A^+, N_B^-\}, \min\{N_A^-, N_B^+\}\}$$

then α_x is one of $M_A^-, M_B^-, M_A^+, M_B^+$ and α_x^* is one of $N_A^-, N_B^-, N_A^+, N_B^+$. We will consider the case when $\alpha_x = M_B^-$ and $\alpha_x^* = N_B^-$ or $\alpha_x = M_B^+$ and $\alpha_x^* = N_B^+$. Similar arguments will work for all remaining cases.

If $\alpha_x = M_B^-$ and $\alpha_x^* = N_B^-$, then

$$M_A^- \leq M_A^+ \leq M_B^- \leq M_B^+ and$$
$$N_A^- \leq N_A^+ \leq N_B^- \leq N_B^+$$

so $\beta_x = M_A^+$ and $\beta_x^* = N_A^+$. Thus,

$$M_B^- = (M_A \cup M_B)^- = \alpha_x > (\alpha_A \wedge \alpha_B) and$$
$$N_A^+ = (N_A \cap N_B)^+ = \beta_x^* < (\beta_A \wedge \beta_B)$$

and, hence,

$$(\alpha_A \wedge \alpha_B) \notin ((M_A \cup M_B)^-, (M_A \cup M_B)^+) and$$
$$(\beta_A \vee \beta_B) \notin ((N_A \cap N_B)^-, (N_A \cap N_B)^+).$$

If $\alpha_x = M_B^+$ and $\alpha_x^* = N_B^+$, then

$$M_A^- \leq M_B^+ \leq M_A^+ and N_A^- \leq N_B^+ \leq N_A^+$$

and so

$$\beta_x = \max\{M_A^-, M_B^-\} and \beta_x^* = \max\{N_A^-, N_B^-\}.$$

Assume that $\beta_x = M_A^-$ and $\beta_x^* = N_A^-$, then

$$M_B^- \leq M_A^+ < (\alpha_A \wedge \alpha_B) < M_B^+ \leq M_A^+ and$$
$$N_B^- \leq N_A^- < (\beta_A \vee \beta_B) \leq N_B^+ \leq N_A^+.$$

We have two cases

Case-1

$$M_B^- \leq M_A^- < (\alpha_A \wedge \alpha_B) < M_B^+ \leq M_A^+ and$$
$$N_B^- \leq N_A^- < (\beta_A \vee \beta_B) < N_B^+ \leq N_A^+.$$

Case-2

$$M_B^- \leq M_A^- = (\alpha_A \wedge \alpha_B) \leq M_B^+ \leq M_A^+ and$$
$$N_B^- \leq N_A^- < (\beta_A \vee \beta_B) = N_B^+ \leq N_A^+.$$

Case-1 contradicts the fact that CISs **A** and **B** are ECISs. From Case-2, it implies that

$$(\alpha_A \wedge \alpha_B) \notin ((M_A \cup M_B)^-, (M_A \cup M_B)^+) and$$

$$(\beta_A \vee \beta_B) \notin ((N_A \cap N_B)^-, (N_A \cap N_B)^+)$$

since

$$(\alpha_A \wedge \alpha_B) = M_A^- = (M_A \cup M_B)^+ and$$
$$(\beta_A \vee \beta_B) = N_B^+ = (N_A \cap N_B)^+.$$

Assume that $\beta_x = M_B^-$ and $\beta_x^* = N_B^-$, then

$$M_A^- \leq M_B^- \leq (\alpha_A \wedge \alpha_B) \leq M_B^+ \leq M_A^+ and$$

$$N_A^- \leq N_B^- < (\beta_A \vee \beta_B) \leq N_B^+ \leq N_A^+.$$

We have two cases
Case-1

$$M_A^- \leq M_B^- < (\alpha_A \wedge \alpha_B) < M_B^+ \leq M_A^+ and$$
$$N_A^- \leq N_B^- < (\beta_A \vee \beta_B) < N_B^+ \leq N_A^+$$

Case-2

$$M_A^- \leq M_B^- = (\alpha_A \wedge \alpha_B) < M_B^+ \leq M_A^+ and$$
$$N_A^- \leq N_B^- < (\beta_A \vee \beta_B) = N_B^+ \leq N_A^+.$$

Case-1 contradicts the fact that CISs **A** and **B** are ECISs. From Case-2, it implies that

$$(\alpha_A \wedge \alpha_B) \notin ((M_A \cup M_B)^-, (M_A \cup M_B)^+) and$$

$$(\beta_A \vee \beta_B) \notin ((N_A \cap N_B)^-, (N_A \cap N_B)^+)$$

since

$$(\alpha_A \wedge \alpha_B) = M_B^- = (M_A \cup M_B)^- and$$
$$(\beta_A \vee \beta_B) = N_B^+ = (N_A \cap N_B)^+.$$

Similar results can be obtained if we assume

$$\beta_x = M_B^- and \beta_x^* = N_A^- or \beta_x = M_A^- and \beta_x^* = N_B^-$$

Hence $\mathbf{A} \cup_R \mathbf{B}$ is ECIS in X. □

Example 8. *Let* $\mathbf{A} = \langle M_A/\alpha_A, N_A/\beta_A \rangle$ *and* $\mathbf{B} = \langle M_B/\alpha_B, N_B/\beta_B \rangle$ *be two ECISs in a set* $X = \{a, b, c\}$ *as shown in Table 3. Then it is easy to see that* **A** *and* **B** *satisfy the conditions*

$$\min\{\max\{M_A^+, M_B^-\}, \max\{M_A^-, M_B^+\}\} = (\alpha_A \wedge \alpha_B)$$

$$> \max\{\min\{M_A^+, M_B^-\}, \min\{M_A^-, M_B^+\}\} and$$
$$\min\{\max\{N_A^+, N_B^-\}, \max\{N_A^-, N_B^+\}\} > (\beta_A \vee \beta_B)$$
$$= \max\{\min\{N_A^+, N_B^-\}, \min\{N_A^-, N_B^+\}\}.$$

However, $\mathbf{A} \cup_R \mathbf{B}$ *is not ECIS because*
$(\alpha_A \wedge \alpha_B)(a) = 0.7 \in [0.6, 0.8] = [(M_A \cup M_B)^-(a), (M_A \cup M_B)^+(a)]$ *and* $(\beta_A \vee \beta_B)(a) = 0.1 \in [0.05, 0.15] = [(N_A \cap N_B)^-(a), (N_A \cap N_B)^+(a)]$.

Table 3. CISs **A** and **B**.

X	M_A/α_A	N_A/β_A	M_B/α_B	N_B/β_B
a	$[0.6, 0.7]/0.7$	$[0.1, 0.15]/0.1$	$[0.5, 0.8]/0.9$	$[0.05, 0.2]/0.03$
b	$[0.1, 0.4]/0.5$	$[0.5, 0.6]/0.5$	$[0.2, 0.3]/0.3$	$[0.55, 0.65]/0.55$
c	$[0.1, 0.2]/0.2$	$[0.45, 0.6]/0.45$	$[0.05, 0.3]/0.4$	$[0.4, 0.5]/0.3$

The following theorems can be easily verified and proved; therefore, we omit the details.

Theorem 9. *Let* $\mathbf{A} = \langle M_A/\alpha_A, N_A/\beta_A \rangle$ *and* $\mathbf{B} = \langle M_B/\alpha_B, N_B/\beta_B \rangle$ *be two ECISs in X, such that*

$$\min\{\max\{M_A^+, M_B^-\}, \max\{M_A^-, M_B^+\}\} \geq (\alpha_A \vee \alpha_B)$$
$$> \max\{\min\{M_A^+, M_B^-\}, \min\{M_A^-, M_B^+\}\} and$$
$$\min\{\max\{N_A^+, N_B^-\}, \max\{N_A^-, N_B^+\}\} > (\beta_A \wedge \beta_B)$$
$$\geq \max\{\min\{N_A^+, N_B^-\}, \min\{N_A^-, N_B^+\}\},$$

then $\mathbf{A} \cap_R \mathbf{B}$ *is also an ECIS in X.*

Theorem 10. *Let* $\mathbf{A} = \langle M_A/\alpha_A, N_A/\beta_A \rangle$ *and* $\mathbf{B} = \langle M_B/\alpha_B, N_B/\beta_B \rangle$ *be two ICISs in X. If*

$$(\alpha_A \wedge \alpha_B) \leq \max\{M_A^-, M_B^-\}$$
$$(\beta_A \vee \beta_B) \geq \min\{N_A^+, N_B^+\},$$

then $\mathbf{A} \cup_R \mathbf{B}$ *is an ECIS in X.*

Theorem 11. *Let* $\mathbf{A} = \langle M_A/\alpha_A, N_A/\beta_A \rangle$ *and* $\mathbf{B} = \langle M_B/\alpha_B, N_B/\beta_B \rangle$ *be two ICISs in X. If*

$$(\alpha_A \vee \alpha_B) \geq \min\{M_A^+, M_B^+\}$$
$$(\beta_A \wedge \beta_B) \leq \max\{N_A^-, N_B^-\},$$

then $\mathbf{A} \cap_R \mathbf{B}$ *is ECIS in X.*

4. MCDM Method Based on Cubic Intuitionistic Sets

In this section, we will apply the proposed operations to deal with the MCDM problems using CISs.

Let $A = \{A_1, A_2, \ldots, A_m\}$ be a set of alternatives, $C = \{C_1, C_2, \ldots, C_n\}$ be a set of criteria, and $E = \{e_1, e_2, \ldots, e_K\}$ be a set of experts. Suppose each alternative $A_i (i = 1, 2, \ldots, m)$ is assessed by the expert $e_k (k = 1, 2, \ldots, K)$ with respect to the criteria $C_j (j = 1, 2, \ldots, n)$ using CISs. The proposed MCDM method is based on the following steps.

Step 1 Construct the decision matrices $R^k = (r_{ij}^k)_{m \times n}$ based on the assessed values of expert $e_k (k = 1, 2, \ldots, K)$ in the form of CISs r_{ij}^k.

Step 2 Calculate the aggregated decision matrix $R = (r_{ij})_{m \times n}$ by using the proposed operations as discussed in Definition 12 where $r_{ij} = \underset{k=1,2,\ldots K}{\cup_P} r_{ij}^k$ or $r_{ij} = \underset{k=1,2,\ldots K}{\cup_R} r_{ij}^k$.

Step 3 Calculate the score value of each r_{ij} of the aggregated decision matrix R by using Definition 9.

Step 4 Calculate the preference values of each alternative $A_i (i = 1, 2, \ldots, m)$ where $P(A_i) = \sum_{i=1}^m \sum_{j=1}^n r_{ij}$.

Step 5 Generate the ranking order of alternatives according to the non-increasing order of the preference values.

An Application Example

Let us suppose that a technical committee composed of three technicians/experts $E = \{e_1, e_2, e_3\}$ wishes to select the best available washing machine on the market. Suppose, there are four types of washing machines $A = \{A_1, A_2, A_3, A_4\}$ available in the market and the experts are requested to select the best one amongst the four with respect to the criteria set $C = \{C_1 = \text{eco-friendly}, C_2 = \text{capacity}, C_3 = \text{price}\}$. Suppose the expert $e_k(k = 1, 2, 3)$ assessed each alternative $A_i(i = 1, 2, \ldots, 4)$ under the criteria $C_j(j = 1, 2, 3)$ by using the CISs. We will now proceed with the following steps.

Step 1 According to the expert's opinion, the individual decision matrices R^1, R^2, R^3 are constructed, which can be seen in Tables 4–6.

Table 4. Decision matrix R^1 provided by expert e_1.

Alt.	C_1	C_2	C_3
A_1	$\langle[0.7, 0.8]/0.35, [0.1, 0.2]/0.6\rangle$	$\langle[0.5, 0.6]/0.5, [0.2, 0.3]/0.2\rangle$	$\langle[0.6, 0.7]/0.4, [0.1, 0.2]/0.7\rangle$
A_2	$\langle[0.2, 0.3]/0.25, [0.3, 0.4]/0.7\rangle$	$\langle[0.3, 0.4]/0.65, [0.5, 0.6]/0.2\rangle$	$\langle[0.2, 0.3]/0.7, [0.4, 0.5]/0.2\rangle$
A_3	$\langle[0.8, 0.9]/0.7, [0.05, 0.1]/0.3\rangle$	$\langle[0.7, 0.8]/0.2, [0.1, 0.2]/0.4\rangle$	$\langle[0.6, 0.7]/0.3, [0.2, 0.3]/0.6\rangle$
A_4	$\langle[0.5, 0.6]/0.5, [0.1, 0.2]/0.3\rangle$	$\langle[0.6, 0.7]/0.3, [0.2, 0.3]/0.4\rangle$	$\langle[0.4, 0.5]/0.6, [0.2, 0.3]/0.2\rangle$

Table 5. Decision matrix R^2 provided by expert e_2.

Alt.	C_1	C_2	C_3
A_1	$\langle[0.6, 0.7]/0.3, [0.1, 0.2]/0.5\rangle$	$\langle[0.45, 0.5]/0.6, [0.25, 0.35]/0.3\rangle$	$\langle[0.5, 0.6]/0.7, [0.2, 0.3]/0.2\rangle$
A_2	$\langle[0.25, 0.4]/0.5, [0.4, 0.5]/0.4\rangle$	$\langle[0.4, 0.5]/0.6, [0.3, 0.4]/0.3\rangle$	$\langle[0.3, 0.4]/0.4, [0.5, 0.6]/0.6\rangle$
A_3	$\langle[0.7, 0.8]/0.8, [0.1, 0.2]/0.1\rangle$	$\langle[0.8, 0.9]/0.7, [0, 0.1]/0.3\rangle$	$\langle[0.5, 0.6]/0.8, [0.1, 0.2]/0.2\rangle$
A_4	$\langle[0.4, 0.5]/0.4, [0.1, 0.2]/0.5\rangle$	$\langle[0.5, 0.6]/0.4, [0.2, 0.3]/0.6\rangle$	$\langle[0.5, 0.6]/0.5, [0.2, 0.3]/0.4\rangle$

Table 6. Decision matrix R^3 provided by expert e_3.

Alt.	C_1	C_2	C_3
A_1	$\langle[0.6, 0.7]/0.7, [0.2, 0.3]/0.2\rangle$	$\langle[0.55, 0.6]/0.8, [0.2, 0.3]/0.1\rangle$	$\langle[0.65, 0.7]/0.6, [0.2, 0.3]/0.3\rangle$
A_2	$\langle[0.2, 0.3]/0.5, [0.4, 0.5]/0.4\rangle$	$\langle[0.3, 0.4]/0.6, [0.4, 0.5]/0.1\rangle$	$\langle[0.25, 0.3]/0.4, [0.5, 0.6]/0.5\rangle$
A_3	$\langle[0.7, 0.85]/0.6, [0.1, 0.15]/0.2\rangle$	$\langle[0.75, 0.8]/0.6, [0.1, 0.2]/0.3\rangle$	$\langle[0.6, 0.7]/0.8, [0.1, 0.2]/0.2\rangle$
A_4	$\langle[0.5, 0.6]/0.7, [0.2, 0.3]/0.2\rangle$	$\langle[0.5, 0.6]/0.5, [0.2, 0.3]/0.4\rangle$	$\langle[0.4, 0.5]/0.7, [0.1, 0.2]/0.1\rangle$

Step 2 The aggregated decision matrix $R = (r_{ij})_{4\times 3}$ is calculated with the help of the proposed operation (P-union) as introduced in Definition 12 where $r_{ij} = \underset{k=1,2,3}{\cup_P} r_{ij}^k$.

The aggregated decision matrix R is shown in Table 7.

Table 7. Aggregated decision matrix R by applying the P-union operation.

Alt.	C_1	C_2	C_3
A_1	$\langle[0.7, 0.8]/0.7, [0.1, 0.2]/0.2\rangle$	$\langle[0.55, 0.6]/0.8, [0.2, 0.3]/0.1\rangle$	$\langle[0.65, 0.7]/0.7, [0.1, 0.2]/0.2\rangle$
A_2	$\langle[0.25, 0.4]/0.5, [0.3, 0.4]/0.4\rangle$	$\langle[0.4, 0.5]/0.65, [0.3, 0.4]/0.1\rangle$	$\langle[0.3, 0.4]/0.7, [0.4, 0.5]/0.2\rangle$
A_3	$\langle[0.8, 0.9]/0.8, [0.05, 0.1]/0.1\rangle$	$\langle[0.8, 0.9]/0.7, [0, 0.1]/0.3\rangle$	$\langle[0.6, 0.7]/0.8, [0.1, 0.2]/0.2\rangle$
A_4	$\langle[0.5, 0.6]/0.7, [0.1, 0.2]/0.2\rangle$	$\langle[0.6, 0.7]/0.5, [0.2, 0.3]/0.4\rangle$	$\langle[0.5, 0.6]/0.7, [0.1, 0.2]/0.1\rangle$

Step 3 By using Definition 9, we will calculate the score value of each r_{ij} of the aggregated decision matrix R. The matrix of the score values of the elements of R is shown in Table 8.

Table 8. Score values of the aggregated decision matrix.

Alt.	C_1	C_2	C_3
A_1	0.5667	0.4500	0.5167
A_2	0.0167	0.2500	0.1000
A_3	0.7500	0.6667	0.5333
A_4	0.4333	0.3000	0.4667

Step 4,5 Finally, the preference value $P(A_i), i = 1, 2, ..., 4$ of each alternative is calculated where $P(A_i) = \sum_{i=1}^{4} \sum_{j=1}^{3} r_{ij}$. The preference values of alternatives by using the P-union operation are given below:

$$P(A_1) = 0.5111, P(A_2) = 0.1222, P(A_3) = 0.6500, P(A_4) = 0.4000.$$

We can see that the ranking order of alternatives according to the non-increasing order of their preference values is $A_3 \succeq A_1 \succeq A_4 \succeq A_2$. Similarly, the preference value of each alternative by using the R-union operation is calculated and given as follows:

$$P(A_1) = 0.2778, P(A_2) = -0.0444, P(A_3) = 0.4389, P(A_4) = 0.2333$$

In this case, the ranking order of alternatives is $A_3 \succeq A_1 \succeq A_4 \succeq A_2$.

We can observe that the ranking order of alternatives by using the R-union operation is exactly the same as that obtained with the help of the P-union operation, which shows the robustness of the proposed approach. We can easily see that by using the P-intersection and R-intersection operations as discussed in Definition 12, the ranking order of alternatives will lead to the reverse order of the raking orders obtained in the P-union and R-union operations, respectively.

5. Conclusions

In this research work, we introduced a new modified form of CIS and discussed some of its related properties. We further introduced two types of CISs, i.e., ICIS and ECIS. The P-(R-) order, P-(R-) union, P-(R-) intersection of CISs, and some useful properties were also discussed with necessary examples. As a supplement, we proved that the P-union and P-intersection of ICISs are also ICISs. Some conditions for the P-(R-) union and P-(R-) intersection of two ECISs to be ICISs were also provided in this paper. We also provided a few conditions for the P-(R-) union and P-(R-) intersection of two ECISs to be ECISs. To check the effectiveness and validity of the proposed operations, we provided an application example at the end by solving a MCDM problem.

In future work, more research can be conducted regarding the intuitionistic cubic soft set and its application in information science and knowledge systems. We intend to apply the intuitionistic cubic soft sets to algebraic structures.

Author Contributions: Conceptualization, S.F., T.R., S.Z., H.S. and W.S.; methodology, S.F., T.R., S.Z., H.S. and W.S.; software, S.F., T.R., S.Z., H.S. and W.S.; validation, S.F., T.R., S.Z., H.S. and W.S.; formal analysis, S.F., T.R., S.Z., H.S. and W.S.; investigation, S.F., T.R., S.Z., H.S. and W.S.; resources, S.F., T.R., S.Z., H.S. and W.S.; data curation, S.F., T.R., S.Z., H.S. and W.S.; writing—original draft preparation, S.F., T.R., S.Z., H.S. and W.S.; writing—review and editing, S.F., T.R., S.Z., H.S. and W.S.; visualization, S.F., T.R., S.Z., H.S. and W.S.; supervision, S.F., T.R., S.Z., H.S. and W.S.; project administration, S.F., T.R., S.Z., H.S. and W.S.; funding acquisition, S.F., T.R., S.Z., H.S. and W.S. All authors have read and agreed to the published version of the manuscript.

Funding: The work was supported by the National Science Centre 2021/41/B/HS4/01296 (W.S.). and 2022/01/4/ST6/00028 (G.S.)

Institutional Review Board Statement: Not applicable.

Informed Consent Statement: Not applicable.

Data Availability Statement: Not applicable.

Acknowledgments: The authors would like to thank the editor and the anonymous reviewers, whose insightful comments and constructive suggestions helped us to significantly improve the quality of this paper.

Conflicts of Interest: The authors declare no conflict of interest.

References

1. Zadeh, L.A. Fuzzy sets. *Inf. Control.* **1965**, *8*, 338–353. [CrossRef]
2. Zadeh, L.A. The concept of linguistic variable and its application to approximate reasoning-I. *Inf. Sci.* **1975**, *8*, 199–249. [CrossRef]
3. Kaklauskas, A. Degree of project utility and investment value assessments. *Int. J. Comput. Commun. Control* **2016**, *11*, 666–683 [CrossRef]
4. Wątróbski, J.; Bączkiewicz, A.; Król, R.; Sałabun, W. Green electricity generation assessment using the CODAS-COMET method. *Ecol. Indic.* **2022**, *143*, 109391. [CrossRef]
5. Atanassov, K. Intuitionistic fuzzy sets. *Fuzzy Set Syst.* **1986**, *20*, 87–96. [CrossRef]
6. Atanassov, K.; Gargov, G. Interval-valued intuitionistic fuzzy sets. *Fuzzy Sets Syst.* **1989**, *31*, 343–349. [CrossRef]
7. Nayagam, V.L.G.; Muralikrishnan, S.; Sivaraman, G. Multi-criteria decision-making method based on interval-valued intuitionistic fuzzy sets. *Expert Syst. Appl.* **2011**, *38*, 464–1467.
8. Torra, V. Hesitant Fuzzy Sets. *Int. J. Intell. Syst.* **2010**, *25*, 529–539. [CrossRef]
9. Rodríguez, R.M.; Martínez, L.; Herrera, F. Hesitant fuzzy linguistic term sets for decision making. *IEEE Trans. Fuzzy Syst.* **2012**, *20*, 109–119. [CrossRef]
10. Molodtsov, D.A. Soft set theory-first results. *Comput. Math. Appl.* **1999**, *37*, 19–31. [CrossRef]
11. Maji, P.K.; Biswas, R.; Roy, A.R. Fuzzy soft sets. *J. Fuzzy Math.* **2001**, *9*, 589–602.
12. Amiri, M.; Hashemi-Tabatabaei, M.; Keshavarz-Ghorabaee, M.; Kaklauskas, A.; Zavadskas, E. K.; Antucheviciene, J. A Fuzzy Extension of Simplified Best-Worst Method (F-SBWM) and Its Applications to Decision-Making Problems. *Symmetry* **2023**, *15*, 81. [CrossRef]
13. Xiao, F.; Wen, J.; Pedrycz, W. Generalized divergence-based decision making method with an application to pattern classification. *IEEE Trans. Knowl. Data Eng.* **2022**. [CrossRef]
14. Xiao, F. Generalized quantum evidence theory. In *Applied Intelligence*; Springer: Berlin/Heidelberg, Germany, 2022; pp. 1–16.
15. Xiao, F. EFMCDM: Evidential fuzzy multicriteria decision making based on belief entropy. *IEEE Trans. Fuzzy Syst.* **2019**, *28*, 1477–1491. [CrossRef]
16. Xiao, F. A distance measure for intuitionistic fuzzy sets and its application to pattern classification problems. *IEEE Trans. Syst. Man Cybern. Syst.* **2019**, *51*, 3980–3992. [CrossRef]
17. Zhang, H.Y.; Zhang, W.X.; Mei, C.L. Entropy of interval-valued fuzzy sets based on distance and its relationship with similarity measure. *Knowl.-Based Syst.* **2009**, *22*, 449–454. [CrossRef]
18. Zeng, W.Y.; Guo, P. Normalized distance, similarity measure, inclusion measure and entropy of interval-valued fuzzy sets and their relationship. *Inf. Sci.* **2008**, *178*, 1334–1342. [CrossRef]
19. Grzegorzewski, P. Distances between intuitionistic fuzzy sets and/or interval-valued fuzzy sets based on the Hausdorff metric. *Fuzzy Set Syst.* **2004**, *148*, 319–328. [CrossRef]
20. Sambuc, R. Functions Φ-Flous, Appication à Ìaide au Diagnostic en Pathologie Thyroidienne. Ph.D. Thesis, University of Marseille, Marseille, France, 1975.
21. Kohout, L.J.; Bandler, W. Fuzzy interval inference utilizing the checklist paradigm and BK-relational products. In *Application of Interval Computations*; Kearfort, R.B., Eds.; Kluwer: Dordrecht, The Netherlands, 1996; pp. 291–335.
22. Gorzałczany, M.B. A method of inference in approximate reasoning based on interval-valued fuzzy sets. *Fuzzy Sets Syst.* **1987**, *21*, 1–17. [CrossRef]
23. Turksen, I.B. Interval-valued fuzzy sets based on normal forms. *Fuzzy Sets Syst.* **1986**, *20*, 191–210. [CrossRef]
24. Turksen, I.B. Interval-valued fuzzy sets and compensatory AND. *Fuzzy Sets Syst.* **1992**, *51*, 295–307. [CrossRef]
25. Turksen, I.B. Interval-valued strict preference with Zadeh triples. *Fuzzy Sets Syst.* **1996**, *78*, 183–195. [CrossRef]
26. Jun, Y.B.; Kim, C.S.; Yang, K.O. Cubic sets. *Ann. Fuzzy Math. Inform.* **2012**, *4*, 83–98.
27. Jun, Y.B. A novel extension of cubic sets and its applications in BCK=BCI-algebras. *Ann. Fuzzy Math. Inform.* **2017**, *14*, 475–486.
28. Jun, Y.B.; Song, S.Z.; Kim, S.J. Cubic interval-valued intuitionistic fuzzy sets and their application in BCK/BCI-algebras. *Axioms* **2018**, *7*, 7. [CrossRef]
29. Ali, M.; Deli, I.; Smarandache, F. The theory of neutrosophic cubic sets and their applications in pattern recognition. *J. Intell. Fuzzy Syst.* **2016**, *30*, 1957–1963. [CrossRef]

30. Kang, J.G.; Kim, C.S. Mappings of cubic sets. *Commun. Korean Math. Soc.* **2016**, *31*, 423–431. [CrossRef]
31. Chinnadurai, V.; Suganya, K.; Swaminathan, A. Some characterizations on cubic sets. *Int. J. Multidiscip. Res. Mod. Educ.* **2016**, *2*, 258–264.
32. Jun, Y.B.; Lee, K.J.; Kang, M.S. Cubic structures applied to ideals of BCI-algebras. *Comput. Math. Appl.* **2011**, *62*, 3334–3342. [CrossRef]

mathematics

MDPI

Article

MemConFuzz: Memory Consumption Guided Fuzzing with Data Flow Analysis

Chunlai Du [1], Zhijian Cui [1], Yanhui Guo [2,*], Guizhi Xu [1] and Zhongru Wang [1,3]

[1] School of Information Science and Technology, North China University of Technology, Beijing 100144, China
[2] Department of Computer Science, University of Illinois Springfield, Springfield, IL 62703, USA
[3] Chinese Academy of Cyberspace Studies, Beijing 100048, China
* Correspondence: yguo56@uis.edu

Abstract: Uncontrolled heap memory consumption, a kind of critical software vulnerability, is utilized by attackers to consume a large amount of heap memory and consequently trigger crashes. There have been few works on the vulnerability fuzzing of heap consumption. Most of them, such as MemLock and PerfFuzz, have failed to consider the influence of data flow. We proposed a heap memory consumption guided fuzzing model named MemConFuzz. It extracts the locations of heap operations and data-dependent functions through static data flow analysis. Based on the data dependency, we proposed a seed selection algorithm in fuzzing to assign more energy to the samples with higher priority scores. The experiment results showed that the MemConFuzz has advantages over AFL, MemLock, and PerfFuzz with more quantity and less time consumption in exploiting the vulnerability of heap memory consumption.

Keywords: fuzzing; memory consumption; data flow; taint analysis

MSC: 90C70

Citation: Du, C.; Cui, Z.; Guo, Y.; Xu, G.; Wang, Z. MemConFuzz: Memory Consumption Guided Fuzzing with Data Flow Analysis. *Mathematics* **2023**, *11*, 1222. https://doi.org/10.3390/math11051222

Academic Editor: Ivan Lorencin

Received: 30 January 2023
Revised: 23 February 2023
Accepted: 27 February 2023
Published: 2 March 2023

1. Introduction

Fuzzing is a kind of random testing technique and is widely used to discover vulnerabilities in computer programs. Blind samples mutation fuzzing models and coverage-guided fuzzing models fail to select interesting seeds and waste testing time. Many fuzzing models are currently guided by exploring ways to improve path coverage. It is believed that the more code blocks that can be covered, the more likely potential vulnerability will be triggered. Many state-of-the-art fuzzing models typically use information from the programs' control flow graph by the program under test (PUT) to determine which samples would be selected as seeds for further mutation. Although there has been a lot of research work on memory overflow vulnerability, most of these methods have mainly exploited memory corruption vulnerabilities, such as stack buffer overflow, use-after-free (UAF), out-of-bounds reading, and out-of-bounds writing, etc. Memory corruption occurs when the contents of memory are overwritten due to malicious instructions or normal instructions with unexpected data beyond the program's original intention. For example, a buffer overflow occurs when a program tries to copy data into a variable whose required memory length is larger than the target. When corrupted memory contents are later used, the program triggers a crash or turns into a shellcode. Most fuzzing models of memory corruption vulnerability depend on the control flow, and seldom on the data semantics.

Memory consumption is a different kind of memory vulnerability in contrast to memory corruption, which is more like a logical vulnerability potentially existing in the action sequence of memory allocation and deallocation. With one goal of making more efficient use of the memory, different code segments in general are stored in different memory areas, among which the stack area and heap area are the two most important types of memory areas. In the process of a program running, the stack area grows up or down by calling subfunctions. It contains local variables, stack register ebp of parent function, return address, and parameters from the parent function. Generally, the heap areas are a series of memory blocks allocated and freed by the programs, which can be used by the pointer of heap blocks. Memory consumption occurs in the process of heap allocation and release. When a program triggers instructions for heap memory allocation enough times without deallocating unused memory in time, it would likely lead to a crash. Uncontrolled heap memory consumption is therefore a critical issue of software security, and can also become an important vulnerability when attackers control execution flow to consume large amounts of memory, and thus, launch denial-of-service attacks.

To solve the problems in vulnerability fuzzing of heap consumption, we propose a heap memory consumption-guided fuzzing model named MemConFuzz in considering the data flow analysis. This paper makes the following contributions:

(1) A novel algorithm is proposed to obtain locations of heap memory operations by taint analysis based on data dependency. The relation of data dependency is deduced from CPG (Code Property Graph). The location serves as an important indicator for seed selection.
(2) A new algorithm for prioritizing seed selection is proposed based on data dependency for discovering memory consumption vulnerability. Input samples covering more heap operations and data-dependent functions will be assigned high scores, and they are chosen as seeds and assigned more energy in the fuzzing loop.
(3) A novel memory consumption guided fuzzing model, MemConFuzz, is proposed. Compared with AFL [1], MemLock [2], and PerfFuzz [3], MemConFuzz has a significant improvement in discovering memory consumption vulnerability with more quantity and lower time cost.

The rest of the paper is organized as follows. Section 2 introduces related work. Section 3 presents the algorithm for extracting locations of heap operations through taint analysis based on data dependency. In Section 4, the proposed MemConFuzz model is described. In Section 5, the experimental process and the results are discussed. Finally, we conclude the paper in Section 6.

2. Related Work

Methods of discovering vulnerability are divided into static techniques and dynamic techniques. Static methods are used to make classification between the target program and known CVE (Common Vulnerabilities and Exposures) code based on structural similarity or statistical similarity by artificial intelligence technology. Dynamic methods include generation fuzzing, coverage-guided fuzzing, and symbolic execution.

Generation fuzzing adopts a generator to create required samples by mapping out all possible fields of the target program. The generator then separately mutates each of these fields to potentially cause crashes. In the generating process, those methods may result in a large number of invalid samples being rejected by the program as they do not follow the correct format. Coverage-guided fuzzing models integrate instrumentation into the target program before tracing the running information. To discover the special target areas in the program, a directed greybox fuzzing is proposed. Symbolic execution analyzes the target program to determine what inputs cause each part of this program to execute. Through symbolic execution, the required samples that execute the constraint code path to reach the target basic block are solved by an SMT (Satisfiability Modulo Theories) solver.

2.1. Static Techniques Based on Artificial Intelligence

During the research of discovering the vulnerability, the bottlenecks are related to how to generate good samples, how to improve path coverage, and how to provide more knowledge support for dynamic methods. Artificial intelligence has been used in the field of vulnerability discovery in recent years.

Machine learning is the most important technology of artificial intelligence, which attains knowledge about features obtained by analyzing an existing vulnerability-related dataset. This knowledge can be used to analyze new objects and thus predict potentially vulnerable locations in static mode. Machine learning methods can be divided into traditional machine learning, deep learning, and reinforcement learning.

Rajpal [4] used neural networks to learn patterns in past samples to highlight useful locations for future mutations, and then improved the AFL approach. Samplefuzz [5] combined learn and fuzz algorithms to leverage learned samples' probability distribution to make the generation of grammar suitable samples by using past samples and a neural network-based statistical machine learning. NEUZZ [6] leveraged neural networks to model the branching behavior of programs, generating interesting seeds by strategically modifying certain bytes of existing samples to trigger edges that had not yet been executed. Angora [7] modeled the target behavior, treated the mutation problem as a search problem, and applied the search algorithm in machine learning, which used a discrete function to represent the path from the beginning of the program to a specific branch under path constraints, and thus used the gradient descent search algorithm to find a set of inputs that satisfied the constraint and make the program go through that particular branch. Cheng [8] used RNNs to predict new paths of the program and then fed these paths into a Seq2Seq model, increasing the coverage of samples in PDF, PNG, and TFF formats. SySeVR [9] proposed a systematic framework for using deep learning to discover vulnerabilities. Based on Syntax, Semantics, and Vector Representations, SySeVR focuses on obtaining program representations that can accommodate syntax and semantic information pertinent to vulnerabilities. VulDeePecker [10] is a deep learning-based vulnerability detection system, which has presented some preliminary principles for guiding the practice of applying deep learning to vulnerability detection. μVulDeePecker [11] proposed a deep learning-based system for multiclass vulnerability detection. It introduced the concept called code attention to learn local features and pinpoint types of vulnerabilities.

However, most of these works are computationally intensive. The cost is very high because deep learning requires a large amount of data and computing power. The quality and quantity of the training data set have a direct impact on the accuracy of the training model, and there is also a key challenge to accurately locate the instructions where the vulnerability occurs.

2.2. Dynamic Execution Fuzzing Technique

Fuzzing has gained popularity as a useful and scalable approach for discovering software vulnerabilities. In the process of dynamic execution, that is, the fuzzing loop, the fuzzer generally uses the seed selection algorithm to select favorable seeds based on the feedback information of PUT execution, and then performs seed mutation according to a series of strategies to generate new samples and explore paths of the target program. Fuzzing is widely used to test application software, libraries, kernel codes, protocols, etc. Furthermore, symbolic execution is another important approach that can create a sample corresponding to a specific constraint path by the SMT solver. The following mainly introduces several popular dynamic technologies and methods in fuzzing.

A. Coverage-guide fuzzing

Coverage-guide greybox fuzzing (CGF) is one of the most effective techniques to discover vulnerabilities. CGF usually uses path coverage information to guide path exploration. In order to improve the coverage of fuzzers, researchers have focused on optimizing the coverage guide engine, which is the main component of fuzzers.

LibFuzzer [12] provided samples into the library through a specific fuzzing entry point, used LLVM's SanitizerCoverage tool to obtain code coverage, and then performed mutations on the sample to maximize coverage. Honggfuzz [13] proposed a genetic algorithm to efficiently mutate seeds. AFL [1] is a coverage-based fuzzing tool that captures basic block transitions by instrumentation and records the path coverage, thereby adjusting the samples to improve the coverage and increase the probability of finding vulnerabilities. OSS-FUZZ [14] was a common platform built by Google to support fuzzing engines in combination with Sanitizers for fuzzing open source programs. GRIMOIRE [15], Superion [16], and Zest [17] leveraged the knowledge in highly structured files to generate well-formed samples and traced the coverage of the program to reach deeper levels of code. Therefore, branch coverage was increased. CollAFL [18] proposed a coverage-sensitive fuzzing scheme to reduce path conflicts and thus improve program branch coverage. TensorFuzz [19] used the activation function as the coverage indicator and leveraged the algorithm of fast-approximate nearest neighbor to check whether the coverage increases to accordingly adjust the neural network. PerfFuzz [3] generated input samples by using multi-dimensional feedback and independently maximizing execution counts for all program locations. Fw-fuzz [20] obtained the code coverage of firmware programs of MIPS, ARM, PPC, and other architectures through dynamic instrumentation of physical devices, and finally implemented a coverage-oriented firmware protocol fuzzing method. T-fuzz [21] used coverage to guide the generation of input, and when the new path could not be accessed, the sanity check was removed to ensure that the fuzzer could continue to discover new paths and vulnerabilities.

Most coverage-based fuzzers treat all codes of a program as equals. However, some vulnerabilities hide in the corners of the code. As a result, the efficiency of CGF suffers and efforts are wasted on bug-free areas of the code.

B. Symbolic execution

Symbolic execution is a technique to systematically explore the paths of a program, which executes programs with symbolic inputs. When used in the field of discovering vulnerabilities, symbolic execution can generate new input samples that have a path reaching target codes from the initial code by solving path constraints with the SMT solver. It can also be said to deduce input from results under constraints.

Driller [22] leveraged fuzzing and selective concolic execution in a complementary manner. Angr [23], which is based on the model popularized and refined by S2E [24] and Mayhem [25], was used by Driller to be a dynamic symbolic execution engine for the concolic execution test. Driller uses selective concolic execution to only explore the paths deemed interesting by the fuzzer and to generate inputs for conditions that the fuzzer cannot satisfy. SAGE [26] is equipped with whitebox fuzzing instead of blackbox fuzzing, with symbolic execution to record path information and constraint solvers to explore different paths. QSYM [27] adopted a symbol execution engine for a greybox fuzzing approach to reach deeper code levels of the program. SAFL [28] augmented the AFL fuzzing approach by additionally leveraging KLEE as the symbolic execution engine.

However, the disadvantage of symbolic execution is that the increased analysis process leads to the program running overhead. In addition, as the depth of the path increases, the path conditions will become more and more complex, which will also pose a great challenge to the constraint solver.

C. Directed greybox fuzzing

Directed Greybox Fuzzing (DGF) is a fuzzing approach based on the target location or the specific program behavior obtained from the characteristics of a vulnerable code. Unlike CGF, which blindly increases path coverage, DGF aims to reach a predetermined set of places in the code (potentially vulnerable parts) and spends most of the time budget getting there, without wasting resources emphasizing irrelevant parts.

AFLgo [29] and Hawkeye [30] used distance metrics in their programs to perform user-specified target sites. A disadvantage of the distance-based approach is that it only focuses on the shortest distance, so when there are multiple paths to the same goal, longer paths may be ignored, resulting in lower efficiency. MemFuzz [31] focused on code regions related to memory access, and further guided the fuzzer by memory access information executed by the target program. UAFuzz [32] and UAFL [33] focused on UAF vulnerability-related code regions, leveraging target sequences to find use-after-free vulnerabilities, where memory operations must be performed in a specific order (e.g., allocate, free then store/write). Memlock [2] mainly focused on memory consumption vulnerabilities, took memory usage as the fitness goal, and searched for uncontrolled memory consumption vulnerabilities, but did not consider the influence of data flow. AFL-HR [34] triggered hard-to-show buffer overflow and integer overflow vulnerabilities through coevolution. IOTFUZZER [35] used a lightweight mechanism based on IoT mobile device APP, and proposed a black-box fuzzing model without protocol specifications to discover memory corruption vulnerabilities of IoT devices.

However, these works focus more on specific measurement strategies. When looking for the optimal path, it is easy to get stuck in local blocks of the program and ignore other paths that may lead to vulnerabilities, thus making the fuzzing results inaccurate.

D. Data flow guided fuzzing

Data flow analysis increases the knowledge set of the fuzzer and semantic information of the PUT by adding data flow information, and thus essentially makes the code characteristics and program behavior clear. Data flow analysis methods, such as taint analysis, can reflect the impact of the mutation on samples that could help optimize seed mutation strategy, input generation, and the seed selection process.

SemFuzz [36] tracked kernel function parameters on which key variables depend through reverse data flow analysis. SeededFuzz [37] proposed a dynamic taint analysis (DTA) approach to identify seed bytes that influence the values of security-sensitive program sites. TIFF [38] proposed a mutation strategy to infer input types through in-memory data structure identification and DTA, which increased the probability of triggering memory corruption vulnerabilities. However, data flow analysis, especially DTA, often increases runtime overhead and slows down the program while obtaining accurate data information of PUT. Fairfuzz [39] and Profuzzer [40] all adopted lightweight taint analysis to find the guiding mutation solution and obtain the variable taint attributes. GREYONE [41] equipped fuzzing with lightweight Fuzzing-Based Taint Inference (FTI) to carry out taint calibration for the branch jump variables of the program control flow. In the process of fuzzing, they mutate the specific bytes of samples and observe the changes of tainted variables to obtain the data dependency relationship between seed bytes and tainted variables.

However, it is impossible to understand the semantics of control flow by simply using data flow for vulnerability discovery, and detailed data flow analysis will increase overhead and reduce fuzzing efficiency. Usually, it can only be used as an important supplementary method of vulnerability discovery based on control flow analysis.

In summary, data flow analysis has become a future research trend, as more additional information of PUT can be obtained for better guidance of fuzzers. Therefore, the performance of the fuzzer can be better played for different vulnerabilities.

3. Enhanced Heap Operation Location Based on Data Semantics

In order to focus on discovering heap vulnerability, we first analyze the program in static mode to identify the locations of heap operation. We not only try to obtain the subsequence of heap operation, but also deduct the relations of heap operation based on data semantics. To achieve this goal, we build CPG including CFG and DDG (Data Dependency Graph). CFG is used to describe the sequence of operations, while DDG is used to point out the relationship between heap pointers. Based on data dependency deduced from CPG, we propose an algorithm to extract the locations of suspected dangerous heap operation code areas.

3.1. Examples of Memory Consumption Vulnerability

If an attacker can control the allocation of limited software resources and use a large number of system resources, the attacker may consume all available resources and then trigger a denial of service attack, which belongs to the category of resource consumption vulnerability CWE-400. This kind of vulnerability may prevent authorized users from accessing the software and have harmful effects on the surrounding memory environment. For example, a memory exhaustion attack could render software or even the operating system unusable. Therefore, we focus on the heap memory consumption vulnerability of code blocks, which is divided into two types named uncontrolled memory allocation and memory leaks.

Definition 1. ***Memory consumption*** *is defined as a vulnerability occupying process memory resources by triggering data storage instructions several times, which affects the normal running of the process and leads to a denial-of-service attack.*

Definition 2. ***Uncontrolled memory allocation*** *is defined as a vulnerability related to heap memory allocation and release, which allocates memory based on untrusted size values, but does not validate or incorrectly validate the size, and allows any amount of memory to be allocated. Its CWE number is CWE-789.*

Definition 3. ***Memory leak*** *is defined as a vulnerability also related to heap memory allocation and release, in which the program does not adequately track and free the allocated memory after allocation, and thus slowly consumes the remaining memory. Its CWE number is CWE-401.*

Compared with non-memory consuming vulnerabilities, uncontrolled memory allocation vulnerability and memory leak vulnerability are more difficult to discover because their conditions of triggering crashes are stricter.

CVE-2019-6988 is a public CVE, and this vulnerability occurs in the *opj_calloc* function. This vulnerability is formed because the program code lacks the detection of the allocated amount and the security mechanism for specially crafted files. In Figure 1, the code snippet related to an uncontrolled memory allocation vulnerability (CVE-2019-6988) exists in the executable program OpenJPEG version 2.3.0. In the source code project, the function *opj_tcd_init_tile* in file tcd.c is called when the OpenJPEG is running to decompress the "specially-crafted" images. This vulnerability allows a remote attacker to attempt too much memory allocation by function *opj_calloc* in the file opj_malloc.c, which calls the system function *calloc* to allocate a large amount of heap memory and ultimately results in a denial-of-service attack due to a lack of enough free heap memory.

```
static INLINE OPJ_BOOL opj_tcd_init_tile(opj_tcd_t *p_tcd, OPJ_UINT32 p_tile_no,
        OPJ_BOOL isEncoder, OPJ_FLOAT32 fraction, OPJ_SIZE_T sizeof_block,
        opj_event_mgr_t* manager){
    ...
    // call opj_tgt_create to creat the file.
    if (! l_current_precinct->incltree) {
        l_current_precinct->incltree = opj_tgt_create(l_current_precinct->cw,
                                       l_current_precinct->ch, manager);
    } else {
        l_current_precinct->incltree = opj_tgt_init(l_current_precinct->incltree,
                                       l_current_precinct->cw, l_current_precinct->ch, manager);
    }
    ...
}

opj_tgt_tree_t *opj_tgt_create(OPJ_UINT32 numleafsh, OPJ_UINT32 numleafsv,
                               opj_event_mgr_t *p_manager){
    ...
    tree = (opj_tgt_tree_t *) opj_calloc(1, sizeof(opj_tgt_tree_t)); // call opj_calloc.
    if (!tree) {
        opj_event_msg(p_manager, EVT_ERROR, "Not enough memory to create Tag-tree\n");
        return 00;
    }
    ...
}
void * opj_calloc(size_t num, size_t size){
    if (num == 0 || size == 0) {
        return NULL;
    }
    return calloc(num, size); // call calloc.
}
```

Figure 1. Code snippet from tcd.c/tgt.c in OpenJPEG v2.3.0.

As shown in Figure 2, the code snippet concerning memory leaks vulnerability exists in a program case of Samate Juliet Test Suite. This case is a memory leak vulnerability caused by allocating heap memory without release. Specifically, the case uses the function *malloc* on line 5 to allocate memory and checks whether the allocation is successful or not on line 7. However, at the end of the function, the allocated memory data is not effectively released, eventually resulting in a heap memory leak.

```
1   void func()
2   {
3       int64_t * data;
4       data = NULL;
5       data = (int64_t *)malloc(100*sizeof(int64_t));
6
7       if (data == NULL){
8           exit(-1);
9           }
10      data[0] = 5LL;
11      printLongLongLine(data[0]);
12  }
```

Figure 2. Code snippet from Samate Juliet Test Suite.

3.2. *Location of Heap Operation Code Based on Data Semantic*

In order to directionally discover heap-memory-consumed vulnerabilities, how to obtain the locations of suspected heap operations is the first essential goal. Once the locations are identified, they will be used as a guided factor to optimize the guidance strategy of vulnerability fuzzing, which is our second essential goal.

We first construct CPG based on the static analysis tool Joern. Then, a scheme is proposed to deduce the explicit and implicit semantic relations between heap pointers based on data flows from CPG. In addition, based on the semantic relations between heap pointers, we analyze the abnormal sequence of heap memory operation concerning allocation and release, and thus demarcate the heap operation code areas with suspected heap consumption. These locations will serve as an important indicator for selecting seeds from input samples during the fuzzing procedure.

3.2.1. Construct CPG

CPG is a graph combining multi-level code information where the information at each level can be related to each other. CPG can be obtained by combining AST (Abstract Syntax Trees), CFG, DDG, and CDG (Control Dependency Graph). Compared with other structures, CPG contains much richer data and relational information, which enables more complex and detailed static analysis of the program source code.

The CPG is composed of nodes and edges. Nodes represent the components of PUT, including functions, variables, etc. Each node has a type, such as a type METHOD representing a method, PARAM representing a parameter, and LOCAL representing a local variable. The directed edges represent the relationship between nodes, and the label is the description of the relationship, such as a label DDG from node A to node B represents B's data dependency on A.

The program files can be parsed using the source code analysis tool Joern to obtain the CPG. In order to show what useful data can be obtained from CPG for data relationship derivation, we analyze OpenJPEG v2.3.0 containing CVE-2019-6988 introduced in Section 3.1. Due to the huge number of codes, we only show the partial CPG shown in Figure 3. Figure 3a is the full CPG of the *opj_calloc* function, in which the *calloc* method is the partial zoom shown as Figure 3b. From Figure 3b, we can find the *calloc* method is dependent on the parament t_nmemb and t_size. We also find the parament t_nmemb and t_size are dependent on the return method. Combined with CFG, we can derive the potential dependent relationship between the *calloc* function and the *calloc* function and the return function.

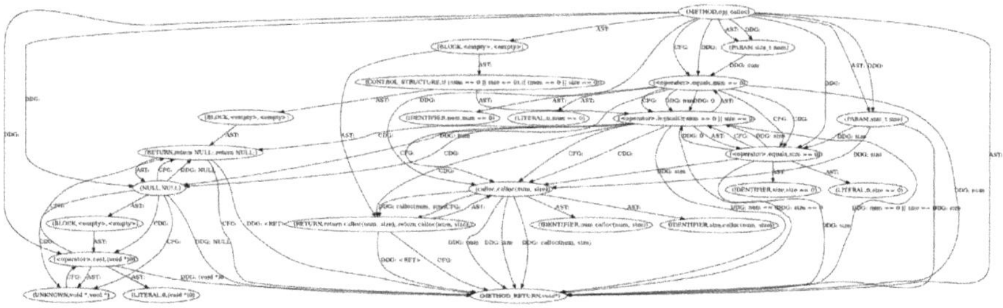

(a) CPG of *opj_calloc* method.

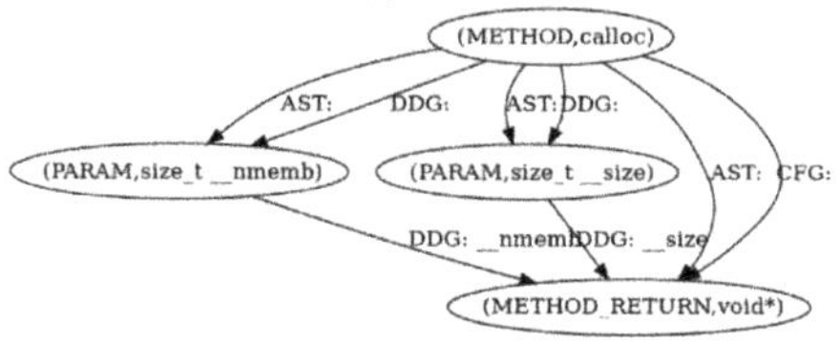

(b) CPG of *calloc* method.

Figure 3. CPG of OpenJPEG v2.3.0 *opj_calloc* and *calloc* methods.

Therefore, after constructing the CPG of the program, we analyze data dependencies using taint analysis on CPG and determine the location of heap operations.

3.2.2. Location Extraction Based on Data Dependency

Current research faces the challenge of finding accurate locations for code areas related to heap operations. In this section, we introduce how to obtain the data dependency by taint analysis. Taint analysis is an effective technology for data flow analysis. In our research work, we use a lightweight static taint analysis method to locate potentially vulnerable code areas.

Because CFG can reflect the jump of code and show all branches, most state-of-the-art fuzzers use CFG as an analysis object. Meanwhile, the data flow can reflect the direct relationship between variables and the function parameters, so some fuzzers also consider data flow as the analysis object. The data flow and the dependence on data semantics can provide positive help for understanding the real behavior of CFG, so we use these advantages to better serve the seed selection for discovering our required types of vulnerabilities. Using CPG for program analysis has many advantages. After using Joern to parse the source code into CPG, it does not need to be further compiled. CPG will be loaded into memory, and we can perform traversal queries, evaluate function leakage problems, perform data flow analysis, etc.

Dynamic taint analysis usually increases program runtime overhead. To this end, we use a lightweight static data flow analysis method to obtain the suspected locations in the target program, thereby reducing the impact on the program runtime overhead. We mainly focus on data dependencies in CPG during static taint analysis. We analyze the CPG to obtain relevant function points that have data dependencies between program input and memory allocation, that is, taint attributes information, and record them. Specifically, we use a static taint analysis approach to obtain the location of functions including heap operation.

Algorithm 1 is proposed to extract location by taint analysis based on data dependency. The static taint analysis is used to track the data flow of heap operation functions such as *malloc*, *calloc*, *realloc*, *free*, *new*, *delete*, and their deformation functions. The source set of the algorithm is the parameter of all the program methods and all the called functions of the program, and the sink set is the function arguments. In the process of data flow analysis, all relevant nodes from source to sink are traversed, and we use Joern's built-in functions, such as *reachableBy* and *reachablebyFlows*, to query the paths from all sources to the sink points in its CPG. Finally, matched functions are obtained, and duplicated ones are removed.

Algorithm 1: Taint analysis approach for locating potentially vulnerable functions

Input: CPG of program under test P
Output: Set of dataflow functions Set_{funcs}

1 $Set_{funcs} = \emptyset$;
2 Target heap functions $funcs \leftarrow \{malloc, calloc, free...\}$;
3 **for** f in $funcs$ **do**
4 $\quad$ $Source$ = called functions, methods' $params$ of P in CPG;
5 $\quad$ $Sink$ = $args$ of f in CPG;
6 $\quad$ **if** $Sink$ $dataFlowReachable$ by $Source$ **then**
7 $\quad\quad$ $Nodes_{related} = \emptyset$; //nodes which are data related;
8 $\quad\quad$ $Nodes_{related} \leftarrow Nodes_{related} \cup Traverse(CPG$ of $P)$;
9 $\quad\quad$ $paths = Query(Nodes_{related}, CPG)$; //dataflow paths of heap opreations;
10 $\quad\quad$ **for** p in $paths$ **do**
11 $\quad\quad\quad$ $(statements, line\ num, funcs_p, source\ locations) \leftarrow regularMatch(p)$;
12 $\quad\quad\quad$ **if** $funcs_p$ not $\emptyset$ **then**
13 $\quad\quad\quad\quad$ Remove duplicate items in $funcs_p$;
14 $\quad\quad\quad\quad$ $Set_{funcs} \leftarrow Set_{funcs} \cup funcs_p$;

15 **return** Set_{funcs};

Figure 4 shows the partial data flow of the heap memory allocation function obtained through static taint analysis in OpenJPEG v2.3.0, which is a data flow path to the parameter value of the standard library function *malloc*. Each path contains four aspects of information. Among them, the tracked column contains the statements in the queried nodes, the lineNumber column contains the line number in the source code file, the method column displays the method names where the statements are located, and the file column displays the locations of the source code file. To construct the source set, we mark the parameters of all methods and call all functions in the CPG into the source set. We find all call-sites for all methods in the graph with the name *malloc* and mark their arguments into the sink set. After identification, we obtain a data flow path to *malloc* in our query of OpenJPEG's CPG. Eventually, we collect dataflow-related functions *jpip_to_jp2*, *fread_jpip*, and *opj_malloc*, which were obtained in the dataflow path after the static taint analysis.

```
| tracked                                               | lineNumber| method      | file                                      |
|=======================================================|===========|=============|===========================================|
| jpip_to_jp2(char *argv[])                             | 45        | jpip_to_jp2 | /src/bin/jpip/opj_jpip_transcode.c        |
| fread_jpip(argv[1], dec)                              | 51        | jpip_to_jp2 | /src/bin/jpip/opj_jpip_transcode.c        |
| fread_jpip(const char fname[], jpip_dec_param_t *dec) | 350       | fread_jpip  | /src/lib/openjpip/openjpip.c              |
| open(fname, O_RDONLY)                                 | 354       | fread_jpip  | /src/lib/openjpip/openjpip.c              |
| open(fname, O_RDONLY)                                 | 354       | fread_jpip  | /src/lib/openjpip/openjpip.c              |
| infd = open(fname, O_RDONLY)                          | 354       | fread_jpip  | /src/lib/openjpip/openjpip.c              |
| get_filesize(infd)                                    | 359       | fread_jpip  | /src/lib/openjpip/openjpip.c              |
| get_filesize(infd)                                    | 359       | fread_jpip  | /src/lib/openjpip/openjpip.c              |
| (Byte8_t)get_filesize(infd)                           | 359       | fread_jpip  | /src/lib/openjpip/openjpip.c              |
| dec->jpiplen = (Byte8_t)get_filesize(infd)            | 359       | fread_jpip  | /src/lib/openjpip/openjpip.c              |
| opj_malloc(dec->jpiplen)                              | 363       | fread_jpip  | /src/lib/openjpip/openjpip.c              |
| opj_malloc(size_t size)                               | 191       | opj_malloc  | /src/lib/openjp2/opj_malloc.c             |
| size == 0U                                            | 193       | opj_malloc  | /src/lib/openjp2/opj_malloc.c             |
| malloc(size)                                          | 196       | opj_malloc  | /src/lib/openjp2/opj_malloc.c             |
```

Figure 4. OpenJPEG v2.3.0 partial data flow path.

In summary, the proposed algorithm 1 analyzes the data flow related to heap memory allocation and release in the program, and obtains the locations, variables, and parameters related to heap operation, which guide seed selection in the following fuzzing process.

4. MemConFuzz Model

After analyzing the CPG in the static analysis stage, we obtain the function locations related to the data flow of the heap memory allocation and release functions and quantitatively record the sizes of the memory block allocated and released by the heap operations. We feed these back to the fuzzer to prioritize the detection of relevant vulnerable code areas in the fuzzing loop.

The prioritizing discovery of consumption-type vulnerabilities is a novel contribution to this paper. Through the calibration of suspected heap memory-consuming instructions, the priority discovering of them is realized. Through investigating the existing vulnerability discovery models, we found that there are few studies on the discovery of heap memory consumption vulnerabilities. At the same time, the existing methods for discovering such vulnerabilities have many deficiencies, such as the lack of the important data flow analysis related to heap operations. Therefore, we propose our model, which has benefits for our purpose of focusing on heap consumption vulnerabilities discovery, while taking into account the discovery of other vulnerabilities.

4.1. Overview

To address problems mentioned in the previous sections, we propose a memory consumption-guided fuzzing model, MemConFuzz, as shown in Figure 5. The main components of MemConFuzz contain a static analyzer, an executor, fuzz loop feedback, a seed selector, and a seed mutator. In MemConFuzz, the static analyzer marks the dataflow-related edges and records the trigger value for each edge by scanning the source code and then inserts code fragments to update the value in the running program. The executor executes the instrumented program. Fuzz loop feedback is used to record and update related information to guide the seed selector after the program execution. The seed selector adopts a priority strategy to select seeds according to the different scores of the seed bank. The seed mutator mutates the selected seed to test the program in the fuzzing loop.

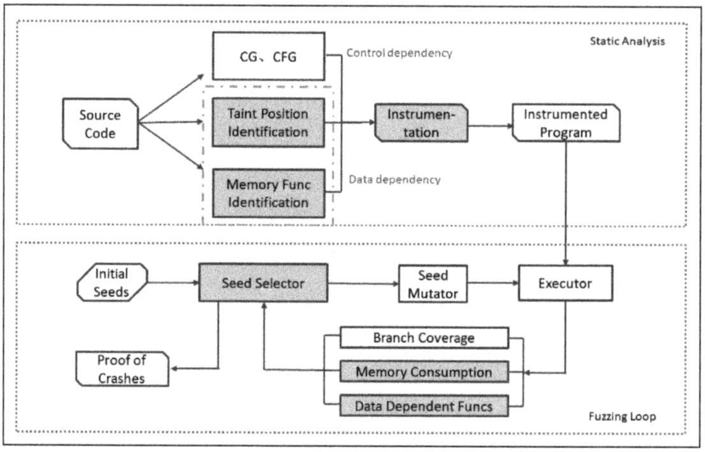

Figure 5. MemConFuzz model.

MemConFuzz contains two main stages: the static analysis stage and the fuzzing loop stage. Dark colors indicate optimized changes to the original AFL approach. Static analysis performs taint position identification and memory function identification for instrumentation. We use lightweight instrumentation to capture basic block transitions, heap memory function locations, and data-flow-dependent function locations at compile-time, while gaining coverage information, heap memory size, and data-flow-dependent information during runtime.

In the static analysis stage, we instrument all the captured target locations and then recompile to obtain an instrumented file. In the main fuzzing loop, the seed is selected for mutation and delivered to the instrumented file for execution. The model continuously tracks the state of the target program and records the cases that cause the program to crash. At the same time, the recorded feedback information is continuously submitted to the seed selector for priority selection, which helps to discover more heap memory consumption vulnerabilities.

4.2. Code Instrumentation at Locations of Suspected Heap Operation

In order to record the execution information in the fuzzing process, the bitmap of AFL records the number of branch executions, and the *perf_bits* of Memlock records the size of the heap allocation. The MemConFuzz also adopts a shared memory and incrementally adds *dataflow_shm* to store the numbers of the data-dependent functions triggered.

The MemConFuzz is derived from AFL. MemConFuzz adds two shared memory areas in AFL and mainly expands the afl-llvm-rt.o.c and afl-llvm-pass.so.cc files for instrumentation. The instrumented contents include branch coverage information, heap memory allocation functions, and data-dependent functions.

The first shared memory *perf_bits* records the size of the memory allocation and release during runtime. In the static analysis stage, we use LLVM [42] to obtain the function Call Graph (CG) and CFG of the program. Through traversing CG and CFG, we search the locations of basic blocks related to heap memory allocation and release functions, including *malloc, calloc, realloc, free, new, delete,* and their variant functions, and locate the call-sites of heap functions for instrumentation. During the fuzzing loop, *perf_bits* records the amount of consumed heap memory.

As shown in Figure 6 below, there are four basic blocks A, B, C and D representing nodes in the CFG of the program. The program will first go to B or C according to a branch condition. Once the branch condition for block C is met, the variable *size* is initialized, and then the memory allocation operation is performed in the block D. We traverse branches of the basic block described by IR language from the beginning of the program. Once we find a match among all our target heap functions, we locate the potential block D and instrument it.

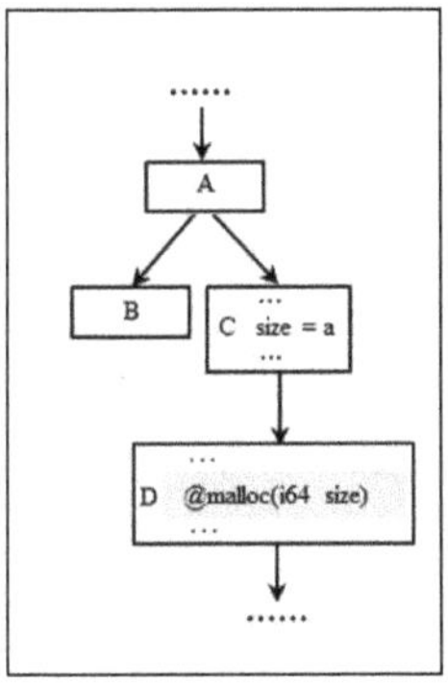

Figure 6. An example of basic block transition.

Meanwhile, we add the second shared memory *dataflow_shm* to record the numbers of data-dependent functions. We traverse the basic blocks of the program to search the locations that belong to Set_{funcs}, and then complete instrumentation. Specifically, after using these locations to analyze the program, we instrument the code of increasing count in *dataflow_shm*. In the fuzzing loop stage, MemConFuzz can increase the count value in *dataflow_shm* corresponding to triggered data-dependent functions when the target program executes an input sample. Thus, we can get the coverage information of heap operation when the execution of an input sample is completed.

In the instrumentation pass file, we declare a pointer variable, *DataflowPtr*, pointing to the shared memory area *dataflow_shm*. Then, the values of *dataflow_shm* are changed based on the number of data-dependent functions triggered. We inject instrumentation codes into the program during compilation. The approximate formulas for instrumentation are shown below, where Formula (1) marks the ID of the current block with a random number *cur_location*, Formula (2) shows that by applying the XOR operation on two IDs of the current block and previous block as the key, the corresponding value in *dataflow_shm* is updated by adding *dataflowfunc_cnt* to self-value, where the *shared_mem[]* array is our *dataflow_shm*, the size of which is 64 Kb, and *dataflowfunc_cnt* is the count of data-dependent functions triggered on this branch, and, in Formula (3), in order to distinguish the paths in different directions between two blocks, the *cur_location* is moved to the right by one bit as the *prev_location* to complete the marking of these two blocks.

$$cur\_location = \text{<COMPILE\_TIME\_RANDOM>} \tag{1}$$

$$shared\_mem[cur\_location \,\hat{}\, prev\_location] += dataflowfunc\_cnt \tag{2}$$

$$prev\_location = cur\_location >> 1 \tag{3}$$

We instrument the program based on static analysis to get the instrumented program. Therefore, we prioritize guidance to the suspected heap operation areas in the fuzzing loop stage to realize our directed fuzzing on the heap consumption vulnerability.

4.3. Strategy of Seed Priority Selection

This model proposes a fine-grained seed priority strategy for discovering heap memory consumption vulnerabilities. Seeds are mainly scored by the following indicators:

(1) We use *dataflow_funcs* as the metric, which is instrumented during the static analysis stage to record the number of data-dependent functions triggered during execution. The more related functions that are triggered, the higher the seed priority.

(2) Like Memlock, we record the size of the allocated heap memory; the larger the heap memory that is allocated, the more power this input sample gets. The input samples with more power at the top of the queue are selected as seeds. We use *new_max_size* as a flag which represents the maximum memory newly consumed on the heap in history. When the flag is triggered, we increase the score of this seed and enlarge its mutation time.

(3) In addition, when no data-dependent function has triggered and the new maximum allocated memory has not been reached, MemConFuzz still retains AFL's path-coverage-based seed prioritization strategy to cover as many program branches as possible.

The first two strategies will help the fuzzer trigger more potential heap memory consumption vulnerabilities.

Principle 1. *During execution, the more data dependent functions that are recorded, the greater the coefficient increase. In the end, the seed score increases and the energy obtained increases.*

Principle 2. *The original scoring strategy of AFL should also be taken into account. The final score of the seed should not be too large, because the execution process may be trapped in local code blocks of the program.*

According to the above two seed selection formula design principles, and in order to evaluate the excellence degree of each input sample, a scoring formula is proposed, in which the more data-dependent functions are recorded, the larger the coefficients are increased, and the higher scores are achieved. In addition, we set the parameters 1.2 and 1/5 according to the design principle, making sure to set the multiplier factor of the increase to the maximum value of 1.2, so that the evaluating strategy does not have too much impact to avoid missing out on other good samples, which are not used for discovering memory consumption vulnerability, but can be used for non-memory consumption vulnerability.

$$Priority\_score(sample_i) = \begin{cases} P_{afl}(sample_i) \cdot \left(1.2 - \frac{1}{5}e^{-dataflow\_funcs}\right), & dataflow\_funcs \,||\, new\_max\_size \\ P_{afl}(sample_i) & , otherwise \end{cases} \tag{4}$$

Equation (4) shows the seed priority strategy adopted by MemConFuzz. Specifically, for each $sample_i$ in the sample queue, when the data-dependent function is triggered or the new maximum memory is reached, we multiply the original AFL score value $P_{afl}(sample_i)$ by our formula and then obtain different seed scores under different numbers of data-dependent functions. The *dataflow_funcs* is the total number of data-dependent functions triggered by the sample during the fuzzing loop. Otherwise, we adopt the original AFL strategy, which is to perform sample scoring according to the execution speed and length of the samples. At last, we choose some samples with high *Priority_score* values from the sample queue as seeds.

In summary, every time the program executes, we detect code coverage, memory usage, and data-dependent functions triggered. For the impact of heap operations, we adopt two equations for different cases. The samples that trigger more data-related functions, allocate larger heap memory sizes, and have higher program path coverage are preferentially given higher power. Furthermore, we set up a maximum time in the havoc mutation phase to prevent wasting too much test time.

4.4. Proposed Model

We implement a directed fuzzing model MemConFuzz to discover heap memory consumption vulnerabilities. Unlike AFL, our model first performs a static analysis approach to analyze program data flow, and then uses the data flow information as a guide in discovering heap memory consumption vulnerabilities. Algorithm 2 describes the workflow of MemConFuzz.

The current vulnerability discovery models are faced with the challenge of not being able to prioritize the discovery of heap memory consumption vulnerabilities, and there is a problem of inaccurate static analysis caused by a lack of data flow information. Algorithm 2 is the pseudo code of our proposed fuzzing model, which is improved based on AFL, and we have optimized and improved the seed selection.

In the seed queue *Queue*, we select a seed q based on our seed priority strategy and then assign energy to the mutation. Meanwhile, we record the hashes, memory size, and data-dependent functions in each running process. If the q' causes the program to crash, add it to the crash set. Otherwise, we select those seeds that can trigger a new path, more heap memory allocation, or trigger more data-dependent functions, set them as interesting samples, and add them to the seed queue for the mutation of the next loop. Finally, the collection set of seeds that trigger heap memory consumption vulnerabilities and cause crashes is obtained.

Algorithm 2: Memory Consumption Fuzzing

Input: Instrumented program P, Initial seed input S
Output: Set of crash outputs Set_{crash}

1 $Set_{crash} = \emptyset$;
2 $Queue \leftarrow S$;
3 **while** *time and resource budget do not expire* **do**
4   **if** $Queue$ not $\emptyset$ **then**
5          $q = ChooseNext(Queue)$; // Our Modifications;
6          $e = AssignEnergy(q)$;
7   **if** i from 1 to e **then**
8              $q' = Mutate(q)$;
9              $(tracebits_i, memory_i, dataflowfuncs_i) \leftarrow Run(q', P)$;
10             $hash_i = Hash(tracebits_i)$;
11   **if** q' triggers crash **then**
12                 $Set_{crash} \leftarrow Set_{crash} \cup q'$;
13   **else**
14   **if** $NewCoverage(q')$ **then**
15                     $Queue \leftarrow Queue \cup q'$;
16   **if** $NewMaxSize(q')$ **then**
17                     $Queue \leftarrow Update(q', memory_i[hash_i])$;
18   **if** $DataflowFuncs(q')$ **then**
19                     $Queue \leftarrow$ Add and Prioritize$(q', dataflowfuncs_i)$;
20 **return** Set_{crash};

5. Experimental Results and Discussions

We implement the MemConFuzz based on the AFL-2.52b framework. We mainly write additional codes for LLVM-mode (based on LLVM v6.0.0) to realize our program static analysis approach related to memory consumption based on data flow and modify *afl-fuzz.c* to support our interaction module with instrumentation information and the fine-grained seed priority selection strategy.

We chose popular open source programs OpenJPEG v2.3.0, jasper v2.0.14, and readelf v2.28 with heap memory consumption vulnerabilities as test datasets, and compared them against AFL, MemLock, and PerfFuzz. Our experiments were performed on Ubuntu LTS 18.04 with a Linux kernel v4.15.0, Intel(R) Xeon(R) CPU E7-4820 processor, and 4GB RAM. The experiment results show that MemConFuzz outperforms the state-of-the-art fuzzing techniques, including AFL, MemLock, and PerfFuzz, in discovering heap memory consumption vulnerabilities. MemConFuzz can discover heap memory consumption CVEs faster and trigger a higher number of heap memory consumption crashes.

5.1. Evaluation Scheme

During the experiment, since the fuzzer heavily relies on random mutations, there may be performance fluctuations between different experiments on our machine, resulting in different experimental results each time. We have taken effective measures to configure experimental parameters and have taken two measures to mitigate the randomness caused by the properties of the fuzzing technology. First, we conduct a uniform long-term test of the experimental process of each PUT performed by each fuzzer until the fuzzer reaches a relatively stable state. Specifically, our stable results are obtained after a uniform 24-h period during every fuzzing execution. Second, we add the -d option to all fuzzers in the experiment to skip the deterministic mutation stage, so that more mutation strategies can be performed in the havoc and splicing stages to discover heap memory consumption vulnerabilities.

Due to factors such as different computer performance and randomness of mutation, the results of each experiment will be different. For the experiments in the comparison model, such as Memlock, we reproduced them on the same machine in order to ensure that each model is based on the same initial experimental conditions. We give the definition of "relatively stable state".

Definition 4. ***Relatively Stable State*** *is defined as a state in which test data smoothly changes. On the same machine, after a certain period of time, the results of multiple experiments are relatively stable compared to the growth rate in the initial stage, and then the test results reach a "relatively stable state".*

Figure 7 below is an experimental record of fuzzing readelf; the ordinate shows the number, and the abscissa shows the time. We mainly focus on the changes in the number of unique crashes. It shows that the growth rate is the fastest in the first 2 h, and the growth rate slows down after about 22 h, which fully meets the definition of a "relatively stable state". The other tests also meet the definition of a "relatively stable state" around 24 h. We consulted a large number of vulnerability discovery studies and methods, and many studies also selected 24 h as the test standard. In addition, MemConFuzz, Memlock, and PerfFuzz are all improved based on AFL, so if the time is too long, when almost no heap consumption vulnerabilities can be found in the end, it will gradually degenerate into AFL's general vulnerability discovering, and the discovery efficiency for heap consumption vulnerability cannot be demonstrated at this time. In order to comprehensively ensure accuracy and efficiency, we uniformly select 24 h as our test standard, which can reflect the ability of vulnerability discovering and also reduce unnecessary time overhead.

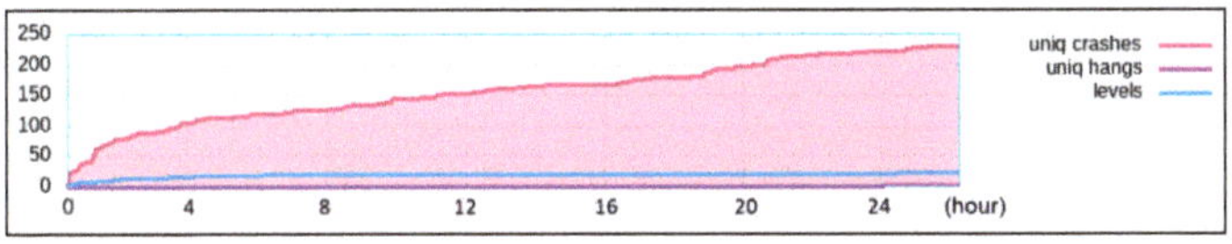

Figure 7. Experimental record of fuzzing readelf.

We enable ASAN [43] compilation of the source program file, and set the *allocator_may_return_null* option so that the program will crash when the heap memory allocation fails due to the allocation of too much memory, which is convenient for us to observe and analyze. In addition, we used LeakSanitizer to detect memory leak vulnerabilities and conduct subsequent analyses.

5.2. Experimental Results and Discussions

We perform fuzzing on the selected real-world program datasets and record the experimental data according to the evaluation metrics.

To demonstrate our work, we compare against some fuzzing techniques, recording the number of triggering heap memory consumption vulnerabilities and the time of triggering real-world CVEs. We select large-scale programs with tens of thousands of lines, which are continuously maintained in the open source community and have high popularity. These programs are from the comparison model. The name and version of the test software are mentioned by Memlock and some other fuzz testing tools. They contain heap consumption vulnerabilities and other types of vulnerabilities as interference items to comprehensively evaluate the models. Because we use the analysis method of source code and semantic heap operation code, other corresponding open-source source codes are difficult to find. There are very few fuzzing research works related to this type of vulnerability. In order to better evaluate the horizontal performance of the model, we choose these programs and ensure that these softwares are publicly available for download. The download link has been added. Additionally, the source code of MemConFuzz will be available for request.

Table 1 shows the crashes related to memory consumption vulnerabilities obtained by fuzzing the programs jasper, readelf, and openjpeg. UA stands for uncontrolled-memory-allocation vulnerabilities, ML stands for memory leak vulnerabilities, and SLoC stands for Source Lines of Code. For each 24-h fuzzing experiment, we use Python to analyze the obtained crashes and automatically reproduce them. We classify the crashes according to the obtained Address Sanitizer function call chain and its output summary information of vulnerability types, and then obtain the memory consumption-related vulnerabilities we need, that is, the number of UA and ML. Among them, most of the crashes triggered by jasper are ML, while the crashes triggered by other programs are UA. The results show that MemConFuzz has an improvement of 43.4%, 13.3%, and 561.2% in the discovery of heap memory consumption vulnerabilities compared with the advanced fuzzing techniques AFL, MemLock, and PerfFuzz, respectively.

Table 1. Number of heap memory consumption vulnerabilities.

Program	Version	SLoC	Type	MemConFuzz	AFL	MemLock	PerfFuzz
jasper	2.0.14	44k	UA	**5**	1	2	0
			ML	208	**212**	190	28
readelf	2.28	1844k	UA	**219**	86	182	39
openjpeg	2.3.0	243k	UA	11	10	**17**	0
Total Unique Crashes (Improvement)				**443**	309 (+43.4%)	391 (+13.3%)	67 (+561.2%)

The test programs [44–46] selected are all historical versions. After our automated crash analysis, the discovered vulnerabilities are all historically reported vulnerabilities. Our experimental comparison mainly focuses on the number and speed of discovering heap memory consumption vulnerabilities. We may consider discovering and analyzing additional new vulnerabilities in future research.

The AFL framework shows that vulnerabilities with the same crash point belong to the same vulnerability. Vulnerabilities are divided into many types. Since we are targeting heap consumption vulnerabilities, the only thing we need to confirm is whether the discovered vulnerabilities belong to heap consumption vulnerabilities. We wrote automated crash analysis scripts and compared the crash function stacks reported by ASAN. Through the ASAN report, the function call relationship, and the location of the crashed code, we spent a lot of time confirming that the vulnerability mentioned in this experiment belonged to the heap consumption vulnerability.

Furthermore, we also recorded the time of triggering real-world CVEs. In order to facilitate experimental comparison, we conducted a 24-h test for each test, and T/O stands for a timeout during the 24-h test. Table 2 shows the time of real-world CVEs triggered after we fuzzed on our dataset. Likewise, we used ASAN to reproduce crashes to detect memory error information. We did not use Valgrind because it slows the program down too much, while ASAN only slows the program down about 2×. We use Python to automatically analyze crashes and search the crash points, and compare the obtained Address Sanitizer function the call chain and crash point with the function location described by the real-world CVE information, therefore gaining the time of the first matching crash. Our experimental results show that MemConFuzz has significant time reduction compared to the state-of-the-art fuzzing techniques AFL, MemLock, and PerfFuzz, respectively. Among them, CVE-2017-12982 has more obvious advantages, which can make the program allocate large heap memory faster and trigger the vulnerability faster. The reason is that the proposed model focuses on the location of functions that are data-dependent on memory consumption, and pays attention to the size of allocated memory, which is more targeted for memory consumption vulnerabilities than other fuzzing models.

Table 2. Trigger time of real-world vulnerability.

Program	Vulnerability	Type	MemConFuzz	AFL	MemLock	PerfFuzz
			Time (h)	Time (h)	Time (h)	Time (h)
jasper	CVE-2016-8886	UA	2.6	10.2	**1.5**	T/O
readelf	CVE-2017-9039	UA	0.1	0.1	0.1	0.1
openjpeg	CVE-2017-12982	UA	**2.2**	12.8	5.5	T/O
	CVE-2019-6988	UA	**12.5**	T/O	14.8	T/O
Average Time Usage (Improvement)			**4.35**	11.78 (2.71×)	5.48 (1.26×)	18.03 (4.14×)

6. Conclusions and Future Work

In this paper, we propose a directed fuzzing approach MemConFuzz model based on data flow analysis of heap operations to discover heap memory consumption vulnerabilities. The MemConFuzz uses the coverage information, memory consumption information, and data dependency information to guide the fuzzing process. The coverage information guides the fuzzer to explore different program paths, the memory consumption information guides the fuzzer to search for program paths that show increasing memory consumption, and the data information guides the fuzzer to explore paths with increasing dependencies on heap memory data flow. Experimental results show that the MemConFuzz outperforms the state-of-the-art fuzzing technologies, AFL, MemLock, and PerfFuzz, in both the number of heap memory vulnerabilities and the time to discovery.

In the future, we plan to enhance the heap memory consumption vulnerability discovery capabilities and vulnerability coverage of our approach with more efficient and more complete data flow analysis. Furthermore, we will add support for binaries to our proposed vulnerability discovery methodology. We will disassemble the binary code to obtain the instruction code set, complete the analysis of the control flow and the data flow, and discover the heap memory consumption vulnerabilities of the binary program more effectively.

Author Contributions: Conceptualization, C.D. and Y.G.; methodology, C.D. and Y.G.; software, Z.C. and G.X.; validation, C.D. and Y.G.; investigation, Z.C. and Z.W.; writing—original draft preparation, Z.C. and Y.G; writing—review and editing, C.D. and Y.G.; visualization, Z.C.; project administration, C.D.; funding acquisition, Z.W. All authors have read and agreed to the published version of the manuscript.

Funding: This research was funded by the National Natural Science Foundation of China grant number 62172006 and the National Key Research and Development Plan of China grant number 2019YFA0706404.

Institutional Review Board Statement: Not applicable.

Informed Consent Statement: Not applicable.

Data Availability Statement: The test results data presented in this study are available on request. The data set can be found in public web sites.

Conflicts of Interest: The authors declare no conflict of interest.

Abbreviations

AFL	America Fuzzy Lop
PUT	Program Under Test
UAF	Use-After-Free
CPG	Code Property Graph
CVE	Common Vulnerabilities and Exposures
SMT	Satisfiability Modulo Theories

CGF	Coverage-guide Greybox Fuzzing
DGF	Directed Greybox Fuzzing
DTA	Dynamic Taint Analysis
FTI	Fuzzing-Based Taint Inference
DDG	Data Dependency Graph
AST	Abstract Syntax Trees
CDG	Control Dependency Graph
CG	Call Graph
CFG	Control Flow Graph

References

1. Zalewski, M. American Fuzzing Lop. Available online: https://lcamtuf.coredump.cx/afl/ (accessed on 31 October 2022).
2. Wen, C.; Wang, H.; Li, Y.; Qin, S.; Liu, Y.; Xu, Z.; Chen, H.; Xie, X.; Pu, G.; Liu, T. Memlock: Memory usage guided fuzzing. In Proceedings of the ACM/IEEE 42nd International Conference on Software Engineering, Seoul, Republic of Korea, 27 June 2020–19 July 2020; pp. 765–777. [CrossRef]
3. Lemieux, C.; Padhye, R.; Sen, K.; Song, D. Perffuzz: Automatically generating pathological inputs. In Proceedings of the 27th ACM SIGSOFT International Symposium on Software Testing and Analysis, Amsterdam, The Netherlands, 16–21 July 2018; pp. 254–265.
4. Rajpal, M.; Blum, W.; Singh, R. Not all bytes are equal: Neural byte sieve for fuzzing. *arXiv* **2017**, arXiv:1711.04596.
5. Godefroid, P.; Peleg, H.; Singh, R. Learn&fuzz: Machine learning for input fuzzing. In Proceedings of the 2017 32nd IEEE/ACM International Conference on Automated Software Engineering (ASE), Urbana, IL, USA, 30 October–3 November 2017; pp. 50–59.
6. She, D.; Pei, K.; Epstein, D.; Yang, J.; Ray, B.; Jana, S. Neuzz: Efficient fuzzing with neural program smoothing. In Proceedings of the 2019 IEEE Symposium on Security and Privacy (SP), San Francisco, CA, USA, 19–23 May 2019; pp. 803–817.
7. Chen, P.; Chen, H. Angora: Efficient fuzzing by principled search. In Proceedings of the 2018 IEEE Symposium on Security and Privacy (SP), San Francisco, CA, USA, 20–24 May 2018; pp. 711–725.
8. Cheng, L.; Zhang, Y.; Zhang, Y.; Wu, C.; Li, Z.; Fu, Y.; Li, H. Optimizing seed inputs in fuzzing with machine learning. In Proceedings of the 2019 IEEE/ACM 41st International Conference on Software Engineering: Companion Proceedings (ICSE-Companion), Montreal, QC, Canada, 25–31 May 2019; pp. 244–245.
9. Li, Z.; Zou, D.; Xu, S.; Jin, H.; Zhu, Y.; Chen, Z. SySeVR: A Framework for Using Deep Learning to Detect Software Vulnerabilities. *IEEE Trans. Dependable Secur. Comput.* **2022**, *19*, 2244–2258. [CrossRef]
10. Li, Z.; Zou, D.; Xu, S.; Ou, X.; Jin, H.; Wang, S.; Deng, Z.; Zhong, Y. Vuldeepecker: A deep learning-based system for vulnerability detection. *arXiv* **2018**, arXiv:1801.01681.
11. Zou, D.; Wang, S.; Xu, S.; Li, Z.; Jin, H. μVulDeePecker: A Deep Learning-Based System for Multiclass Vulnerability Detection. *IEEE Trans. Dependable Secur. Comput.* **2021**, *18*, 2224–2236. [CrossRef]
12. LibFuzzer—A Library for Coverage-Guided Fuzz Testing. Available online: http://llvm.org/docs/LibFuzzer.html (accessed on 31 October 2022).
13. Honggfuzz. Available online: http://honggfuzz.com/ (accessed on 31 October 2022).
14. Serebryany, K. OSS-Fuzz-Google's Continuous Fuzzing Service for Open Source Software. In Proceedings of the USENIX Security symposium, Vancouver, BC, Canada, 16–18 August 2017.
15. Blazytko, T.; Bishop, M.; Aschermann, C.; Cappos, J.; Schlögel, M.; Korshun, N.; Abbasi, A.; Schweighauser, M.; Schinzel, S.; Schumilo, S. GRIMOIRE: Synthesizing structure while fuzzing. In Proceedings of the 28th USENIX Security Symposium (USENIX Security 19), Santa Clara, CA, USA, 14–16 August 2019; pp. 1985–2002.
16. Wang, J.; Chen, B.; Wei, L.; Liu, Y. Superion: Grammar-aware greybox fuzzing. In Proceedings of the 2019 IEEE/ACM 41st International Conference on Software Engineering (ICSE), Montreal, QC, Canada, 25–31 May 2019; pp. 724–735.
17. Padhye, R.; Lemieux, C.; Sen, K.; Papadakis, M.; Le Traon, Y. Semantic fuzzing with zest. In Proceedings of the 28th ACM SIGSOFT International Symposium on Software Testing and Analysis, Beijing, China, 15–19 July 2019; pp. 329–340.
18. Gan, S.; Zhang, C.; Qin, X.; Tu, X.; Li, K.; Pei, Z.; Chen, Z. Collafl: Path sensitive fuzzing. In Proceedings of the 2018 IEEE Symposium on Security and Privacy (SP), San Francisco, CA, USA, 21–23 May 2018; pp. 679–696.
19. Odena, A.; Olsson, C.; Andersen, D.; Goodfellow, I. Tensorfuzz: Debugging neural networks with coverage-guided fuzzing. In Proceedings of the International Conference on Machine Learning, Long Beach, CA, USA, 10–15 July 2019; pp. 4901–4911.
20. Gao, Z.; Dong, W.; Chang, R.; Wang, Y.J.C.; Practice, C. Experience. Fw-fuzz: A code coverage-guided fuzzing framework for network protocols on firmware. *Concurr. Comput. Pract. Exp.* **2022**, *34*, e5756. [CrossRef]
21. Peng, H.; Shoshitaishvili, Y.; Payer, M. T-Fuzz: Fuzzing by program transformation. In Proceedings of the 2018 IEEE Symposium on Security and Privacy (SP), San Francisco, CA, USA, 20–24 May 2018; pp. 697–710. [CrossRef]
22. Stephens, N.; Grosen, J.; Salls, C.; Dutcher, A.; Wang, R.; Corbetta, J.; Shoshitaishvili, Y.; Kruegel, C.; Vigna, G. Driller: Augmenting fuzzing through selective symbolic execution. In Proceedings of the NDSS, San Diego, CA, USA, 21–24 February 2016; pp. 1–16.
23. Shoshitaishvili, Y.; Wang, R.; Hauser, C.; Kruegel, C.; Vigna, G. Firmalice-automatic detection of authentication bypass vulnerabilities in binary firmware. In Proceedings of the NDSS, San Diego, CA, USA, 7 February 2015; pp. 1.1–8.1.

24. Chipounov, V.; Kuznetsov, V.; Candea, G. S2E: A platform for in-vivo multi-path analysis of software systems. *Acm Sigplan Not.* **2011**, *46*, 265–278. [CrossRef]
25. Cha, S.K.; Avgerinos, T.; Rebert, A.; Brumley, D. Unleashing mayhem on binary code. In Proceedings of the 2012 IEEE Symposium on Security and Privacy, San Francisco, CA, USA, 20–23 May 2012; pp. 380–394.
26. Godefroid, P.; Levin, M.Y.; Molnar, D. SAGE: Whitebox fuzzing for security testing. *Commun. ACM* **2012**, *55*, 40–44. [CrossRef]
27. Yun, I.; Lee, S.; Xu, M.; Jang, Y.; Kim, T. QSYM: A practical concolic execution engine tailored for hybrid fuzzing. In Proceedings of the 27th USENIX Security Symposium (USENIX Security 18), Baltimore, MD, USA, 15–17 August 2018; pp. 745–761.
28. Wang, M.; Liang, J.; Chen, Y.; Jiang, Y.; Jiao, X.; Liu, H.; Zhao, X.; Sun, J. SAFL: Increasing and accelerating testing coverage with symbolic execution and guided fuzzing. In Proceedings of the 40th International Conference on Software Engineering: Companion Proceeedings, Gothenburg, Sweden, 27 May 2018—-3 June 2018; pp. 61–64.
29. Böhme, M.; Pham, V.-T.; Nguyen, M.-D.; Roychoudhury, A. Directed greybox fuzzing. In Proceedings of the 2017 ACM SIGSAC Conference on Computer and Communications Security, Dallas, TX, USA, 30 October 30–3 November 2017; pp. 2329–2344.
30. Chen, H.; Xue, Y.; Li, Y.; Chen, B.; Xie, X.; Wu, X.; Liu, Y. Hawkeye: Towards a desired directed grey-box fuzzer. In Proceedings of the 2018 ACM SIGSAC Conference on Computer and Communications Security, Toronto, ON, Canada, 15–19 October 2018; pp. 2095–2108.
31. Coppik, N.; Schwahn, O.; Suri, N. Memfuzz: Using memory accesses to guide fuzzing. In Proceedings of the 2019 12th IEEE Conference on Software Testing, Validation and Verification (ICST), Xi'an, China, 22–27 April 2019; pp. 48–58. [CrossRef]
32. Nguyen, M.-D.; Bardin, S.; Bonichon, R.; Groz, R.; Lemerre, M. Binary-level Directed Fuzzing for Use-After-Free Vulnerabilities. In Proceedings of the RAID, San Sebastian, Spain, 14–15 October 2020; pp. 47–62.
33. Wang, H.; Xie, X.; Li, Y.; Wen, C.; Li, Y.; Liu, Y.; Qin, S.; Chen, H.; Sui, Y. Typestate-guided fuzzer for discovering use-after-free vulnerabilities. In Proceedings of the ACM/IEEE 42nd International Conference on Software Engineering, Seoul, Republic of Korea, 27 June 202–19 July 2020; pp. 999–1010.
34. Medicherla, R.K.; Komondoor, R.; Roychoudhury, A. Fitness guided vulnerability detection with greybox fuzzing. In Proceedings of the IEEE/ACM 42nd International Conference on Software Engineering Workshops, Seoul, Republic of Korea, 27 June 202–19 July 2020; pp. 513–520.
35. Chen, J.; Diao, W.; Zhao, Q.; Zuo, C.; Lin, Z.; Wang, X.; Lau, W.C.; Sun, M.; Yang, R.; Zhang, K. IoTFuzzer: Discovering Memory Corruptions in IoT Through App-based Fuzzing. In Proceedings of the NDSS, San Diego, CA, USA, 18–21 February 2018.
36. You, W.; Zong, P.; Chen, K.; Wang, X.; Liao, X.; Bian, P.; Liang, B. Semfuzz: Semantics-based automatic generation of proof-of-concept exploits. In Proceedings of the 2017 ACM SIGSAC Conference on Computer and Communications Security, Abu Dhabi, United Arab Emirates, 2–6 April 2017; pp. 2139–2154.
37. Wang, W.; Sun, H.; Zeng, Q. Seededfuzz: Selecting and generating seeds for directed fuzzing. In Proceedings of the 2016 10th International Symposium on Theoretical Aspects of Software Engineering (TASE), Shanghai, China, 17–19 July 2016; pp. 49–56.
38. Jain, V.; Rawat, S.; Giuffrida, C.; Bos, H. TIFF: Using input type inference to improve fuzzing. In Proceedings of the 34th Annual Computer Security Applications Conference, San Juan, PR, USA, 3–7 December 2018; pp. 505–517.
39. Lemieux, C.; Sen, K. Fairfuzz: A targeted mutation strategy for increasing greybox fuzz testing coverage. In Proceedings of the 33rd ACM/IEEE International Conference on Automated Software Engineering, Virtual Event, 18–21 February 2018; pp. 475–485.
40. You, W.; Wang, X.; Ma, S.; Huang, J.; Zhang, X.; Wang, X.; Liang, B. Profuzzer: On-the-fly input type probing for better zero-day vulnerability discovery. In Proceedings of the 2019 IEEE symposium on security and privacy (SP), San Francisco, CA, USA, 19–23 May 2019; pp. 769–786.
41. Gan, S.; Zhang, C.; Chen, P.; Zhao, B.; Qin, X.; Wu, D.; Chen, Z. GREYONE: Data Flow Sensitive Fuzzing. In Proceedings of the USENIX Security Symposium, Boston, MA, USA, 12–14 August 2020; pp. 2577–2594.
42. Lattner, C.; Adve, V. LLVM: A compilation framework for lifelong program analysis & transformation. In Proceedings of the International Symposium on Code Generation and Optimization, San Jose, CA, USA, 20–24 March 2004; pp. 75–86.
43. Serebryany, K.; Bruening, D.; Potapenko, A.; Vyukov, D. AddressSanitizer: A fast address sanity checker. In Proceedings of the Usenix Conference on Technical Conference, Boston, MA, USA, 13–15 June 2012.
44. Openjpeg. An Open-Source JPEG 2000 Codec Written in C Language. Available online: https://github.com/uclouvain/openjpeg (accessed on 19 February 2023).
45. Jasper. Image Processing/Coding Tool Kit. Available online: https://www.ece.uvic.ca/~frodo/jasper (accessed on 19 February 2023).
46. GNU Binutils. Acollection of Binary Tools. Available online: https://www.gnu.org/software/binutils/ (accessed on 19 February 2023).

mathematics

MDPI

Article

A Hybrid MCDM Approach Based on Fuzzy MEREC-G and Fuzzy RATMI

Anas A. Makki [1,*] and Reda M. S. Abdulaal [2]

[1] Department of Industrial Engineering, Faculty of Engineering—Rabigh, King Abdulaziz University, Jeddah 21589, Saudi Arabia
[2] Department of Industrial Engineering, College of Applied Sciences, AlMaarefa University, Ad Diriyah 13713, Saudi Arabia; rabdelaal@um.edu.sa
* Correspondence: nhmakki@kau.edu.sa

Abstract: Multi-criteria decision-making (MCDM) assists in making judgments on complex problems by evaluating several alternatives based on conflicting criteria. Several MCDM methods have been introduced. However, real-world problems often involve uncertain and ambiguous decision-maker inputs. Therefore, fuzzy MCDM methods have emerged to handle this problem using fuzzy logic. Most recently, the method based on the removal effects of criteria using the geometric mean (MEREC-G) and ranking the alternatives based on the trace to median index (RATMI) were introduced. However, to date, there is no fuzzy extension of the two novel methods. This study introduces a new hybrid fuzzy MCDM approach combining fuzzy MEREC-G and fuzzy RATMI. The fuzzy MEREC-G can accept linguistic input terms from multiple decision-makers and generates consistent fuzzy weights. The fuzzy RATMI can rank alternatives according to their fuzzy performance scores on each criterion. The study provides the algorithms of both fuzzy MEREC-G and fuzzy RATMI and demonstrates their application in adopted real-world problems. Correlation and scenario analyses were performed to check the new approach's validity and sensitivity. The new approach demonstrates high accuracy and consistency and is sufficiently sensitive to changes in the criteria weights, yet not too sensitive to produce inconsistent rankings.

Keywords: fuzzy MEREC-G; fuzzy RATMI; fuzzy logic; hybrid; MCDM

MSC: 03E72; 90B50

Citation: Makki, A.A.; Abdulaal, R.M.S. A Hybrid MCDM Approach Based on Fuzzy MEREC-G and Fuzzy RATMI. *Mathematics* **2023**, *11*, 3773. https://doi.org/10.3390/math11173773

Academic Editors: Yanhui Guo and Jun Ye

Received: 27 July 2023
Revised: 30 August 2023
Accepted: 31 August 2023
Published: 2 September 2023

1. Introduction

Multi-criteria decision-making (MCDM), a major subdiscipline of the operations research domain, assists in making judgments in complex real-world challenges. It allows for formulating problems comprising several alternatives in a structured format to find the best ranking or select the best alternative based on multiple conflicting criteria. The criteria are conflicting in the sense of being benefit criteria and non-benefit criteria to reflect their roles in maximizing or minimizing the alternatives, respectively. Moreover, the criteria are weighted to represent the problem better and make the best decision on the alternatives. Several MCDM methods have emerged, with different characteristics and purposes, with broad applications in many disciplines [1,2]. The two primary components of MCDM are weighing the criteria and ranking the alternatives.

The first component of MCDM, weighting the criteria, entails designating importance or preference values to each criterion. Depending on whether the weights are based on quantified qualitative inputs from the decision-maker's judgments using a predefined scale (i.e., subjective data) [3–5], based on quantitative data (i.e., objective data) [6–10], or a combination of both (i.e., a mix of subjective and objective data) [11–13], there are various MCDM methods for weighting criteria. Methods like the analytic hierarchy process (AHP), analytic network process (ANP), and best-worst method (BWM) are examples of subjective methodologies for finding the weights of criteria [4,5]. These pairwise-based methods compare criteria using a scale of preferences to quantify qualitative inputs. Entropy and criteria importance through inter-criteria correlation (CRITIC) are examples of objective methods [14]. These data-based methods use mathematical algorithms to calculate the weights based on the information entropy, the correlation coefficients, or the compromise ranking of the alternatives. However, fuzzy AHP, fuzzy ANP, and fuzzy BWM accept a combination of subjective and objective data for finding the criteria weights. These methods base the calculations of weights in a fuzzy environment to account for uncertainty and ambiguity in decision-makers' inputs [15].

The second component of MCDM, ranking the alternatives, entails the performance scoring of each alternative on each criterion and finding the best ranking or choice accordingly. Various techniques for ranking alternatives based on multiple criteria have been developed. Such methods include outranking algorithms like "élimination et choix traduisant la realité" (ELECTRE), which translates to elimination and choice translating reality, and the preference ranking organization method for enrichment evaluations (PROMETHEE) [16–18], to mention two. These methods compare alternatives pair-wisely using measures of concordance and discordance between them on each criterion.

However, fuzzy MCDM alternative ranking methods have been developed and applied to enable them to handle the uncertainty and ambiguity of decision-makers' subjective scoring inputs. Such methods are the fuzzy BWM [19–26], fuzzy additive ratio assessment (ARAS) [27–29], fuzzy measurement alternatives and ranking according to compromise solution (MARCOS) [30–32], fuzzy technique for order preference by similarity to ideal solution (TOPSIS) [24,33,34], fuzzy multi-attributive border approximation area comparison (MABAC) [35–38], fuzzy VlseKriterijumska Optimizacija I Kompromisno Resenje (VIKOR) [39–42], fuzzy multi-attributive ideal–real comparative analysis (MAIRCA) [43–47], and, most recently, the fuzzy multiple criteria ranking by alternative trace (MCRAT) [48]. Several investigators applied the two components of MCDM in different fields [49–63].

Two of the most recent MCDM methods for weighting the criteria and ranking the alternatives are the method based on the removal effects of criteria (MEREC) [64–66] and ranking the alternatives based on the trace to median index (RATMI) techniques [67]. The MEREC was developed as an objective method for weighting the criteria. In 2023, an updated and enhanced version of the MEREC, labeled as the method for removal effects of criteria with a geometric mean (MEREC-G), was developed to enable it to process objective and subjective data [65]. Also, fuzzy extension and modification of the MEREC method were recently developed, enabling it to process subjective data using linguistic term judgments by decision-makers [68,69]. However, to date, there is no fuzzy extension to the enhanced MEREC-G. Additionally, in 2022, the RATMI was developed as an alternative ranking method. RATMI bases the ranking algorithm on the trace to median index, which combines ranking alternatives based on median similarity (RAMS), and the MCRAT methods, using a majority index and the concept of the VIKOR method [67]. In addition, despite this, the RATMI method is a relatively new alternative ranking method; it has proven its efficacy in real-world applications [70,71]. However, to date, there is no fuzzy extension to the RATMI method.

Therefore, this study aims to first develop a fuzzy MEREC-G as a weighting criteria method and a fuzzy RATMI as an alternative ranking method. Secondly, it proposes a new hybrid MCDM approach based on the developed fuzzy MEREC-G and fuzzy RATMI. The proposed new hybrid MCDM approach will provide advancements in that the fuzzy MEREC-G can accept linguistic input terms from multiple decision-makers, handle their ambiguous judgments on a complex problem, and produce consistent fuzzy weights of the criteria when converted to crisp values. This, in turn, will enable the use of the produced fuzzy weights from the fuzzy MEREC-G in the fuzzy RATMI, which will be able to accept and process fuzzy ranking scores of each alternative for each criterion and rank them accordingly.

The new proposed hybrid MCDM approach is provided in the following section. In the subsequent sections, along with a discussion, a numerical application of the proposed approach is provided to compare its results with other fuzzy MCDM methods to check its validity and sensitivity. Finally, the last section of this paper provides a conclusion to the proposed approach and some future research directions.

2. Preliminaries of Fuzzy Sets

Definition 1 ([69]). *$\tilde{a} = (k,\ l,\ m)$ is a representation of a triangular fuzzy number (TFN). The $\mu_{\tilde{a}}(z)$ membership function of a TEN, $\tilde{a}$, has the definition given by Equation (1).*

$$\mu_{\tilde{a}}(z) = \begin{cases} 0, & if\ z < k, \\ \frac{z-k}{l-k}, & if\ k \le z < l, \\ \frac{m-z}{m-l}, & if\ l \le z \le m, \\ 0, & if\ z > m, \end{cases} \quad (1)$$

Definition 2 ([72]). *Let $\tilde{x} = (a_1,\ b_1,\ c_1)$ and $\tilde{y} = (a_2,\ b_2,\ c_2)$ be two non-negative TFNs. According to the extension principle, the arithmetic operations are defined as follows:*

- $\tilde{x} \oplus \tilde{y} = (a_1 + a_2,\ b_1 + b_2,\ c_1 + c_2);$
- $\tilde{x} \ominus \tilde{y} = (a_1 - c_2,\ b_1 - b_2,\ c_1 - a_2);$
- $\alpha \odot \tilde{x} = (\alpha.a_1,\ \alpha.b_1,\ \alpha.c_1);$
- $\tilde{x}^{-1} \cong \left(\frac{1}{c_1}, \frac{1}{b_1}, \frac{1}{a_1}\right);$
- $\tilde{x} \otimes \tilde{y} \cong (a_1 \times a_2,\ b_1 \times b_2,\ c_1 \times c_2);$
- $\tilde{x} \oplus \tilde{y} \cong (a_1 / c_2,\ b_1 / b_2,\ c_1 / a_2).$

3. The Proposed Hybrid Fuzzy MEREC-G and Fuzzy RATMI Methods

Figure 1 illustrates the proposed fuzzy MEREC-G and fuzzy RATMI methods in three main phases. The first phase involves defining the problem under study by specifying the alternatives and criteria with their objective. The decision-maker invites the experts who will provide their initial fuzzy decision matrices between the alternatives and criteria. The second phase applies the fuzzy MEREC-G method to assign weights to each criterion based on the information from the first phase. The third step uses the fuzzy RATMI method to rank the alternatives according to the weighted fuzzy criteria obtained in the second phase. The following sections explain these phases in more detail.

3.1. Phase 1: Formulate the Problem Using the MCDM Model

Step 1.1: The decision-maker identifies "m" possible alternatives, "n" relevant criteria, and the nature of each criterion (i.e., whether it is a benefit criterion that should be maximized or a non-benefit criterion that should be minimized) for the problem at hand.

Step 1.2: The decision-maker determines "k" experts who have knowledge and experience about the problem to participate in the decision-making process by providing either subjective or objective input data represented by triangular fuzzy numbers (TFNs).

Step 1.3: The experts, $E = \{E_1,\ E_2,\ \ldots,\ E_k\}$, will provide a realistic evaluation of each alternative in $A = \{A_1,\ A_2,\ \ldots,\ A_m\}$ based on each criterion in $C = \{C_1,\ C_2,\ \ldots,\ C_n\}$, which is represented by the fuzzy number $x_{ij}^u = \left(a_{ij}^u,\ b_{ij}^u, c_{ij}^u\right)$, $i = 1, \ldots,\ m; j = 1, \ldots,\ n; u = 1, \ldots,\ k$. The fuzzy decision matrix, X^u, for each expert, "u", can be constructed using Equation (2).

$$X^u = \left[x_{ij}^u\right]_{mxn} = \begin{bmatrix} A/C & C_1 & C_2 & \cdots & C_n \\ A_1 & x_{11}^u & x_{12}^u & \cdots & x_{1n}^u \\ A_2 & x_{21}^u & x_{22}^u & \cdots & x_{2n}^u \\ \vdots & \vdots & \vdots & \ddots & \vdots \\ A_m & x_{m1}^u & x_{m2}^u & \cdots & x_{mn}^u \end{bmatrix} \quad (2)$$

Step 1.4: Construct the combined fuzzy decision matrix, $\tilde{X}$, using Equation (3).

$$\tilde{X} = \left[\tilde{x}_{ij}\right]_{mxn} \quad (3)$$

where

$\tilde{x}_{ij} = \left(a_{ij},\, b_{ij},\, c_{ij}\right)$, $a_{ij} = min_k\left(a_{ij}^k\right)$, $b_{ij} = \frac{1}{k}\left(\sum_{u=1}^{k} b_{ij}^u\right)$, *and* $c_{ij} = max_k\left(c_{ij}^k\right)$.

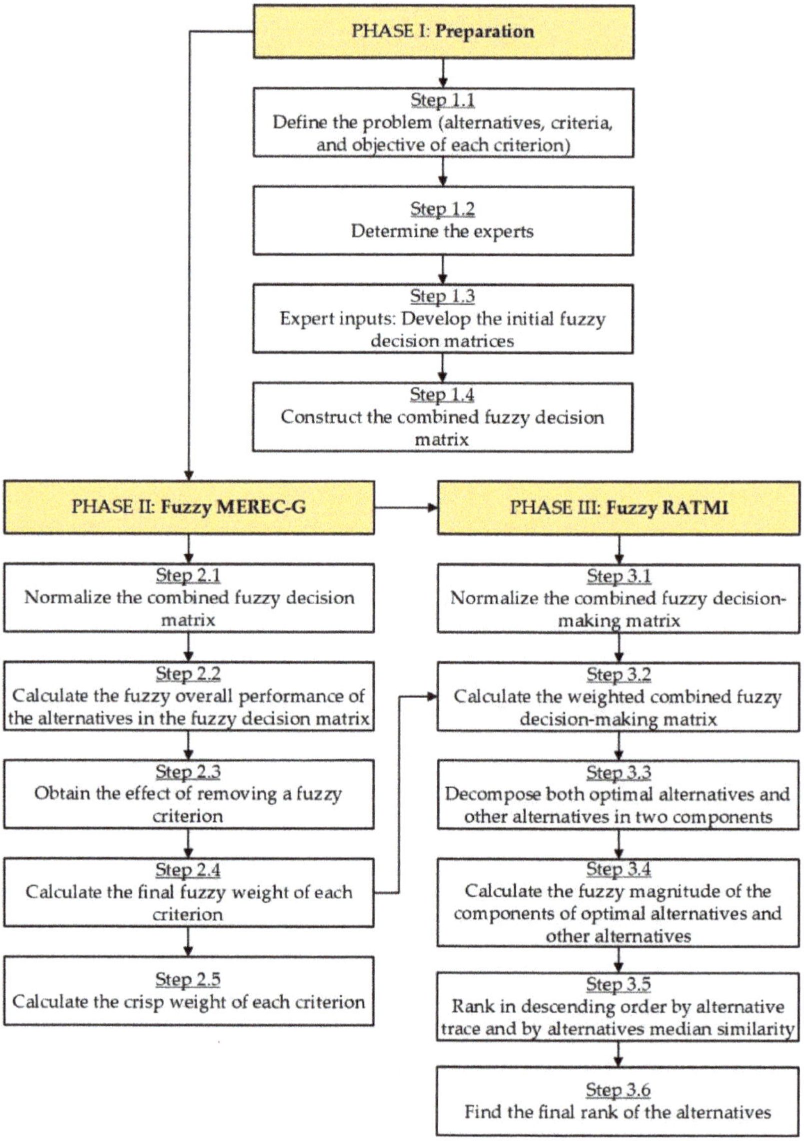

Figure 1. The framework of the proposed hybrid fuzzy MEREC-G and fuzzy RATMI methods.

3.2. Phase 2: Fuzzy MEREC-G Method

Step 2.1: Normalize the combined fuzzy decision matrix to reduce the disparity between the magnitude of alternatives and dimensions, with a normalized value within

[0, 1]. The component of a normalized matrix, $\tilde{e}_{ij}$, will be produced by the triangular fuzzy number (TFN) according to [69] using Equation (4) for benefit criteria and Equation (5) for non-benefit criteria.

$$\tilde{e}_{ij} = \left(r_{ij}^{l}, r_{ij}^{m}, r_{ij}^{u}\right) = \left(\frac{a_{ij}}{c_j^{max}}, \frac{b_{ij}}{c_j^{max}}, \frac{c_{ij}}{c_j^{max}}\right) \quad \forall\, i \in [1, \ldots, m]\ , \forall\, j \in [1, \ldots, n] \tag{4}$$

$$\tilde{e}_{ij} = \left(r_{ij}^{l}, r_{ij}^{m}, r_{ij}^{u}\right) = \left(\frac{a_j^{min}}{c_{ij}}, \frac{a_j^{min}}{b_{ij}}, \frac{a_j^{min}}{a_{ij}}\right) \quad \forall\, i \in [1, \ldots, m]\ , \forall\, j \in [1, \ldots, n] \tag{5}$$

Step 2.2: Calculate the fuzzy overall performance value, $\tilde{P}_i$, of the alternatives using the geometric mean of the fuzzy normalized matrix, as presented by Equation (6).

$$\tilde{P}_i = \left(\sqrt[n]{\prod_{j=1}^{n} r_{ij}^{l}}, \sqrt[n]{\prod_{j=1}^{n} r_{ij}^{m}}, \sqrt[n]{\prod_{j=1}^{n} r_{ij}^{u}}\right) \quad \forall\, i \in [1, \ldots, m] \tag{6}$$

Step 2.3: This step considers the core of the classical MEREC-G [65], in which the changes in the overall performance value of the alternatives will be calculated by removing the effect of each criterion from the overall performance. This step can be calculated for the fuzzy MEREC-G using Equation (7) to find the changes represented by the fuzzy number, $\tilde{t}_{ij}$.

$$\tilde{t}_{ij} = \left(\sqrt[n]{\frac{\prod_{j=1}^{n} r_{ij}^{l}}{r_{ik}^{l}}}, \sqrt[n]{\frac{\prod_{j=1}^{n} r_{ij}^{m}}{r_{ik}^{m}}}, \sqrt[n]{\frac{\prod_{j}^{n} r_{ij}^{u}}{r_{ik}^{u}}}\right) \quad \forall\, i \in [1, \ldots, m]\ , k \neq j \tag{7}$$

Step 2.4: Find the removal effect, $\tilde{E}_j$, using Equation (8) to obtain the final fuzzy weights, $\tilde{w}_j$, of each criterion using Equation (9) and Equation (10).

$$\tilde{E}_j = \left(\sum_{i=1}^{m} \tilde{t}_{ij}^{l}, \sum_{i=1}^{m} \tilde{t}_{ij}^{m}, \sum_{i=1}^{m} \tilde{t}_{ij}^{u}\right) \quad \forall\, j \in [1, \ldots, n] \tag{8}$$

$$\tilde{w}_j = \left(\frac{\sum_{i=1}^{m} \tilde{t}_{ij}^{l}}{\sum_{j=1}^{n} \tilde{E}_j^{u}}, \frac{\sum_{i=1}^{m} \tilde{t}_{ij}^{m}}{\sum_{j=1}^{n} \tilde{E}_j^{m}}, \frac{\sum_{i=1}^{m} \tilde{t}_{ij}^{u}}{\sum_{j=1}^{n} \tilde{E}_j^{l}}\right) \quad \forall\, j \in [1, \ldots, n] \tag{9}$$

$$\tilde{w}_j = \left(w_j^{l}, w_j^{m}, w_j^{u}\right) \quad \forall\, j \in [1, \ldots, n] \tag{10}$$

Step 2.5: To obtain the crisp weights, w_j^*, of the criteria, the obtained fuzzy weights, $\tilde{w}_j$, are converted using Equation (11). The sum of the crisp weights equals one.

$$w_j^* = \frac{w_j^{l} + 4w_j^{m} + w_j^{u}}{6} \tag{11}$$

3.3. Phase 3: Fuzzy RATMI Method

Step 3.1: The values in the combined fuzzy decision-making matrix will be normalized by the Equations (4) and (5) that are used for the fuzzy MEREC-G technique.

Step 3.2: The fuzzy weights of the criteria are multiplied by the fuzzy normalized values to obtain fuzzy weighted normalized values using Equation (12).

$$\tilde{g}_{ij} = \left(g_{ij}^{l}, g_{ij}^{m}, g_{ij}^{u}\right) = \tilde{w}_j \times \tilde{e}_{ij} = \left(w_j^{l} \times r_{ij}^{l}, w_j^{m} \times r_{ij}^{m}, w_j^{l} \times r_{ij}^{u}\right) \tag{12}$$

Step 3.3: Determine the fuzzy optimal alternative using Equations (13) and (14). Then, decompose the fuzzy optimal alternative into two components using Equations (15) and (16), followed by decomposing the other alternatives into two components using Equations (17) and (18).

$$\tilde{q}_j = max\left(\tilde{g}_{ij} | 1 \leq j \leq n\right) \tag{13}$$

$$\tilde{Q} = \{\tilde{q}_1,\ \tilde{q}_2,\ \ldots,\ \tilde{q}_n\} \tag{14}$$

$$\tilde{Q} = \tilde{Q}^{max} \cup \tilde{Q}^{min} \tag{15}$$

$$\tilde{Q} = \{\tilde{q}_1,\ \tilde{q}_2,\ \ldots,\ \tilde{q}_k\} \cup \{\tilde{q}_1,\ \tilde{q}_2,\ \ldots,\ \tilde{q}_h\}; k+h=j \tag{16}$$

$$\tilde{V} = \tilde{V}^{max} \cup \tilde{V}^{min} \tag{17}$$

$$\tilde{V} = \{\tilde{v}_1,\ \tilde{v}_2,\ \ldots,\ \tilde{v}_k\} \cup \{\tilde{v}_1,\ \tilde{v}_2,\ \ldots,\ \tilde{v}_h\}; k+h=j \tag{18}$$

Step 3.4: Calculate the fuzzy magnitude of optimal alternative components using Equations (19) and (20) and the fuzzy magnitude of other alternative components using Equations (21) and (22).

$$\tilde{Q}_k = \left(q_k^l,\ q_k^m,\ q_k^u\right) = \left(\sqrt{\left(q_1^l\right)^2 + \left(q_2^l\right)^2 + \ldots + \left(q_k^l\right)^2},\ \sqrt{\left(q_1^m\right)^2 + \left(q_2^m\right)^2 + \ldots + \left(q_k^m\right)^2},\ \sqrt{\left(q_1^u\right)^2 + \left(q_2^u\right)^2 + \ldots + \left(q_k^u\right)^2}\right) \tag{19}$$

$$\tilde{Q}_h = \left(q_h^l,\ q_h^m,\ q_h^u\right) = \left(\sqrt{\left(q_1^l\right)^2 + \left(q_2^l\right)^2 + \ldots + \left(q_h^l\right)^2},\ \sqrt{\left(q_1^m\right)^2 + \left(q_2^m\right)^2 + \ldots + \left(q_h^m\right)^2},\ \sqrt{\left(q_1^u\right)^2 + \left(q_2^u\right)^2 + \ldots + \left(q_h^u\right)^2}\right) \tag{20}$$

$$\tilde{V}_k = \left(v_k^l,\ v_k^m,\ v_k^u\right) = \left(\sqrt{\left(v_1^l\right)^2 + \left(v_2^l\right)^2 + \ldots + \left(v_k^l\right)^2},\ \sqrt{\left(v_1^m\right)^2 + \left(v_2^m\right)^2 + \ldots + \left(v_k^m\right)^2},\ \sqrt{\left(v_1^u\right)^2 + \left(v_2^u\right)^2 + \ldots + \left(v_k^u\right)^2}\right) \tag{21}$$

$$\tilde{V}_h = \left(v_h^l,\ v_h^m,\ v_h^u\right) = \left(\sqrt{\left(v_1^l\right)^2 + \left(v_2^l\right)^2 + \ldots + \left(v_h^l\right)^2},\ \sqrt{\left(v_1^m\right)^2 + \left(v_2^m\right)^2 + \ldots + \left(v_h^m\right)^2},\ \sqrt{\left(v_1^u\right)^2 + \left(v_2^u\right)^2 + \ldots + \left(v_h^u\right)^2}\right) \tag{22}$$

Step 3.5: In this step, the alternatives will be ranked twice. The first uses the fuzzy MCRAT [48], and the second uses fuzzy RAMS as a part of the proposed fuzzy RATMI. Ranking by fuzzy MCRAT uses the following sub-steps:

Step 3.5.1: Create the matrix, $\tilde{Y}$, composed of the optimal alternative component, as shown in Equation (23).

$$\tilde{Y} = \begin{bmatrix} \tilde{Q}_k & 0 \\ 0 & \tilde{Q}_h \end{bmatrix} \tag{23}$$

Step 3.5.2: Create the matrix, $\tilde{B}_i$, composed of the alternative's component using Equation (24).

$$\tilde{B}_i = \begin{bmatrix} \tilde{V}_{ik} & 0 \\ 0 & \tilde{V}_{ih} \end{bmatrix} \tag{24}$$

Step 3.5.3: Create the matrix, $\tilde{Z}_i$, using Equation (25).

$$\tilde{Z}_i = \tilde{Y} \times \tilde{B}_i = \begin{bmatrix} \tilde{z}_{11;i} & 0 \\ 0 & \tilde{z}_{22;i} \end{bmatrix} \tag{25}$$

Step 3.5.4: Then, the fuzzy trace of the matrix, $\tilde{Z}_i$, can be obtained using Equation (26).

$$tr\left(\tilde{Z}_i\right) = \tilde{z}_{11;i} + \tilde{z}_{22;i} = \left(z_{11,i}^l + z_{22,i}^l,\ z_{11,i}^m + z_{22,i}^m,\ z_{11,i}^u + z_{22,i}^u\right) \tag{26}$$

In Equation (26), $tr\left(\tilde{Z}_i\right) = \left(Z_i^l,\ Z_i^m,\ Z_i^u\right)$ indicates the fuzzy trace of the Z_i matrix, and the value is defuzzied to obtain $tr(Z_i)$ by using Equation (27). Here, rank the alternatives in descending order of the $tr(Z_i)$ values.

$$Z_i = \frac{Z_j^l + 4Z_j^m + Z_j^u}{6} \tag{27}$$

Ranking by fuzzy alternatives median similarity (RAMS) uses the following sub-steps:

Step 3.5.5: Determine the fuzzy median of similarity of the optimal alternative using Equation (28).

$$\tilde{D} = \left(d^l,\ d^m,\ d^u\right) = \left(\sqrt{\tilde{Q}_k^2 + \tilde{Q}_h^2}\right)/2 \tag{28}$$

Step 3.5.6: Determine the fuzzy median of similarity of the alternatives using Equation (29).

$$\tilde{D}_i = \left(d_i^l,\ d_i^m,\ d_i^l\right) = \left(\sqrt{\tilde{V}_{ik}^2 + \tilde{V}_{ih}^2}\right)/2 \tag{29}$$

Step 3.5.7: Calculate the fuzzy median similarity, $ms(\tilde{M}_i)$, which represents the ratio between the perimeter of each alternative and the optimal alternative using Equation (30).

$$ms(\tilde{M}_i) = \frac{\tilde{D}_i}{\tilde{D}} = \left(\frac{d_i^l}{d^u},\ \frac{d_i^m}{d^m},\ \frac{d_i^u}{d^l}\right) \tag{30}$$

In Equation (30), $ms(\tilde{M}_i) = \left(M_i^l,\ M_i^m,\ M_i^u\right)$ indicates the median similarity of the M_i matrix, and the value is defuzzied to obtain $ms(M_i)$ by using Equation (31). Here, rank the alternatives in descending order of the $ms(M_i)$ values.

$$M_i = \frac{M_j^l + 4M_j^m + M_j^u}{6} \tag{31}$$

Step 3.6: If v is the weight of fuzzy MCRAT's strategy, and $(1 - v)$ is the weight of RAMS's strategy, then the majority index, E_i, between the two strategies can be calculated using Equation (32). Then, find the final rank of the alternatives in descending order of E_i.

$$E_i = v\frac{(tr(Z_i) - tr^*)}{(tr^- - tr^*)} + (1 - v)\frac{(ms(M_i) - ms^*)}{(ms^- - ms^*)} \tag{32}$$

where
$tr^* = \min(tr(Z_i),\ \forall i \in [1,\ 2,\ \ldots,\ m]);$
$tr^- = \max\ (tr(Z_i), \forall i \in [1,\ 2,\ \ldots,\ m]);$
$ms^* = \min\ (ms(M_i), \forall i \in [1,\ 2,\ \ldots,\ m]);$
$ms^- = \max(ms(M_i),\ \forall i \in [1,\ 2,\ \ldots,\ m]);$
v is a value from 0 to 1. Here, $v = 0.5$.

4. Applications and Results

This section applies the proposed hybrid fuzzy MEREC-G and fuzzy RATMI methods using the data from Ulutaş et al. [48] to purchase a forklift that laborers can use in the warehouse. The following is an application of the three phases previously mentioned to rank the alternatives based on weighted criteria.

4.1. Phase 1: Formulate the Problem Using the MCDM Model

Following step 1.1, the decision-maker determined eight criteria and six forklifts as alternatives. The criteria for assessment of the forklifts were C_1 (purchasing price), C_2 (lifting height), C_3 (lowering speed), C_4 (loading capacity), C_5 (lifting speed), C_6 (movement area requirement), C_7 (image of the manufacturer company), and C_8 (supply of spare parts). Only two criteria (C_1 and C_6) were non-benefit, and the others were benefit criteria. Using steps 1.2, 1.3, and 1.4, the decision maker determined six experts to evaluate the performance of the forklifts under each criterion using the linguistic phrases shown in Stanković et al. [31]. The experts' assessments were transformed into fuzzy values using those linguistic phrases and aggregated using Equation (3). The combined fuzzy decision matrix, as given by Ulutaş et al. [48], is presented in Table 1.

Table 1. The combined fuzzy decision matrix [48].

Alternatives	C_1	C_2	C_3	C_4
A1	(4.0000, 5.6670, 6.0000)	(5.3330, 6.3330, 7.3330)	(2.0000, 3.0000, 4.0000)	(5.6670, 6.6670, 7.6670)
A2	(5.0000, 5.6670, 7.0000)	(5.3330, 6.3330, 7.3330)	(3.6670, 5.0000, 5.6670)	(4.0000, 5.0000, 6.0000)
A3	(5.6670, 7.3330, 7.6670)	(6.6670, 7.3330, 8.6670)	(4.0000, 5.6670, 6.0000)	(5.6670, 6.6670, 7.6670)
A4	(5.6670, 7.3330, 7.6670)	(5.6670, 6.6670, 7.6670)	(4.0000, 5.6670, 6.0000)	(5.6670, 6.6670, 7.6670)
A5	(5.0000, 6.0000, 7.0000)	(5.6670, 6.6670, 7.6670)	(4.0000, 5.6670, 6.0000)	(4.0000, 5.0000, 6.0000)
A6	(5.0000, 6.0000, 7.0000)	(5.6670, 6.3330, 7.6670)	(4.0000, 5.6670, 6.0000)	(4.0000, 5.0000, 6.0000)
Alternatives	C_5	C_6	C_7	C_8
A1	(4.3330, 5.3330, 6.3330)	(5.3330, 6.3330, 7.3330)	(5.3330, 6.0000, 7.3330)	(4.6670, 5.6670, 6.6670)
A2	(4.3330, 5.3330, 6.3330)	(6.3330, 7.3330, 8.3330)	(6.0000, 6.6670, 8.0000)	(5.6670, 6.0000, 7.6670)
A3	(6.0000, 7.0000, 8.0000)	(6.3330, 7.3330, 8.3330)	(6.0000, 7.0000, 8.0000)	(5.0000, 6.0000, 7.0000)
A4	(6.0000, 7.0000, 8.0000)	(6.3330, 7.3330, 8.3330)	(5.3330, 6.0000, 7.3330)	(4.6670, 5.6670, 6.6670)
A5	(4.3330, 6.0000, 6.3330)	(5.3330, 6.3330, 7.3330)	(5.6670, 6.0000, 7.6670)	(5.0000, 6.0000, 7.0000)
A6	(4.3330, 5.6670, 6.3330)	(5.0000, 5.6670, 7.0000)	(5.0000, 5.6670, 7.0000)	(5.6670, 6.3330, 7.6670)

4.2. Phase 2: Application and Results of the Fuzzy MEREC-G Method

Equations (4) and (5) of step 2.1 have been used to determine the fuzzy decision matrix with normalization. Table 2 presents the results obtained from this step.

Table 2. The normalized fuzzy decision matrix.

Alternatives	C_1	C_2	C_3	C_4
A1	(0.6667, 0.7058, 1.0000)	(0.6153, 0.7307, 0.8461)	(0.3333, 0.5000, 0.6667)	(0.7391, 0.8696, 1.0000)
A2	(0.5714, 0.7058, 0.8000)	(0.6153, 0.7307, 0.8461)	(0.6112, 0.8333, 0.9445)	(0.5217, 0.6521, 0.7826)
A3	(0.5217, 0.5455, 0.7058)	(0.7692, 0.8461, 1.0000)	(0.6667, 0.9445, 1.0000)	(0.7391, 0.8696, 1.0000)
A4	(0.5217, 0.5455, 0.7058)	(0.6539, 0.7692, 0.8846)	(0.6667, 0.9445, 1.0000)	(0.7391, 0.8696, 1.0000)
A5	(0.5714, 0.6667, 0.8000)	(0.6539, 0.7692, 0.8846)	(0.6667, 0.9445, 1.0000)	(0.5217, 0.6521, 0.7826)
A6	(0.5714, 0.6667, 0.8000)	(0.6539, 0.7307, 0.8846)	(0.6667, 0.9445, 1.0000)	(0.5217, 0.6521, 0.7826)
Alternatives	C_5	C_6	C_7	C_8
A1	(0.5416, 0.6666, 0.7916)	(0.6818, 0.7895, 0.9376)	(0.6666, 0.7500, 0.9166)	(0.6087, 0.7391, 0.8696)
A2	(0.5416, 0.6666, 0.7916)	(0.6000, 0.6818, 0.7895)	(0.7500, 0.8334, 1.0000)	(0.7391, 0.7826, 1.0000)
A3	(0.7500, 0.8750, 1.0000)	(0.6000, 0.6818, 0.7895)	(0.7500, 0.8750, 1.0000)	(0.6521, 0.7826, 0.9130)
A4	(0.7500, 0.8750, 1.0000)	(0.6000, 0.6818, 0.7895)	(0.6666, 0.7500, 0.9166)	(0.6087, 0.7391, 0.8696)
A5	(0.5416, 0.7500, 0.7916)	(0.6818, 0.7895, 0.9376)	(0.7084, 0.7500, 0.9584)	(0.6521, 0.7826, 0.9130)
A6	(0.5416, 0.7084, 0.7916)	(0.7143, 0.8823, 1.0000)	(0.6250, 0.7084, 0.8750)	(0.7391, 0.8260, 1.0000)

Steps 2.2 and 2.3 have been applied with the help of Equations (6) and (7), respectively, to calculate the overall performance of alternatives in the fuzzy decision matrix and then calculate the changes in this overall performance by removing each fuzzy number. Table 3 shows the results of Equation (7) of step 2.3.

Table 3. The changes in the overall performance of alternatives.

Alternatives	C_1	C_2	C_3	C_4
A1	(0.6231, 0.7428, 0.8718)	(0.6294, 0.7396, 0.8902)	(0.6795, 0.7755, 0.9171)	(0.6151, 0.7237, 0.8718)
A2	(0.6585, 0.7653, 0.8892)	(0.6524, 0.7620, 0.8830)	(0.6530, 0.7496, 0.8709)	(0.6660, 0.7729, 0.8917)
A3	(0.7331, 0.8544, 0.9599)	(0.6984, 0.8088, 0.9190)	(0.7110, 0.7977, 0.9190)	(0.7019, 0.8060, 0.9190)
A4	(0.7018, 0.8223, 0.9294)	(0.6823, 0.7877, 0.9035)	(0.6806, 0.7677, 0.8898)	(0.6719, 0.7757, 0.8898)
A5	(0.6662, 0.7981, 0.9049)	(0.6551, 0.7840, 0.8936)	(0.6535, 0.7641, 0.8800)	(0.6738, 0.8003, 0.9074)
A6	(0.6701, 0.7981, 0.9122)	(0.6589, 0.7890, 0.9008)	(0.6573, 0.7641, 0.8871)	(0.6777, 0.8003, 0.9147)

Table 3. *Cont.*

Alternatives	C_5	C_6	C_7	C_8
A1	(0.5999, 0.7178, 0.8839)	(0.5805, 0.7006, 0.8628)	(0.5824, 0.7058, 0.8656)	(0.5900, 0.7073, 0.8721)
A2	(0.6251, 0.7427, 0.8757)	(0.6160, 0.7403, 0.8761)	(0.5967, 0.7194, 0.8470)	(0.5979, 0.7259, 0.8470)
A3	(0.6659, 0.7808, 0.9080)	(0.6874, 0.8092, 0.9392)	(0.6659, 0.7808, 0.9080)	(0.6793, 0.7934, 0.9199)
A4	(0.6335, 0.7474, 0.8751)	(0.6540, 0.7745, 0.9051)	(0.6442, 0.7640, 0.8860)	(0.6526, 0.7656, 0.8927)
A5	(0.6335, 0.7599, 0.8934)	(0.6130, 0.7544, 0.8721)	(0.6096, 0.7599, 0.8694)	(0.6169, 0.7553, 0.8754)
A6	(0.6377, 0.7661, 0.9017)	(0.6130, 0.7425, 0.8721)	(0.6248, 0.7661, 0.8889)	(0.6100, 0.7495, 0.8721)

Equations (8)–(10) from step 2.4 have been used to calculate the fuzzy criteria weight of each criterion. Then, Equation (11) from step 2.5 was used to calculate the crisp value of each criterion. Table 4 shows the results of these calculations.

Table 4. Resulting effect and weights of the fuzzy MEREC-G.

Removal effect	$\tilde{E}_1$	$\tilde{E}_2$	$\tilde{E}_3$	$\tilde{E}_4$
	(4.0527, 4.7810, 5.4674)	(3.9764, 4.6710, 5.3902)	(4.0348, 4.6188, 5.3640)	(4.0064, 4.6790, 5.3944)
	$\tilde{E}_5$	$\tilde{E}_6$	$\tilde{E}_7$	$\tilde{E}_8$
	(3.7955, 4.5147, 5.3378)	(3.7639, 4.5214, 5.3273)	(3.7236, 4.4961, 5.2648)	(3.7467, 4.4970, 5.2792)
Fuzzy weights	$\tilde{w}_1$	$\tilde{w}_2$	$\tilde{w}_3$	$\tilde{w}_4$
	(0.0946, 0.1300, 0.1758)	(0.0929, 0.1270, 0.1733)	(0.0942, 0.1256, 0.1725)	(0.0936, 0.1272, 0.1735)
	$\tilde{w}_5$	$\tilde{w}_6$	$\tilde{w}_7$	$\tilde{w}_8$
	(0.0886, 0.1228, 0.1716)	(0.0879, 0.1229, 0.1713)	(0.0869, 0.1222, 0.1693)	(0.0875, 0.1223, 0.1697)
Crisp weights	$\overset{*}{w}_1$	$\overset{*}{w}_2$	$\overset{*}{w}_3$	$\overset{*}{w}_4$
	0.1317	0.1290	0.1282	0.1293
	$\overset{*}{w}_5$	$\overset{*}{w}_6$	$\overset{*}{w}_7$	$\overset{*}{w}_8$
	0.1252	0.1252	0.1242	0.1244

4.3. Phase 3: Application and Results of the Fuzzy RATMI Method

The fuzzy MEREC-G method is used to determine the fuzzy criteria weights, which are then combined with the decision matrix to form the decision-making matrix. The fuzzy RATMI method is applied to this matrix to rank the alternatives. From step 3.1, the fuzzy decision-making matrix is normalized using Equations (4) and (5), which are the same as those used in the fuzzy MEREC-G. The fuzzy weighted decision-making matrix is obtained using Equation (12) from step 3.2 and shown in Table 5.

First, the fuzzy optimal alternatives are determined using Equations (13) and (14), and then they are decomposed into their components using Equations (15) and (16). Next, Equations (17) and (18) are used to decompose the alternatives into their components. Finally, the fuzzy magnitude of the components is calculated using Equations (19) and (20). The values of the fuzzy magnitude of components are shown in Table 6.

The same process is performed for the alternatives using Equations (21) and (22). Then, with Equations (23)–(25), the values of $\tilde{z}_{11;i}$ and $\tilde{z}_{22;i}$, which are the elements of the $\tilde{Z}_i$, are found. Equation (26) is used to obtain the fuzzy trace, $tr(\tilde{Z}_i)$, of the matrix, $\tilde{Z}_i$. Finally, this fuzzy value is defuzzified using Equation (27). Table 7 shows these values and the results of the fuzzy MCRAT method.

Table 5. The fuzzy weighted decision-making matrix.

Alternatives	C_1	C_2	C_3	C_4
A1	(0.0631, 0.0918, 0.1758)	(0.0571, 0.0928, 0.1466)	(0.0314, 0.0628, 0.1150)	(0.0691, 0.1106, 0.1735)
A2	(0.0541, 0.0918, 0.1406)	(0.0571, 0.0928, 0.1466)	(0.0576, 0.1047, 0.1629)	(0.0488, 0.0830, 0.1357)
A3	(0.0494, 0.0709, 0.1241)	(0.0714, 0.1075, 0.1733)	(0.0628, 0.1186, 0.1725)	(0.0691, 0.1106, 0.1735)
A4	(0.0494, 0.0709, 0.1241)	(0.0607, 0.0977, 0.1533)	(0.0628, 0.1186, 0.1725)	(0.0691, 0.1106, 0.1735)
A5	(0.0541, 0.0867, 0.1406)	(0.0607, 0.0977, 0.1533)	(0.0628, 0.1186, 0.1725)	(0.0488, 0.0830, 0.1357)
A6	(0.0541, 0.0867, 0.1406)	(0.0607, 0.0928, 0.1533)	(0.0628, 0.1186, 0.1725)	(0.0488, 0.0830, 0.1357)
Alternatives	C_5	C_6	C_7	C_8
A1	(0.0480, 0.0818, 0.1359)	(0.0599, 0.0971, 0.1606)	(0.0580, 0.0917, 0.1552)	(0.0533, 0.0904, 0.1476)
A2	(0.0480, 0.0818, 0.1359)	(0.0527, 0.0838, 0.1352)	(0.0652, 0.1019, 0.1693)	(0.0647, 0.0957, 0.1697)
A3	(0.0665, 0.1074, 0.1716)	(0.0527, 0.0838, 0.1352)	(0.0652, 0.1070, 0.1693)	(0.0571, 0.0957, 0.1550)
A4	(0.0665, 0.1074, 0.1716)	(0.0527, 0.0838, 0.1352)	(0.0580, 0.0917, 0.1552)	(0.0533, 0.0904, 0.1476)
A5	(0.0480, 0.0921, 0.1359)	(0.0599, 0.0971, 0.1606)	(0.0616, 0.0917, 0.1622)	(0.0571, 0.0957, 0.1550)
A6	(0.0480, 0.0870, 0.1359)	(0.0628, 0.1085, 0.1713)	(0.0543, 0.0866, 0.1481)	(0.0647, 0.1010, 0.1697)

Table 6. The fuzzy magnitude of components' values.

Components	Magnitude
$\tilde{Q}_k$	(0.1633, 0.2665, 0.4205)
$\tilde{Q}_h$	(0.0890, 0.1421, 0.2455)

Table 7. Results of the fuzzy MCRAT method.

Alternatives	$\tilde{V}_{ik}$	$\tilde{V}_{ih}$	$\tilde{z}_{11;i}$	$\tilde{z}_{22;i}$
A1	(0.1324, 0.2192, 0.3594)	(0.0870, 0.1336, 0.2381)	(0.0216, 0.0584, 0.1511)	(0.0077, 0.0190, 0.0584)
A2	(0.1404, 0.2295, 0.3774)	(0.0755, 0.1243, 0.1951)	(0.0229, 0.0612, 0.1587)	(0.0067, 0.0177, 0.0479)
A3	(0.1605, 0.2646, 0.4147)	(0.0722, 0.1098, 0.1835)	(0.0262, 0.0705, 0.1744)	(0.0064, 0.0156, 0.0451)
A4	(0.1517, 0.2529, 0.3983)	(0.0722, 0.1098, 0.1835)	(0.0248, 0.0674, 0.1675)	(0.0064, 0.0156, 0.0451)
A5	(0.1392, 0.2378, 0.3748)	(0.0807, 0.1301, 0.2135)	(0.0227, 0.0634, 0.1576)	(0.0072, 0.0185, 0.0524)
A6	(0.1395, 0.2341, 0.3754)	(0.0829, 0.1388, 0.2216)	(0.0228, 0.0624, 0.1578)	(0.0074, 0.0197, 0.0544)
Alternatives	$tr(\tilde{Z}_i)$	$tr(Z_i)$	**Rankings**	
A1	(0.0294, 0.0774, 0.2095)	0.0914	6	
A2	(0.0296, 0.0788, 0.2066)	0.0919	5	
A3	(0.0326, 0.0861, 0.2194)	0.0994	1	
A4	(0.0312, 0.0830, 0.2125)	0.0960	2	
A5	(0.0299, 0.0819, 0.2100)	0.0946	4	
A6	(0.0302, 0.0821, 0.2122)	0.0952	3	

Another ranking will be obtained by the fuzzy RAMS method. In this method, the alternatives are ranked based on the median similarity between the optimal alternatives and other alternatives by applying Equations (28)–(31). This was followed by finding the majority index between the fuzzy MCRAT and fuzzy RAMS methods using Equation (32) with $v = 0.5$. The results of these calculations are shown in Tables 8 and 9, along with the alternative rankings according to the fuzzy RATMI method.

Table 8. Results of the fuzzy RAMS technique.

Alternatives	Max	Min	Median	Median similarity
	$\tilde{Q}_k$	$\tilde{Q}_h$	$\tilde{D}$	$ms(\tilde{M}_i)$
	(0.1633, 0.2665, 0.4205)	(0.0890, 0.1421, 0.2455)	(0.0930, 0.1510, 0.2434)	
	$\tilde{V}_{ik}$	$\tilde{V}_{ih}$	$\tilde{D}_i$	
A1	(0.1324, 0.2192, 0.3594)	(0.0870, 0.1336, 0.2381)	(0.0792, 0.1284, 0.2155)	(0.3254, 0.8500, 2.3175)
A2	(0.1404, 0.2295, 0.3774)	(0.0755, 0.1243, 0.1951)	(0.0797, 0.1305, 0.2124)	(1.4031, 2.2837, 15.0859)
A3	(0.1605, 0.2646, 0.4147)	(0.0722, 0.1098, 0.1835)	(0.0880, 0.1432, 0.2268)	(1.5398, 2.4381, 16.5554)
A4	(0.1517, 0.2529, 0.3983)	(0.0722, 0.1098, 0.1835)	(0.0840, 0.1379, 0.2193)	(1.4822, 2.3577, 15.9359)
A5	(0.1392, 0.2378, 0.3748)	(0.0807, 0.1301, 0.2135)	(0.0804, 0.1355, 0.2157)	(1.4571, 2.3188, 15.6662)
A6	(0.1395, 0.2341, 0.3754)	(0.0829, 0.1388, 0.2216)	(0.0811, 0.1361, 0.2180)	(1.4634, 2.3434, 15.7337)
Alternatives	$ms(M_i)$	**Rankings**		
A1	1.0071	6		
A2	1.0113	5		
A3	1.0989	1		
A4	1.0591	2		
A5	1.0398	4		
A6	1.0470	3		

Table 9. Alternatives rankings according to the fuzzy RATMI method.

Alternatives	Fuzzy MCRAT	Fuzzy RAMS	Majority Index	Rankings
	$tr^* = 0.0914$	$ms^* = 1.0071$		
	$tr^- = 0.0094$	$ms^- = 1.0989$	E_i	
	$tr(Z_i)$	$ms(M_i)$		
A1	0.0914	1.0071	0.0000	6
A2	0.0919	1.0113	0.0538	5
A3	0.0994	1.0989	1.0000	1
A4	0.0960	1.0591	0.5670	2
A5	0.0946	1.0398	0.3742	4
A6	0.0952	1.0470	0.4502	3

Another application of the proposed fuzzy MCDM approach was conducted using two other problems [61,62] that are demonstrated in Table 10. The computations of these two examples are attached in the Supplementary Materials as Table S1 for Example 1 and Table S2 for Example 2.

Table 10. Details of the selected problems and comparisons with the proposed approaches.

Prob No.	Ref. No.	Problem Field	Objective of the Study	Fuzzy MCDM Tool(s) Used	Comparison of Results with the Proposed and Used Fuzzy Approaches
1	[61]	Food security	This study examined the various supplier selection approaches to determine Jordan's primary wheat suppliers and rank them according to specified criteria. The fuzzy VIKOR approach assessed, selected, and ranked the best wheat suppliers in Jordan.	Fuzzy VIKOR with the following characteristics: • 12 experts • seven criteria • five wheat suppliers as alternatives (Romania, Ukraine, Syria, Russia, and Australia)	• The used approach found Romania is the best supplier, followed by Ukraine. • The proposed approach found that Ukraine is a better supplier than Romania. • The Spearman's *rho* and Kendall's *tau_b* correlations between the alternative rankings of the two methods are 60% and 40%, respectively. • The used approach used crisp weights as input to the fuzzy VIKOR matrix, while the proposed approach used fuzzy weights created by fuzzy MEREC-G to be an input to the fuzzy RATMI decision matrix.
2	[62]	Waste management	This study used a fuzzy TOPSIS to evaluate the performance of five waste disposal locations in Park Avenue, Vijayashanti apartments in Chennai, Tamil Nadu (India)	Fuzzy TOPSIS with the following characteristics: • one expert • five criteria • five garbage disposal places as alternatives	• The used approach ranked the five disposal sites in the order S5, S4, S3, S1, and S2. • The proposed approach ranked the fice disposal sites in the order S5, S4, S3, S2, and S1. • The Spearman's *rho* and Kendall's *tau_b* correlations between the alternative rankings of the two methods are 90% and 78%, respectively. • The used approach applied, given fuzzy weights as input to the fuzzy TOPSIS matrix, while the proposed approach used fuzzy weights obtained from fuzzy MEREC-G.

5. Discussion

The numerical application of the proposed hybrid MCDM approach based on fuzzy MEREC-G and fuzzy RATMI methods in this research study showed that it can generate alternative rankings. However, ensuring its validity and checking how those generated alternative rankings compare with rankings of other fuzzy MCDM methods is essential. Moreover, it is also necessary to check the sensitivity of the proposed model. Therefore, the validity and sensitivity analyses are provided in the following subsections.

5.1. Validity Analysis of the Proposed Approach

The validity of the resulting alternative rankings from the fuzzy MCRAT, fuzzy RAMS, and fuzzy RATMI methods presented in Tables 7–9, respectively, are checked. This was done by comparing these rankings from the proposed methods in this study with those resulting from multiple fuzzy MCDM methods presented in Table 11. Those other MCDM methods are the fuzzy ARAS, fuzzy MARCOS, fuzzy TOPSIS, fuzzy MABAC, fuzzy VIKOR, and fuzzy MAIRCA. It is worth mentioning that the researchers who created these fuzzy MCDM methods applied criteria with established fuzzy weights. In contrast, in this research study, the fuzzy weights were unknown and determined by the proposed MEREC-G method. The nonparametric correlation coefficients of ranked data, Spearman's *rho*, and Kendall's *tau_b*, which might be better for smaller samples [73], were found as shown in Tables 12 and 13, respectively. The correlation analyses show high correlations with statistical significance levels between the resulting alternative rankings from the fuzzy MCRAT, fuzzy RAMS, and fuzzy RATMI methods and those resulting from the other fuzzy MCDM methods. This result indicates high accuracy and consistency between the alternative rankings of the proposed hybrid MCDM approach based on fuzzy MEREC-G and fuzzy RATMI methods in this research study and the other fuzzy MCDM methods. Therefore, the proposed approach is deemed valid.

Table 11. Alternative rankings resulted from multiple fuzzy MCDM methods.

Alternatives	Fuzzy ARAS *	Fuzzy MARCOS *	Fuzzy TOPSIS *	Fuzzy MABAC *	Fuzzy VIKOR *	Fuzzy MAIRCA *	Fuzzy MCRAT **	Fuzzy RAMS **	Fuzzy RATMI **
A1	5	6	6	5	6	5	6	6	6
A2	6	5	5	6	5	6	5	5	5
A3	1	1	1	1	1	1	1	1	1
A4	2	2	2	2	2	2	2	2	2
A5	4	4	4	4	4	4	4	4	4
A6	3	3	3	3	3	3	3	3	3

* Alternative ranking adopted from [48]. ** Alternative ranking based on Tables 7–9.

Table 12. Spearman's *rho* correlation coefficients between alternative rankings resulted from multiple fuzzy MCDM methods.

	Fuzzy ARAS	Fuzzy MARCOS	Fuzzy TOPSIS	Fuzzy MABAC	Fuzzy VIKOR	Fuzzy MAIRCA	Fuzzy MCRAT	Fuzzy RAMS	Fuzzy RATMI
Fuzzy ARAS		0.943	0.943	1.000	0.943	1.000	0.943	0.943	0.943
Fuzzy MARCOS			1.000	0.943	1.000	0.943	1.000	1.000	1.000
Fuzzy TOPSIS				0.943	1.000	0.943	1.000	1.000	1.000
Fuzzy MABAC					0.943	1.000	0.943	0.943	0.943
Fuzzy VIKOR						0.943	1.000	1.000	1.000
Fuzzy MAIRCA							0.943	0.943	0.943
Fuzzy MCRAT								1.000	1.000
Fuzzy RAMS									1.000
Fuzzy RATMI									

Note: All Spearman's *rho* correlation coefficients are significant at the $p \leq 0.01$ level (2-tailed).

Table 13. Kendall's *tau_b* correlation coefficients between alternative rankings resulted from multiple fuzzy MCDM methods.

	Fuzzy ARAS	Fuzzy MARCOS	Fuzzy TOPSIS	Fuzzy MABAC	Fuzzy VIKOR	Fuzzy MAIRCA	Fuzzy MCRAT	Fuzzy RAMS	Fuzzy RATMI
Fuzzy ARAS		0.867 *	0.867 *	1.000 **	0.867 *	1.000 **	0.867 *	0.867 *	0.867 *
Fuzzy MARCOS			1.000 **	0.867 *	1.000 **	0.867 *	1.000 **	1.000 **	1.000 **
Fuzzy TOPSIS				0.867 *	1.000 **	0.867 *	1.000 **	1.000 **	1.000 **
Fuzzy MABAC					0.867 *	1.000 **	0.867 *	0.867 *	0.867 *
Fuzzy VIKOR						0.867 *	1.000 **	1.000 **	1.000 **
Fuzzy MAIRCA							0.867 *	0.867 *	0.867 *
Fuzzy MCRAT								1.000 **	1.000 **
Fuzzy RAMS									1.000 **
Fuzzy RATMI									

* Correlation is significant at the $p \leq 0.05$ level (2-tailed). ** Correlation is significant at the $p \leq 0.01$ level (2-tailed).

5.2. Sensitivity Analysis of the Proposed Approach

The sensitivity of the proposed MCDM approach in this study is checked by analyzing the effect of different criteria weights on the resulting rankings of alternatives (A1–A6) from the fuzzy RATMI. The sensitivity analysis was performed by calculating different fuzzy criteria weights of each of the eight criteria (C1–C8) based on a range of 10% to 90% with 10% increments and equally distributing the remainder of the 100% on the reset of criteria in each scenario. This has created a total of 72 run scenarios of the fuzzy RATMI algorithm (i.e., nine sets of criteria weights × eight criteria = 72 run scenarios). This procedure enabled comparing the effect of different weights of each criterion on the resulting alternative rankings.

Figure 2 shows the resulting alternative rankings from the sensitivity analysis. As shown in Figure 2a, criterion C1 demonstrated its sensitivity in most of the alternative rankings in the 10% and 20% scenarios and provided consistent rankings for the 30% to 90% scenarios. Figure 2b shows that criterion C2 changed the rankings of the alternatives A3 and A4 only in the 10% scenario and showed consistent alternative rankings in the 20% to 90% scenarios. For criterion C3, the analysis shows that it gave consistent alternative rankings for the whole range of scenarios from 10% to 90%, as presented in Figure 2c, indicating that changing its weight does not influence the decision-making problem. Figure 2d shows that criterion C4 changed the rankings of the alternatives in the 10%, 80%, and 90% scenarios and gave consistent alternative rankings in the 20% to 70% scenarios. Figure 2e shows that criterion C5 changed the rankings of the alternatives A2, A3, and A4 only in the 10% scenario and showed consistent alternative rankings in the 20% to 90% scenarios. Figure 2f shows that criterion C6 changed the rankings of the alternatives in the 10% and 20% scenarios while giving consistent alternative rankings in the 30% to 90% scenarios. Figure 2g shows that criterion C7 changed the rankings of the alternatives in the 10%, 20%, and 70% scenarios while giving consistent alternative rankings in the other scenarios. Finally, Figure 2h shows that criterion C8 changed the rankings of the alternatives in the 10%, 20%, and 30% scenarios and gave consistent alternative rankings in the 40% to 90% scenarios. These results indicate that the proposed approach is sensitive enough to changes in the criteria weights and reflects those changes on the alternative rankings, yet not too sensitive and capable of producing consistent rankings based on alternatives' performance scoring.

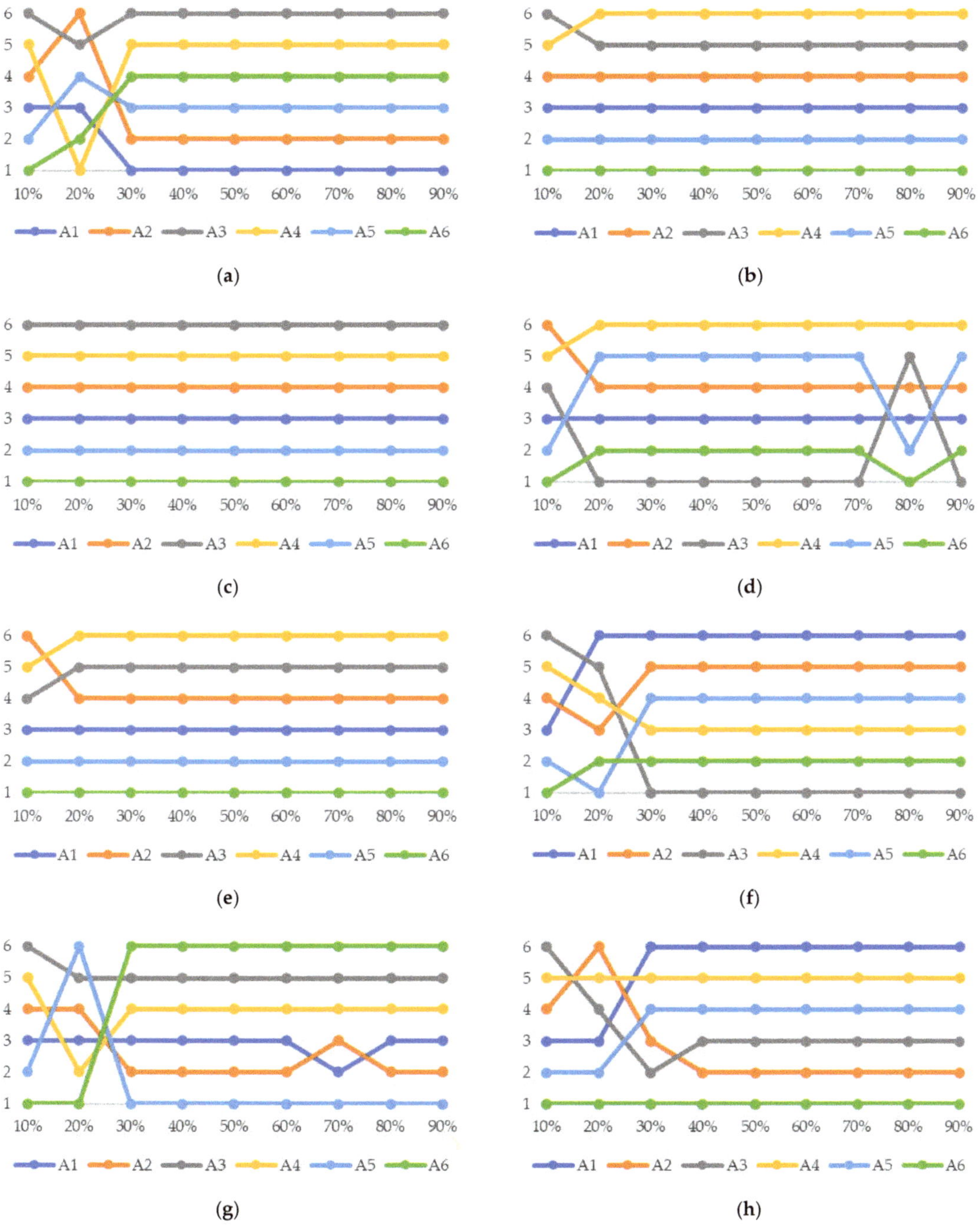

Figure 2. Sensitivity analysis of alternative rankings resulted from using different criteria weight percentages for (**a**) C1; (**b**) C2; (**c**) C3; (**d**) C4; (**e**) C5; (**f**) C6; (**g**) C7; and (**h**) C8.

6. Conclusions

Decision-making can be challenging when faced with multiple conflicting criteria and uncertain or vague information. Fuzzy logic can model the uncertainty and ambiguity in the decision process and provide a framework for fuzzy MCDM methods. These methods help decision-makers assign weights to the criteria and rank the alternatives systematically. This paper introduces a new hybrid fuzzy MCDM approach that combines two novel methods: fuzzy MEREC-G for criteria weighting and fuzzy RATMI for alternative rankings. The new approach was tested with real-world problem data adopted from Ulutaş et al. [48] and compared with other MCDM methods: fuzzy ARAS, fuzzy MARCOS, fuzzy TOPSIS, fuzzy MABAC, fuzzy VIKOR, and fuzzy MAIRCA, fuzzy MCRAT, and fuzzy RAMS. The validity and sensitivity of the proposed hybrid MCDM approach were evaluated. The validity was measured using the nonparametric Spearman's *rho* and Kendall's *tau_b* correlation coefficients of ranked data. The correlation coefficients were 0.943 and 1.00 using Spearman's *rho* methodology, while they were 0.867 and 1.00 using Kendall's *tau_b* methodology. These figures indicate that the proposed approach was valid and can be applied to different real problems with fuzzy data, such as supplier selection [49,52] and selecting pandemic hospital sites [55]. The sensitivity was checked by analyzing how different criteria weights affected the alternative rankings from the fuzzy RATMI, which showed that the approach was sensitive enough to reflect the changes in the criteria weights on the alternative rankings, but not too sensitive and able to produce consistent rankings based on the alternatives' performance scorings. Therefore, this study's new hybrid fuzzy approach is deemed valid.

There are always opportunities for further studies in any new approach. The following are possible future directions to extend the study on the proposed hybrid fuzzy MEREC-G and fuzzy RATMI approach:

- Using the proposed fuzzy hybrid approach for different problems in multi-disciplines can further ensure its effectiveness in solving research and industrial decision-making problems.
- Conduct comparative studies between the new hybrid fuzzy approach and different hybrid fuzzy methods in the literature or to be developed in the future.
- Study the efficacy of the proposed fuzzy hybrid approach when the number of decision criteria increases.
- Apply other variations and extensions of traditional fuzzy set theory, such as intuitionistic, hesitant, and Pythagorean fuzzy, in the developed method, which might better handle the uncertainty and vagueness of inputs in decision-making problems.
- For further comparative analyses, the proposed fuzzy hybrid approach could apply to other studies, such as the recent study presented by Görçün et al. [63].

Supplementary Materials: The following supporting information can be downloaded at: https://www.mdpi.com/article/10.3390/math11173773/s1, Table S1: Example 1; Table S2: Example 2.

Author Contributions: Conceptualization, A.A.M. and R.M.S.A.; data curation, A.A.M. and R.M.S.A.; formal analysis, A.A.M. and R.M.S.A.; investigation, A.A.M. and R.M.S.A.; methodology, A.A.M. and R.M.S.A.; project administration, A.A.M. and R.M.S.A.; resources, A.A.M. and R.M.S.A.; software, A.A.M. and R.M.S.A.; supervision, A.A.M. and R.M.S.A.; validation, A.A.M. and R.M.S.A.; visualization, A.A.M. and R.M.S.A.; writing—original draft, A.A.M. and R.M.S.A.; writing—review and editing, A.A.M. and R.M.S.A. All authors have read and agreed to the published version of the manuscript.

Funding: This research received no external funding.

Data Availability Statement: Not applicable.

Conflicts of Interest: The authors declare no conflict of interest.

References

1. Azhar, N.A.; Radzi, N.A.; Wan Ahmad, W.S.H.M. Multi-criteria decision making: A systematic review. *Recent Adv. Electr. Electron. Eng. Former. Recent Pat. Electr. Electron. Eng.* **2021**, *14*, 779–801. [CrossRef]
2. Taherdoost, H.; Madanchian, M. Multi-Criteria Decision Making (MCDM) Methods and Concepts. *Encyclopedia* **2023**, *3*, 77–87. [CrossRef]
3. Robert, M.X.; Yongwen, W. Which objective weight method is better: PCA or entropy? *Sci. J. Res. Rev.* **2022**, *3*, 1–4. [CrossRef]
4. Singh, M.; Pant, M. A review of selected weighing methods in MCDM with a case study. *Int. J. Syst. Assur. Eng. Manag.* **2021**, *12*, 126–144. [CrossRef]
5. Odu, G.O. Weighting methods for multi-criteria decision-making technique. *J. Appl. Sci. Environ. Manag.* **2019**, *23*, 1449–1457. [CrossRef]
6. Mukhametzyanov, I. Specific character of objective methods for determining weights of criteria in MCDM problems: Entropy, CRITIC and SD. *Decis. Mak. Appl. Manag. Eng.* **2021**, *4*, 76–105. [CrossRef]
7. Keshavarz-Ghorabaee, M.; Amiri, M.; Zavadskas, E.K.; Turskis, Z.; Antucheviciene, J. Determination of Objective Weights Using a New Method Based on the Removal Effects of Criteria (MEREC). *Symmetry* **2021**, *13*, 525. [CrossRef]
8. Beed, R.S.; Sarkar, S.; Roy, A. Hierarchical Bayesian approach for improving weights for solving multi-objective route optimization problem. *Int. J. Inf. Technol.* **2021**, *13*, 1331–1341. [CrossRef]
9. Krishnan, A.R.; Kasim, M.M.; Hamid, R.; Ghazali, M.F. A Modified CRITIC Method to Estimate the Objective Weights of Decision Criteria. *Symmetry* **2021**, *13*, 973. [CrossRef]
10. Xing, J.; Wenshuo, Z. The optimization of objective weighting method based on relative importance. In Proceedings of the 2020 5th International Conference on Mechanical, Control and Computer Engineering (ICMCCE), Harbin, China, 25–27 December 2020; pp. 1234–1237. [CrossRef]
11. Chang, K.-H. Integrating Subjective–Objective Weights Consideration and a Combined Compromise Solution Method for Handling Supplier Selection Issues. *Systems* **2023**, *11*, 74. [CrossRef]
12. Paramanik, A.R.; Sarkar, S.; Sarkar, B. OSWMI: An objective-subjective weighted method for minimizing inconsistency in multi-criteria decision making. *Comput. Ind. Eng.* **2022**, *169*, 108138. [CrossRef]
13. Şahin, M. A comprehensive analysis of weighting and multi-criteria methods in the context of sustainable energy. *Int. J. Environ. Sci. Technol.* **2021**, *18*, 1591–1616. [CrossRef]
14. Adalı, E.A.; Işık, A.T. CRITIC and MAUT methods for the contract manufacturer selection problem. *Eur. J. Multidiscip. Stud.* **2017**, *2*, 88–96. [CrossRef]
15. Kaya, I.; Çolak, M.; Terzi, F. A comprehensive review of fuzzy multi criteria decision making methodologies for energy policy making. *Energy Strategy Rev.* **2019**, *24*, 207–228. [CrossRef]
16. Akram, M.; Garg, H.; Zahid, K. Extensions of ELECTRE-I and TOPSIS methods for group decision-making under complex Pythagorean fuzzy environment. *Iran. J. Fuzzy Syst.* **2020**, *17*, 147–164. [CrossRef]
17. Jayant, A.; Sharma, J. A comprehensive literature review of MCDM techniques ELECTRE, PROMETHEE, VIKOR and TOPSIS applications in business competitive environment. *Int. J. Curr. Res.* **2018**, *10*, 65461–65477.
18. Sari, F.; Kandemir, İ.; Ceylan, D.A.; Gül, A. Using AHP and PROMETHEE multi-criteria decision making methods to define suitable apiary locations. *J. Apic. Res.* **2020**, *59*, 546–557. [CrossRef]
19. Guo, S.; Zhao, H. Fuzzy best-worst multi-criteria decision-making method and its applications. *Knowl.-Based Syst.* **2017**, *121*, 23–31. [CrossRef]
20. Gul, M.; Ak, M.F. Assessment of occupational risks from human health and environmental perspectives: A new integrated approach and its application using fuzzy BWM and fuzzy MAIRCA. *Stoch. Environ. Res. Risk Assess.* **2020**, *34*, 1231–1262. [CrossRef]
21. Khan, S.; Haleem, A.; Khan, M.I. Assessment of risk in the management of Halal supply chain using fuzzy BWM method. *Supply Chain Forum Int. J.* **2021**, *22*, 57–73. [CrossRef]
22. Amiri, M.; Hashemi-Tabatabaei, M.; Ghahremanloo, M.; Keshavarz-Ghorabaee, M.; Zavadskas, E.K.; Banaitis, A. A new fuzzy BWM approach for evaluating and selecting a sustainable supplier in supply chain management. *Int. J. Sustain. Dev. World Ecol.* **2021**, *28*, 125–142. [CrossRef]
23. Gan, J.; Zhong, S.; Liu, S.; Yang, D. Resilient supplier selection based on fuzzy BWM and GMo-RTOPSIS under supply chain environment. *Discret. Dyn. Nat. Soc.* **2019**, *2019*, 2456260. [CrossRef]
24. Gupta, H. Assessing organizations performance on the basis of GHRM practices using BWM and Fuzzy TOPSIS. *J. Environ. Manag.* **2018**, *226*, 201–216. [CrossRef]
25. Mei, M.; Chen, Z. Evaluation and selection of sustainable hydrogen production technology with hybrid uncertain sustainability indicators based on rough-fuzzy BWM-DEA. *Renew. Energy* **2021**, *165*, 716–730. [CrossRef]
26. Ecer, F.; Pamucar, D. Sustainable supplier selection: A novel integrated fuzzy best worst method (F-BWM) and fuzzy CoCoSo with Bonferroni (CoCoSo'B) multi-criteria model. *J. Clean. Prod.* **2020**, *266*, 121981. [CrossRef]
27. Rostamzadeh, R.; Esmaeili, A.; Sivilevičius, H.; Nobard, H.B.K. A fuzzy decision-making approach for evaluation and selection of third party reverse logistics provider using fuzzy ARAS. *Transport* **2020**, *35*, 635–657. [CrossRef]
28. Karagöz, S.; Deveci, M.; Simic, V.; Aydin, N. Interval type-2 Fuzzy ARAS method for recycling facility location problems. *Appl. Soft Comput.* **2021**, *102*, 107107. [CrossRef]

29. Mavi, R.K. Green supplier selection: A fuzzy AHP and fuzzy ARAS approach. *Int. J. Serv. Oper. Manag.* **2015**, *22*, 165–188. [CrossRef]
30. Bakır, M.; Atalık, Ö. Application of fuzzy AHP and fuzzy MARCOS approach for the evaluation of e-service quality in the airline industry. *Decis. Mak. Appl. Manag. Eng.* **2021**, *4*, 127–152. [CrossRef]
31. Stanković, M.; Stević, Ž.; Das, D.K.; Subotić, M.; Pamučar, D. A new fuzzy MARCOS method for road traffic risk analysis. *Mathematics* **2020**, *8*, 457. [CrossRef]
32. Pamucar, D.; Ecer, F.; Deveci, M. Assessment of alternative fuel vehicles for sustainable road transportation of United States using integrated fuzzy FUCOM and neutrosophic fuzzy MARCOS methodology. *Sci. Total Environ.* **2021**, *788*, 147763. [CrossRef] [PubMed]
33. Wang, C.-N.; Pan, C.-F.; Nguyen, H.-P.; Fang, P.-C. Integrating Fuzzy AHP and TOPSIS Methods to Evaluate Operation Efficiency of Daycare Centers. *Mathematics* **2023**, *11*, 1793. [CrossRef]
34. Pompilio, G.G.; Sigahi, T.F.A.C.; Rampasso, I.S.; Moraes, G.H.S.M.d.; Ávila, L.V.; Leal Filho, W.; Anholon, R. Innovation in Brazilian Industries: Analysis of Management Practices Using Fuzzy TOPSIS. *Mathematics* **2023**, *11*, 1313. [CrossRef]
35. Jiang, Z.; Wei, G.; Guo, Y. Picture fuzzy MABAC method based on prospect theory for multiple attribute group decision making and its application to suppliers selection. *J. Intell. Fuzzy Syst.* **2022**, *42*, 3405–3415. [CrossRef]
36. Komatina, N.; Tadić, D.; Aleksić, A.; Jovanović, A.D. The assessment and selection of suppliers using AHP and MABAC with type-2 fuzzy numbers in automotive industry. *Proc. Inst. Mech. Eng. Part O J. Risk Reliab.* **2022**, *237*, 836–852. [CrossRef]
37. Tan, J.; Liu, Y.; Senapati, T.; Garg, H.; Rong, Y. An extended MABAC method based on prospect theory with unknown weight information under Fermatean fuzzy environment for risk investment assessment in B&R. *J. Ambient Intell. Humaniz. Comput.* **2022**, *14*, 13067–13096. [CrossRef]
38. Salimian, S.; Mousavi, S.M.; Antuchevičienė, J. Evaluation of infrastructure projects by a decision model based on RPR, MABAC, and WASPAS methods with interval-valued intuitionistic fuzzy sets. *Int. J. Strateg. Prop. Manag.* **2022**, *26*, 106–118. [CrossRef]
39. Lam, W.S.; Lam, W.H.; Jaaman, S.H.; Liew, K.F. Performance evaluation of construction companies using integrated entropy fuzzy VIKOR model. *Entropy* **2021**, *23*, 320. [CrossRef]
40. Wang, C.N.; Nguyen, N.A.T.; Dang, T.T.; Lu, C.M. A compromised decision-making approach to third-party logistics selection in sustainable supply chain using fuzzy AHP and fuzzy VIKOR methods. *Mathematics* **2021**, *9*, 886. [CrossRef]
41. Poormirzaee, R.; Hosseini, S.; Taghizadeh, R. Smart mining policy: Integrating fuzzy-VIKOR technique and the Z-number concept to implement industry 4.0 strategies in mining engineering. *Resour. Policy* **2022**, *77*, 102768. [CrossRef]
42. Deveci, M.; Gokasar, I.; Pamucar, D.; Zaidan, A.A.; Wen, X.; Gupta, B.B. Evaluation of Cooperative Intelligent Transportation System scenarios for resilience in transportation using type-2 neutrosophic fuzzy VIKOR. *Transp. Res. Part A Policy Pract.* **2023**, *172*, 103666. [CrossRef]
43. Ecer, F. An extended MAIRCA method using intuitionistic fuzzy sets for coronavirus vaccine selection in the age of COVID-19. *Neural Comput. Appl.* **2022**, *34*, 5603–5623. [CrossRef] [PubMed]
44. García Mestanza, J.; Bakhat, R. A fuzzy ahp-mairca model for overtourism assessment: The case of Malaga province. *Sustainability* **2021**, *13*, 6394. [CrossRef]
45. Ecer, F.; Böyükaslan, A.; Hashemkhani Zolfani, S. Evaluation of cryptocurrencies for investment decisions in the era of Industry 4.0: A borda count-based intuitionistic fuzzy set extensions EDAS-MAIRCA-MARCOS multi-criteria methodology. *Axioms* **2022**, *11*, 404. [CrossRef]
46. Hezam, I.M.; Vedala, N.R.D.; Kumar, B.R.; Mishra, A.R.; Cavallaro, F. Assessment of Biofuel Industry Sustainability Factors Based on the Intuitionistic Fuzzy Symmetry Point of Criterion and Rank-Sum-Based MAIRCA Method. *Sustainability* **2023**, *15*, 6749. [CrossRef]
47. Haq, R.S.U.; Saeed, M.; Mateen, N.; Siddiqui, F.; Ahmed, S. An interval-valued neutrosophic based MAIRCA method for sustainable material selection. *Eng. Appl. Artif. Intell.* **2023**, *123*, 106177. [CrossRef]
48. Ulutaş, A.; Topal, A.; Karabasevic, D.; Balo, F. Selection of a Forklift for a Cargo Company with Fuzzy BWM and Fuzzy MCRAT Methods. *Axioms* **2023**, *12*, 467. [CrossRef]
49. Dang, T.-T.; Nguyen, N.-A.-T.; Nguyen, V.-T.-T.; Dang, L.-T.-H. A Two-Stage Multi-Criteria Supplier Selection Model for Sustainable Automotive Supply Chain under Uncertainty. *Axioms* **2022**, *11*, 228. [CrossRef]
50. Wang, C.-N.; Yang, F.-C.; Vo, N.T.M.; Nguyen, V.T.T. Enhancing Lithium-Ion Battery Manufacturing Efficiency: A Comparative Analysis Using DEA Malmquist and Epsilon-Based Measures. *Batteries* **2023**, *9*, 317. [CrossRef]
51. Ayağ, Z. A comparison study of fuzzy-based multiple-criteria decision-making methods to evaluating green concept alternatives in a new product development environment. *Int. J. Intell. Comput. Cybern.* **2021**, *14*, 412–438. [CrossRef]
52. Afrasiabi, A.; Tavana, M.; Di Caprio, D. An extended hybrid fuzzy multi-criteria decision model for sustainable and resilient supplier selection. *Environ. Sci. Pollut. Res.* **2022**, *29*, 37291–37314. [CrossRef]
53. Hien, D.N.; Thanh, N.V. Optimization of Cold Chain Logistics with Fuzzy MCDM Model. *Processes* **2022**, *10*, 947. [CrossRef]
54. Bekesiene, S.; Vasiliauskas, A.V.; Hošková-Mayerová, Š.; Vasilienė-Vasiliauskienė, V. Comprehensive Assessment of Distance Learning Modules by Fuzzy AHP-TOPSIS Method. *Mathematics* **2021**, *9*, 409. [CrossRef]
55. Al Mohamed, A.A.; Al Mohamed, S.; Zino, M. Application of fuzzy multicriteria decision-making model in selecting pandemic hospital site. *Futur. Bus. J.* **2023**, *9*, 14. [CrossRef]

56. Kwok, C.P.; Tang, Y.M. A fuzzy MCDM approach to support customer-centric innovation in virtual reality (VR) metaverse headset design. *Adv. Eng. Inform.* **2023**, *56*, 101910. [CrossRef]
57. Lo, H.W. A data-driven decision support system for sustainable supplier evaluation in the Industry 5.0 era: A case study for medical equipment manufacturing. *Adv. Eng. Inform.* **2023**, *56*, 101998. [CrossRef]
58. Siddiqui, Z.A.; Haroon, M. Research on significant factors affecting adoption of blockchain technology for enterprise distributed applications based on integrated MCDM FCEM-MULTIMOORA-FG method. *Eng. Appl. Artif. Intell.* **2023**, *118*, 105699. [CrossRef]
59. Sotoudeh-Anvari, A. The applications of MCDM methods in COVID-19 pandemic: A state of the art review. *Appl. Soft Comput.* **2022**, *126*, 109238. [CrossRef] [PubMed]
60. Pamucar, D.; Žižović, M.; Biswas, S.; Božanić, D. A new logarithm methodology of additive weights (LMAW) for multi-criteria decision-making: Application in logistics. *Facta Univ. Ser. Mech. Eng.* **2021**, *19*, 361–380. [CrossRef]
61. Magableh, G.M. Evaluating Wheat Suppliers Using Fuzzy MCDM Technique. *Sustainability* **2023**, *15*, 10519. [CrossRef]
62. Vadivel, S.M.; Sakthivel, V.; Praveena, L.; Chandana, V. Apartments Waste Disposal Location Evaluation Using TOPSIS and Fuzzy TOPSIS Methods. In *Innovations in Bio-Inspired Computing and Applications*; IBICA 2022. Lecture Notes in Networks and Systems; Abraham, A., Bajaj, A., Gandhi, N., Madureira, A.M., Kahraman, C., Eds.; Springer: Cham, Switzerland, 2023; Volume 649. [CrossRef]
63. Görçün, Ö.F.; Pamucar, D.; Biswas, S. The blockchain technology selection in the logistics industry using a novel MCDM framework based on Fermatean fuzzy sets and Dombi aggregation. *Inf. Sci.* **2023**, *635*, 345–374. [CrossRef]
64. Ayan, B.; Abacıoğlu, S.; Basilio, M.P. A Comprehensive Review of the Novel Weighting Methods for Multi-Criteria Decision-Making. *Information* **2023**, *14*, 285. [CrossRef]
65. Keleş, N. Measuring performances through multiplicative functions by modifying the MEREC method: MEREC-G and MEREC-H. *Int. J. Ind. Eng. Oper. Manag.* **2023**, *5*, 181–199. [CrossRef]
66. Pala, O. A new objective weighting method based on robustness of ranking with standard deviation and correlation: The ROCOSD method. *Inf. Sci.* **2023**, *636*, 118930. [CrossRef]
67. Abdulaal, R.; Bafail, O.A. Two New Approaches (RAMS-RATMI) in Multi-Criteria Decision-Making Tactics. *J. Math.* **2022**, *2022*, 6725318. [CrossRef]
68. Narang, M.; Kumar, A.; Dhawan, R. A fuzzy extension of MEREC method using parabolic measure and its applications. *J. Decis. Anal. Intell. Comput.* **2023**, *3*, 33–46. [CrossRef]
69. Saidin, M.S.; Lee, L.S.; Marjugi, S.M.; Ahmad, M.Z.; Seow, H.-V. Fuzzy Method Based on the Removal Effects of Criteria (MEREC) for Determining Objective Weights in Multi-Criteria Decision-Making Problems. *Mathematics* **2023**, *11*, 1544. [CrossRef]
70. Makki, A.A.; Alqahtani, A.Y.; Abdulaal, R.M.S. An Mcdm-Based Approach to Compare the Performance of Heuristic Techniques for Permutation Flow-Shop Scheduling Problems. *Int. J. Ind. Eng. Theory Appl. Pract.* **2023**, *30*, 728–749. [CrossRef]
71. Makki, A.A.; Alqahtani, A.Y.; Abdulaal, R.M.S.; Madbouly, A.I. A Novel Strategic Approach to Evaluating Higher Education Quality Standards in University Colleges Using Multi-Criteria Decision-Making. *Educ. Sci.* **2023**, *13*, 577. [CrossRef]
72. Nădăban, S.; Dzitac, S.; Dzitac, I. Fuzzy TOPSIS: A general view. *Procedia Comput. Sci.* **2016**, *91*, 823–831. [CrossRef]
73. Field, A. Chapter 8: Correlation. In *Discovering Statistics Using IBM SPSS Statistics*, 5th ed.; Sage Publications: Thousand Oaks, CA, USA, 2018; pp. 334–367.

MDPI
St. Alban-Anlage 66
4052 Basel
Switzerland
www.mdpi.com

Mathematics Editorial Office
E-mail: mathematics@mdpi.com
www.mdpi.com/journal/mathematics

www.ingramcontent.com/pod-product-compliance
Lightning Source LLC
LaVergne TN
LVHW070155120526
838202LV00013BA/1143
* 9 7 8 3 0 3 6 5 9 7 2 0 1 *